Undergraduate Lecture Notes in Physics

Series Editors

Neil Ashby, University of Colorado, Boulder, CO, USA

William Brantley, Department of Physics, Furman University, Greenville, SC, USA

Matthew Deady, Physics Program, Bard College, Annandale-on-Hudson, NY, USA

Michael Fowler, Department of Physics, University of Virginia, Charlottesville, VA, USA

Morten Hjorth-Jensen, Department of Physics, University of Oslo, Oslo, Norway

Michael Inglis, Department of Physical Sciences, SUNY Suffolk County Community College, Selden, NY, USA

T0171795

Undergraduate Lecture Notes in Physics (ULNP) publishes authoritative texts covering topics throughout pure and applied physics. Each title in the series is suitable as a basis for undergraduate instruction, typically containing practice problems, worked examples, chapter summaries, and suggestions for further reading.

ULNP titles must provide at least one of the following:

- An exceptionally clear and concise treatment of a standard undergraduate subject.
- A solid undergraduate-level introduction to a graduate, advanced, or non-standard subject.
- A novel perspective or an unusual approach to teaching a subject.

ULNP especially encourages new, original, and idiosyncratic approaches to physics teaching at the undergraduate level.

The purpose of ULNP is to provide intriguing, absorbing books that will continue to be the reader's preferred reference throughout their academic career.

More information about this series at http://www.springer.com/series/8917

Nicola Manini

Introduction to the Physics of Matter

Basic Atomic, Molecular, and Solid-State Physics

Second Edition

 Springer

Nicola Manini
Department of Physics
University of Milan
Milan, Italy

ISSN 2192-4791 ISSN 2192-4805 (electronic)
Undergraduate Lecture Notes in Physics
ISBN 978-3-030-57242-6 ISBN 978-3-030-57243-3 (eBook)
https://doi.org/10.1007/978-3-030-57243-3

This Springer imprint is published by the registered company Springer Nature Switzerland AG
The registered company address is: Gewerbestrasse 11, 6330 Cham, Switzerland

Preface

This book fulfills a twofold purpose: to provide a pedagogic panorama of the current microscopic understanding of the basic physics of matter and to help students to acquire a quantitative feeling of the typical orders of magnitude of the main physical quantities (energy, time, temperature, length) involved in the specific conditions relevant for "matter" in its atomic, molecular, and condensed forms. Both tasks are favored by keeping structurally and conceptually well distinct the analysis of the adiabatically separate motions of electrons and of atoms. This distinct treatment is organized in close parallel for molecules and for solids.

While keeping different degrees of freedom well distinct, formal likeness is noted whenever useful, following the standard strategy of *similar solutions for similar equations*. Noteworthy examples of this approach include the spherically symmetric motion of electrons in atoms and of nuclei in diatomic molecules, as well as applications of the Fermi-gas model to electrons in metals and to fluid ^{3}He.

This book includes several detailed derivations, when they are useful to understand the physical reasons of certain facts. These details deserve being understood and then forgotten. In contrast, a number of mathematical relations summarize essential physical information and results, and are therefore worth retaining. As a guide for the reader, a gray background highlights these *essential equations*.

Numbers and orders of magnitude are important, at least as much as mathematical derivations, probably more. A broad selection of numerical problems invites the reader to familiarize with the conceptually simple but often practically intricate numerical evaluations and unit conversions required to reach quantitatively correct estimates in real-life or laboratory conditions.

At variance with many textbooks in this field, the present one adopts SI units throughout. This choice does not only follow the international recommendation, but also helps to compare all results with the output of instruments. The one indulgence to non-SI units is a frequent quotation of energies in eV, which represents a practical unit for most atomic-scale phenomena and converts easily to Joule.

As a basic textbook, this one focuses on what is currently well under control: a selection of well-established systems, phenomena, experimental techniques, and conceptual schemes. From this panorama, the reader should be warned against gathering the false impression that the last word has been said about the physics of matter. On the contrary, physicists, chemists, materials scientists, and biologists currently investigate matter in its multi-faceted forms and collect a wealth of experimental data, for which understanding is often only qualitative and partial. Creativity and insight help scientists to develop novel approximate schemes, or models, to interpret these data and achieve a better understanding of the intimate structure and dynamics of matter.

Not everything is equally important. For a reader wishing to focus on a bare minimum content, while still grasping the basics of the physics of matter, the author suggests to skip the following non-essential topics:

- the spectral-broadening mechanisms—Sect. 1.2,
- the hyperfine structure of H—Sect. 2.1.8,
- perturbation theory applied to the 2-electron atom—Sect. 2.2.4,
- the details of the variational calculation for H_2^+ in Sect. 3.2.1,
- the electronic molecular transitions discussed in Sect. 3.3,
- the density-operator formalism—Sect. 4.1.2,
- the foundation of the canonical ensemble—Sect. 4.2 before Eq. (4.10),
- the connection of entropy and statistics—Sect. 4.2.2,
- the laser—Sect. 4.4.1,
- the tight-binding and plane-waves models for the electron bands of crystals— Sect. 5.2.1,
- the extrinsic semiconductors and their technological applications, from page 214 to the end of Sect. 5.2.2.

On the other hand, a reader who wishes to broaden her/his view of the field beyond the topics covered by the present textbook can follow extended introductory treatments, for example, in Refs. [1, 2, 3, 4]. Specialized texts focus on advanced approaches closer to the frontiers of research, covering both the theoretical and the experimental side. The reader is invited to browse in particular Refs. [5, 6, 7] for atomic physics, Refs. [6, 8, 9] for molecules, and Refs. [10, 11, 12] for solid-state physics. Finally, this textbook focuses on the present-day understanding of the physics of matter, omitting most of the fundamental experiments and conceptual steps through which the scientific community reached this understanding. Reference [13] provides an insider's view of the historical evolution of the basic concepts in this field.

The present volume draws its initial inspiration from Luciano Reatto's course delivered in the 1990s at the University of Milano. In the past, the bulk of this volume was made available as lecture notes, first released on January 15, 2004. Precious feedback and suggestions from Giovanni Onida, other colleagues, and several students in the Milan Physics Diploma course prompted numerous corrections and improvements to the second edition. Compared to the first edition, the second edition also benefits from the addition of an Appendix about lighting applications, a few new problems, a periodic table, and an index. The problems included in this book, plus many more, are available in Italian at the web site http:// materia.fisica.unimi.it/manini/dida/archive_exam.html.

The author acknowledges the warm feedback and interest from the students in physics at the University of Milano; this book was originally written for them. Last but not least, the patient care and love of the author's family was a primary ingredient in nurturing the present textbook.

Milan, Italy   Nicola Manini
June 2020

Contents

Chapter 1
Introductory Concepts

1.1 Basic Ingredients

A substantial body of experimental evidence accumulated mainly through the late 19th and early 20th century convinced the community of physicists and chemists that any piece of matter, e.g., a helium-gas sample, a block of solid ice, a metal wire, a smartphone, a block of wood, a bee,... consists of two kinds of ingredients: *electrons* and *atomic nuclei*.

Electrons are *bonafide* elementary point-like particles. An electron is characterized by a mass $m_e \simeq 0.911 \times 10^{-30}$ kg, and a negative charge $= -q_e$, where the elementary charge $q_e \simeq 1.60 \times 10^{-19}$ C.

Atomic nuclei come in numerous flavors, with complicated inner structures, involving length scales $\simeq 10^{-15}$ m and excitation energies $\simeq 10^{-13}$ J. Many intimate nuclear properties are largely irrelevant to the structure of ordinary matter: for most practical purposes one can model nuclei as approximately structureless point-like particles. A nucleus composed of Z protons and $A - Z$ neutrons, has positive charge $= Z\, q_e$ and mass $M \simeq A$ a.m.u. $= A \times 1.66 \times 10^{-27}$ kg $\gg m_e$.

The only interactions responsible for binding together these elementary building blocks and form pieces of structured matter are the *electromagnetic* ones. Gravity is extremely weak, and only starts to become relevant in shaping very big chunks of matter, such as planets. Nuclear forces are extremely short-ranged, and only play a role in ensuring the existence and stability of the nuclei.

The non-relativistic motion of all nuclei and electrons in a matter sample is governed by the following (Hamiltonian) energy operator:

$$H_{\text{tot}} = T_n + T_e + V_{ne} + V_{nn} + V_{ee} . \tag{1.1}$$

Here

$$T_n = \frac{1}{2} \sum_\alpha \frac{|\mathbf{P}_{\mathbf{R}_\alpha}|^2}{M_\alpha} \tag{1.2}$$

© The Editor(s) (if applicable) and The Author(s), under exclusive license to Springer Nature Switzerland AG 2020
N. Manini, *Introduction to the Physics of Matter*, Undergraduate Lecture Notes in Physics, https://doi.org/10.1007/978-3-030-57243-3_1

is the kinetic energy of the nuclei ($\mathbf{P}_{\mathbf{R}_\alpha}$ is the conjugate momentum to the position $\mathbf{R}_\alpha$ of the αth nucleus);

$$T_e = \frac{1}{2m_e} \sum_i \mathbf{P}_{\mathbf{r}_i}{}^2 \tag{1.3}$$

is the kinetic energy of the electrons ($\mathbf{P}_{\mathbf{r}_i}$ is the conjugate momentum to the position $\mathbf{r}_i$ of the ith electron);

$$V_{ne} = -\frac{q_e^2}{4\pi\epsilon_0} \sum_\alpha \sum_i \frac{Z_\alpha}{|\mathbf{R}_\alpha - \mathbf{r}_i|} \tag{1.4}$$

is the potential energy describing the attraction between nuclei and electrons;

$$V_{nn} = \frac{q_e^2}{4\pi\epsilon_0} \frac{1}{2} \sum_\alpha \sum_{\beta \neq \alpha} \frac{Z_\alpha Z_\beta}{|\mathbf{R}_\alpha - \mathbf{R}_\beta|} \tag{1.5}$$

measures the total nucleus-nucleus repulsion energy; and finally

$$V_{ee} = \frac{q_e^2}{4\pi\epsilon_0} \frac{1}{2} \sum_i \sum_{j \neq i} \frac{1}{|\mathbf{r}_i - \mathbf{r}_j|} \tag{1.6}$$

measures the electron-electron repulsion energy. While the structure of H_{tot} is always the same, the distinction between, e.g., a steel key and a bottle of beer is the result of their different "ingredients", namely the number of electrons and the number and types of nuclei (charge numbers Z_α and masses M_α) involved.

A state ket $|\xi\rangle$ containing all quantum-mechanical information describing the motion of all nuclei and electrons evolves according to Schrödinger's equation

$$i\hbar \frac{d}{dt} |\xi(t)\rangle = H_{\text{tot}} |\xi(t)\rangle . \tag{1.7}$$

This equation, based on Hamiltonian (1.1), is apparently simple and universal. This simplicity and universality indicates that in principle it is possible to understand the observable behavior of any isolated macroscopic object in terms of its microscopic interactions. In practice, however, exact solutions of Eq. (1.7) are available for few simple and idealized cases only. If one attempts an approximate numerical solution of Eq. (1.7), (s)he usually faces the problem that the information contents of a N-particles ket increases exponentially with N, and soon exceeds the capacity of any computer. To describe even a relatively basic material such as a pure rarefied molecular gas, or an elemental solid, nontrivial approximations to the solution of Eq. (1.7) are called for.

The application of smart approximations to Eq. (1.7) to understand observed properties and to make *first-principles* predictions of new properties of material systems is a refined art. These approximations often represent important conceptual tools linking the macroscopic properties of matter to the underlying microscopic interactions. The present textbook proposes a panoramic view of selected observed phenomena in the physics of matter, introducing a few standard conceptual tools for their understanding. The proposed schemes of approximation represent a pedagogical selection of rather primitive idealizations: the bibliography at the end suggests directions to expand the reader's conceptual toolbox to approach today's state of the art in research. We should be aware of the limitations of state-of-the-art tools based on Eq. (1.7): even an experienced physicist of matter risks to predict such a basic property as the electric conductivity of a pure material of known composition and structure in stark conflict with experiment. For more complex systems (e.g., biological matter), quantitative and often even qualitative predictions based purely on Eq. (1.7) lay far beyond the capability of today's modeling and computing power.

1.1.1 Typical Scales

The motions described by Hamiltonian H_{tot} involve several characteristic dimensional scales, dictated by the physical constants [14] present in H_{tot}. Firstly, observe that in H_{tot} the elementary charge q_e and electromagnetic constant ϵ_0 appear everywhere in the fixed combination

$$e^2 \equiv \frac{q_e^2}{4\pi \epsilon_0} = 2.3071 \times 10^{-28} \text{ J m},$$

with dimension energy × length. A unique combination of e^2, Planck's constant $\hbar$, and electron mass m_e yields the length

$$a_0 = \frac{\hbar^2}{m_e e^2} = 0.529177 \times 10^{-10} \text{ m} \tag{1.8}$$

named *Bohr radius*. a_0 sets the characteristic length scale of electronic motions. Consequently, a_0 determines the natural spacings of most microscopic structures and patterns formed by atoms in matter. These spacings can be probed e.g., by means of scanning microscopes. These instruments slide a very sharp tip over a solid surface: the atomic force microscope (AFM) maps the tiny forces that the surface atomic corrugations exert as they come into contact with the tip; the scanning tunneling microscope (STM) maps an electronic tunneling current between the tip and the surface as they are kept a fraction of nm apart. This class of experiments, and many others, provide consistent evidence that indeed in matter atoms are spaced by a fraction of nm, typically in the range $2 - 10 \times a_0$, see Fig. 1.1.

The interaction energy of two elementary point charges at the typical distance a_0

$$E_{Ha} = \frac{e^2}{a_0} = \frac{m_e e^4}{\hbar^2} = 4.35974 \times 10^{-18} \text{ J} = 27.2114 \text{ eV}, \tag{1.9}$$

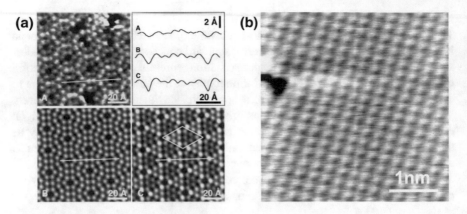

Fig. 1.1 **a** The AFM [A] and STM [B, C] topography images of a clean silicon (111) surface. The STM map visualizes empty [B] and filled [C] electronic states. Reprinted figure with permission from R. Erlandsson, L. Olsson, and P. Mårtensson, http://link.aps.org/abstract/PRB/v54/pR8309 Phys. Rev. B **54**, R8309 (1996). Copyright (1996) by the American Physical Society. **b** The AFM topography of a clean copper (100) surface. Reprinted figure with permission from Ch. Loppacher, M. Bammerlin, M. Guggisberg, S. Schär, R. Bennewitz, A. Baratoff, E. Meyer, and H.-J. Güntherodt, http://link.aps.org/abstract/PRB/v62/p16944 Phys. Rev. B **62**, 16944 (2000). Copyright (2000) by the American Physical Society

is named *Hartree energy*. E_{Ha} sets a natural energy scale for phenomena involving one electron in ordinary matter. In practice, the eV unit is quite popular precisely due to its vicinity to this natural energy scale (1 eV $\simeq 1.60 \times 10^{-19}$ J $\simeq 0.037\, E_{Ha}$). The nuclear charge factors $Z_\alpha \lesssim 10^2$ can scale the e^2 electron-nucleus coupling constant up by $\lesssim 10^2$, the electron-nucleus distance down by $\gtrsim 10^{-2}$, and therefore increase the binding energies of individual electrons by up to 4 orders of magnitude to $\sim 10^4\, E_{Ha} \simeq$ 300 keV. On the opposite, small-energy end, delicate balances can sometimes yield electronic excitation energies as small as 1 meV. The motions of the nuclei are usually associated to slower velocities and smaller energies ($\sim 10^{-4} - 10^{-3}\, E_{Ha}$) than electronic motions, because of the at least 1836 times larger mass at the denominator of the kinetic term of Eq. (1.2).

The typical timescale of electronic motions is inversely proportional to its energy scale:

$$t_0 = \frac{\hbar}{E_{Ha}} = \frac{\hbar^3}{m_e e^4} = 2.4189 \times 10^{-17} \text{ s.} \tag{1.10}$$

Oscillations with period $2\pi t_0$ have a frequency $\nu_0 = \omega_0/(2\pi) = E_{Ha}/(2\pi \hbar) = 6.5797 \times 10^{15}$ Hz.

The typical electron speed is then set by the ratio

$$v_0 = \frac{a_0}{t_0} = \left(\frac{E_{Ha}}{m_e}\right)^{1/2} = \frac{e^2}{\hbar} \simeq 2.1877 \times 10^6 \text{ m/s.} \tag{1.11}$$

v_0 is approximately 1% of the speed of light c. whose absence in H_{tot} is noteworthy. The small ratio v_0/c justifies *a posteriori* the initial neglect of relativity. In detail, the dimensionless ratio

$$\alpha = \frac{v_0}{c} = \sqrt{\frac{E_{Ha}}{m_e c^2}} = \frac{e^2}{\hbar c} \simeq 7.29735 \times 10^{-3} \simeq \frac{1}{137.036}, \qquad (1.12)$$

called *fine-structure constant*, measures the relative importance of relativistic effects in the electronic dynamics. As anticipated, compared to electrons, nuclei usually move far more slowly: therefore relativistic effects are usually entirely negligible for the motions of nuclei.

Figure 1.2 compares the typical length and frequency scales of electrons in matter to the wavelengths and frequencies of the electromagnetic waves, that are used to investigate matter itself, as discussed below mainly in Sects. 1.2, 4.4, 5.1.3, and 5.2.3. The radiation wavelength matches the typical interatomic distances ($\sim 10^{-10}$ m) in the X-rays region. On the other hand, radiation frequencies (and thus radiation energy quanta, or *photons*, see Sect. 4.3.2.2) match the typical electronic frequency scale v_0 (and thus the energy scale E_{Ha}) in the ultraviolet (UV) region: in this spectral region, the typical electromagnetic wavelength is hundreds of nm, i.e. three orders of magnitude larger than the typical atomic spacing $\sim a_0$. Typical frequencies (~ 5 THz) and energies (~ 10 meV) associated to the motion of the nuclei in matter belong to the infrared (IR) region of the electromagnetic spectrum.

1.1.2 Perspectives on the Structure of Matter (Contains Spoilers)

In ordinary matter, electrons, driven by the attraction V_{ne}, Eq. (1.4), tend to lump around the strongest positive charges around, namely the nuclei. Despite the diverging attraction, due to the kinetic energy T_e, Eq. (1.3), and Heisenberg's uncertainty relation (see Sect. C.2.1), electrons fail to collapse inside the nuclei, and rather form *atoms* with a finite size $\approx a_0$, as discussed in Chap. 2. Atoms, in turn, act as the building blocks of matter, in its gaseous and condensed phases. Ultimately, the interaction among atoms, governing these motions, is driven by the quantum dynamics of charged electrons and nuclei described by Eqs. (1.1) and (1.7), usually modeled within the *adiabatic separation* scheme (Sect. 3.1). Understanding the dynamics of a finite number of electrons in the field of two or several nuclei, and the motion of these nuclei themselves (Chap. 3) provides the basics of interatomic *bonding*, namely the mechanism granting the very existence of *molecules* and of condensed states of matter. The methods and approximations developed for few-atom systems and for large statistical ensembles (Chap. 4) lead eventually to collective properties such as elasticity, heat transport, heat capacity, and features, such as transparency or opacity, associated to the macroscopic size of extended objects such as *crystalline solids* (Chap. 5).

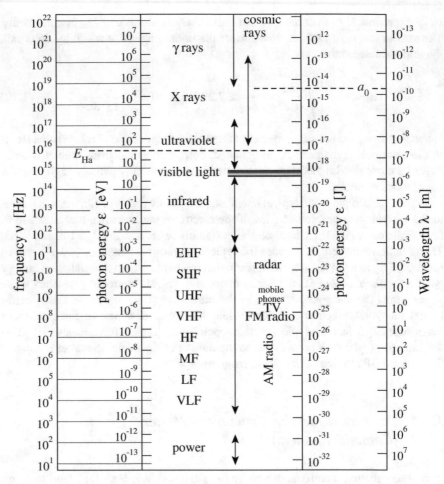

Fig. 1.2 The spectrum of electromagnetic radiation, characterized by frequency ν, photon energy $\mathcal{E} = \hbar\omega = 2\pi\hbar\nu$, and wavelength $\lambda = c/\nu$. Visible light is highlighted with the appropriate colors. *Dashed lines*: the characteristic atomic length scale a_0, Eq. (1.8), and energy scale E_{Ha}, Eq. (1.9)

1.2 Spectra and Broadening

Starting from the late 19th century, physicists have developed and employed all sorts of techniques to investigate the intimate excitations of matter. A rich body of evidence is gathered through *spectroscopies*, i.e. experiments where some property characterizing the interaction of radiation and matter is recorded as a function of the radiation frequency ν (or equivalently angular frequency ω, or photon energy $\mathcal{E} = \hbar\omega$, or wavelength $\lambda = c/\nu$). Out of the many spectroscopies employed routinely in the lab, we focus on absorption and emission, due to their fundamental importance and conceptual simplicity.

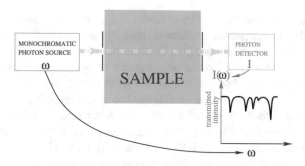

Fig. 1.3 A conceptual scheme of the experimental setup for absorption spectroscopy. The sample is in a "cold" state, essentially at equilibrium with the containing sample cell. Atoms/molecules that the probing radiation excites are a negligibly small minority

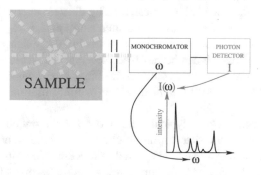

Fig. 1.4 A conceptual scheme of the experimental setup for emission spectroscopy. The sample is brought to a state of high excitation, e.g., by heating it with a flame or by bombarding it with accelerated electrons. A spectrum analyzer separates and counts the emitted photons

- *Absorption*: a collimated beam of monochromatic (not necessarily visible) light crosses the sample; if the light frequency matches a transition specific of that sample, numerous photons are absorbed; otherwise the beam traverses the sample remaining essentially unchanged. The transmitted beam intensity $I(\omega)$ is recorded as a function of the radiation frequency/energy/wavelength, as sketched in Fig. 1.3. Alternatively, the absorbed intensity $I_0 - I(\omega)$ is often recorded.[1]
- *Emission*: the sample is brought to excited states, e.g., by heating it in a flame or by bombarding it with high-energy electrons or photons. The light that the sample subsequently emits in the deexcitation transitions is collected, frequency separated, and its intensity $I(\omega)$ is recorded as above, see Fig. 1.4.

---

[1]The transmitted intensity $I(\omega)$ can be converted into an absorption coefficient $\alpha(\omega)$ by inverting the Beer-Lambert attenuation law $I = I_0 \exp(-\Delta x\, \alpha\, N/V)$. Here I_0 is the incoming radiation intensity, Δx is the probed sample thickness, and N/V is the number density of absorbing atoms/molecules. The power removed from the original beam is transformed, e.g., into heat, or re-irradiated in random directions, mostly different from the original beam direction.

Scientists record routinely spectra of both kinds in the IR, visible, UV and X-ray ranges, to probe the properties of gaseous (both atomic and molecular), liquid, and solid samples. Even the color analysis that the human eyes carry out for the light emitted by light sources (e.g., lamps or stars) can be qualified as a rough emission-spectroscopy experiment. We live in a world full of colors precisely because atoms, molecules, and—to a lesser extent—solids tend to absorb/emit light preferentially at certain frequencies characteristic of the transitions between their quantum states.

The absorption/emission concepts are useful for other routine lab techniques, e.g., nuclear magnetic resonance and UV/visible fluorescence. However, several other spectroscopies, e.g., Raman, Auger, and photoemission, are completely unrelated to absorption and to emission spectroscopies.

Atomic and molecular spectra are often characterized by sharp monochromatic peaks (also called "lines") associated to resonant *transitions* between an initial quantum state $|i\rangle$ and a final one $|f\rangle$, characterized by discrete energy levels E_i and E_f respectively. The angular frequency of one such sharp peak is $\omega_{if} = |E_i - E_f|/\hbar$.

In practice, no absorption/emission peak is ever infinitely sharp. Broadening limits the spectral detail which can be resolved, see Fig. 1.5. At least 3 simultaneous effects combine to broaden the spectra: (i) the finite *experimental resolution* of the spectrometer, (ii) a *"natural" broadening* due to finite lifetime, (iii) the *Doppler broadening*.

Experimental resolution is typically limited by the monochromator resolving power, noise in the photon detector and electronics, inhomogeneities of the sample or of some optional external field applied to it. Resolution broadening is usually generated by several concurrent random effects, and determines therefore a Gaussian line shape. It has no fundamental nature, thus it can be (and often is) reduced by means of technical improvements.

Elementary Schrödinger theory predicts all quantum eigenstates to be stationary states, because this theory neglects the interaction of matter with the ever-present fluctuating radiation field. In reality, the ground state is the one really stationary state, while that interaction makes all excited states decay spontaneously to lower-energy

Fig. 1.5 The same 3-peaks spectrum broadened by increasing spectral width. Broadening limits the detail observable in a spectrum: when the width of neighboring peaks exceeds their energy separation, as in the *dot-dashed curve*, they may appear as a single bump. As a result, the "fine" detail of near features may get lost

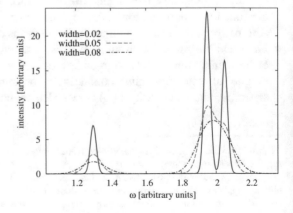

eigenstates. A given individual excited atom decays at an unpredictable random future time. However the statistic decay of a large number N_0 of initially excited atoms is predictable. After a time t, an average number of atoms

$$[N](t) = N_0 \, e^{-t\gamma} = N_0 \, e^{-t/\tau}, \tag{1.13}$$

remains in that excited level. This decay law defines the *lifetime* τ and decay rate $\gamma = 1/\tau$ of that atomic level.[2] This decay rate γ is constant in time: an atom becomes no more or less likely to decay at the next instant because it has already spent some time in an excited level. Equation (1.13) defines a uniform random decay. In practice, very few atoms last in an excited state longer than a few times its characteristic τ. Accordingly, τ sets the typical duration of any spectroscopy experiment involving that excited state. Due to the time-energy uncertainty, the energy of an atomic level cannot be measured with better precision than

$$\Delta E \simeq \frac{\hbar}{\tau} = \hbar\gamma \,. \tag{1.14}$$

This energy uncertainty of the quantum levels produces therefore a broadening of the otherwise infinitely sharp spectral lines. This natural broadening due to the finite lifetime manifests itself in the spectrum as a Lorentzian peak profile

$$I(\omega) = I_0 \, \frac{\gamma^2}{(\omega - \omega_{if})^2 + \gamma^2} \,, \tag{1.15}$$

around the original line position ω_{if}. Atomic excited states are characterized by typical lifetimes of several ns, thus by natural spectral broadening $\hbar\gamma$ of a fraction of μeV.

The random thermal motion in a gas-phase sample introduces an extra source of line broadening: the Doppler effect. When viewed from the lab frame, the intrinsic transition frequencies of atoms/molecules moving randomly toward or away from the detector with a velocity v_x are blue- or red-shifted relative to those at rest. Since the thermal (molecular center-mass) velocities are non-relativistic, the Doppler frequency shift is given by the simple expression

$$\omega = \omega_{if} \left(1 \pm \frac{v_x}{c} \right) \qquad \omega_{if} = \text{angular frequency at rest.} \tag{1.16}$$

---

[2]The total decay rate γ is the sum of the individual decay rates to all lower-lying states. For example, the excited 3p state of hydrogen decays at a rate $\gamma_{3p \to 1s} = 1.67 \times 10^8$ s^{-1} to the ground state 1s, and at a rate $\gamma_{3p \to 2s} = 2.25 \times 10^7$ s^{-1} to the excited 2s state, while decay to excited 2p state is negligible because it is dipole forbidden, see Sect. 2.1.9. Overall the number of 3p excited atoms decays at a total rate $\gamma_{3p} = \gamma_{3p \to 1s} + \gamma_{3p \to 2s} = 1.90 \times 10^8$ s^{-1}.

The molecular velocities are randomly distributed [see Eq. (4.50) in Sect. 4.3.1] depending on the gas temperature T: the average number of atoms/molecules with velocity component v_x in the direction of detection is

$$\frac{dN(v_x)}{dv_x} = N \sqrt{\frac{M}{2\pi k_B T}} \exp\left(-\frac{M v_x^2}{2k_B T}\right),$$ (1.17)

where M is the atomic/molecular mass, and N is the total number of atoms/molecules. Radiation intensity then spreads around the rest frequency ω_{if} as

$$I(\omega) = I_0 \exp\left[-\frac{Mc^2}{2k_B T}\left(\frac{\omega - \omega_{if}}{\omega_{if}}\right)^2\right].$$ (1.18)

This represents a Gaussian broadening with full width at half-maximum

$$\Delta\omega_{\text{Doppler}} = \omega_{if} \sqrt{8\ln(2)\frac{k_B T}{Mc^2}}.$$ (1.19)

As $\lambda = 2\pi c/\omega$, the relative broadening in terms of wavelength is the same as for frequency:

$$\frac{\Delta\lambda_{\text{Doppler}}}{\lambda_{if}} = \sqrt{8\ln(2)\frac{k_B T}{Mc^2}}.$$ (1.20)

For a given temperature, lighter atoms/molecules move faster, thus producing larger Doppler broadening than heavier ones, as accounted by the $M^{-1/2}$ dependence of the Doppler width, Eqs. (1.19) and (1.20). For this reason, the spectra of atomic hydrogen is especially affected by Doppler broadening. For example, at $T = 300$ K all lines of gas-phase H suffer a broadening with fractional width $\Delta\omega_{\text{Doppler}}/\omega_{if} \simeq 3 \times 10^{-6}$. For the H$\alpha$ line (introduced in Sect. 2.1.1) at 1.89 eV, this corresponds to an absolute width $\hbar\Delta\omega_{\text{Doppler}} \simeq 10 \ \mu$eV.

According to the laws of statistics, the individual sources of broadening combine into a total width $\Delta\omega_{\text{tot}} = \left[(\Delta\omega_{\text{exp}})^2 + \gamma^2 + (\Delta\omega_{\text{Doppler}})^2\right]^{1/2}$.

Chapter 2
Atoms

The importance of the spectroscopy of atoms and ions for the understanding of the whole physics of matter cannot be overestimated. The study of atoms starts off naturally from the exact dynamics of a single electron in the central field of one positively charged nucleus (Sect. 2.1) because, beside the intrinsic interest of this system, the notation and concepts developed here are at the basis of the language of all atomic physics. In Sect. 2.2 we then apply this language to the spectroscopy of many-electron atoms, which are clusters of 2 to $\sim 10^2$ electrons repelling each other, but trapped together in the attractive central field of one nucleus.

2.1 One-Electron Atom/Ions

The one-electron atom is one of the few quantum problems whose Schrödinger equation (1.7) can be solved analytically. Comparison of this exact solution with experiments allows physicists to evaluate the limits of validity and predictive power of the quantum mechanical model Eqs. (1.1–1.6). When relativistic effects are included (Sect. 2.1.7), this model is found in almost perfect agreement with extremely accurate experimental data, all the tiny discrepancies being satisfactorily accounted for by a perturbative treatment of residual interactions (Sect. 2.1.8).

The solution of the Schrödinger equation (1.7), or rather its time-independent counterpart (C.30), for the one-electron atom is a basic exercise in quantum mechanics (QM). Both the V_{ee} and V_{nn} terms in the Hamiltonian (1.1) vanish, and only the nuclear and electronic kinetic energies plus the Coulomb attraction V_{ne} are relevant. Many textbooks [5, 15, 16] report detailed solutions of the one-electron atom problem. Here we recall the general strategy and main results.

- **Separation of the center-of-mass motion**. Like in the solution of the classical Kepler-Newton two-body planetary problem, the position operators for the nucleus $\mathbf{R}$ (which has mass $M \gg m_e$) and for the electron $\mathbf{r}_e$ are usefully substituted by the following combinations:

© The Editor(s) (if applicable) and The Author(s), under exclusive license
to Springer Nature Switzerland AG 2020
N. Manini, *Introduction to the Physics of Matter*, Undergraduate Lecture Notes
in Physics, https://doi.org/10.1007/978-3-030-57243-3_2

$$\mathbf{R}_{cm} = \frac{M\mathbf{R} + m_e \mathbf{r}_e}{M + m_e} \quad \text{and} \quad \mathbf{r} = \mathbf{r}_e - \mathbf{R}. \tag{2.1}$$

In these new coordinates, the Hamiltonian separates as the sum of two parts. One is a purely kinetic term for the translation of the atomic center of mass $\mathbf{R}_{cm}$:

$$H_{cm} = -\frac{\hbar^2}{2(M + m_e)} \nabla^2_{\mathbf{R}_{cm}}. \tag{2.2}$$

The global free translational motion for $\mathbf{R}_{cm}$ is described trivially in terms of plane-wave solutions, see Sect. C.7.1. The remaining part is a central-force Hamiltonian for the Coulombic dynamics of the relative coordinate $\mathbf{r}$:

$$H_{Coul} = -\frac{\hbar^2}{2\mu} \nabla^2_{\mathbf{r}} - \frac{Ze^2}{|\mathbf{r}|}. \tag{2.3}$$

Here

$$\mu = \frac{Mm_e}{M + m_e} \tag{2.4}$$

is the *reduced mass* of this 2-particles system. The resulting internal atomic dynamics is that of a single particle with mass μ, moving in the same Coulombic central field as the original nucleus-electron attraction.

- **Separation in spherical coordinates**. To exploit the rotational invariance of the potential, the Schrödinger equation is conveniently rewritten in spherical coordinates r, θ, φ. By factorizing the total wavefunction $\psi(r, \theta, \varphi) = R(r)\,\Theta(\theta)\,\Phi(\varphi)$, the variables separate, and the original three-dimensional (3D) equation splits into three independent second-order equations for the r, θ, and φ motions:

$$-\frac{d^2\Phi}{d\varphi^2} = \eta\,\Phi, \tag{2.5}$$

$$-\frac{1}{\sin\theta} \frac{d}{d\theta} \left(\sin\theta \frac{d\Theta}{d\theta}\right) + \frac{\eta}{\sin^2\theta}\,\Theta = \lambda\,\Theta, \tag{2.6}$$

$$-\frac{\hbar^2}{2\mu} \frac{1}{r^2} \frac{d}{dr} \left(r^2 \frac{dR}{dr}\right) + \left[U(r) + \frac{\hbar^2}{2\mu} \frac{\lambda}{r^2}\right] R = E\,R. \tag{2.7}$$

Here we put a general function $U(r)$ in place of the potential energy $-Ze^2/r$, to emphasize that this same technique can be applied to any central potential (e.g. in Sect. 3.3 below).

- **Solution** of the separate eigenvalue problems: the differential equations are solved under the appropriate boundary conditions for $R(r)$, $\Theta(\theta)$ and $\Phi(\varphi)$. The eigenvalues η, λ, and E can assume only certain "quantized" values, compatible with the boundary conditions[1]:

$$\eta = \eta_{m_l} = m_l^2, \quad m_l = 0, \pm 1, \pm 2, \ldots \tag{2.8}$$

$$\lambda = \lambda_l = l(l+1), \quad l = |m_l|, |m_l|+1, |m_l|+2, \ldots \tag{2.9}$$

$$E = \mathcal{E}_n = -\frac{\mu Z^2 e^4}{2\hbar^2 n^2} = -\frac{\mu}{m_e} \frac{E_{\text{Ha}}}{2} \frac{Z^2}{n^2}, \quad n = l+1, l+2, l+3, \ldots \tag{2.10}$$

The integer quantum numbers m_l (magnetic q.n.), l (azimuthal q.n.), and n (principal q.n.) parameterize the eigenvalues as indicated above; likewise they affect the corresponding eigenfunctions as follows:

$$\Phi_{m_l}(\varphi) = \frac{1}{\sqrt{2\pi}} e^{im_l\varphi} \tag{2.11}$$

$$\Theta_{l m_l}(\theta) = (-1)^{\frac{|m_l|+m_l}{2}} \sqrt{\frac{2l+1}{2} \frac{(l-|m_l|)!}{(l+|m_l|)!}} P_l^{|m_l|}(\cos\theta) \tag{2.12}$$

$$R_{nl}(r) = k^{3/2} \sqrt{\frac{(n-l-1)!}{2n[(n+l)!]^3}} (kr)^l L_{n+l}^{2l+1}(kr) e^{-kr/2}, \tag{2.13}$$

where k is a shorthand for $2Z/(an)$, and a is a mass-corrected Bohr radius $a = a_0 m_e/\mu = \hbar^2/(\mu e^2)$. For nonnegative integers l and m, the associated Legendre functions $P_l^m(x)$ in Eq. (2.12) are defined by

$$P_l^m(x) = (1-x^2)^{m/2} \frac{d^m}{dx^m} P_l(x), \quad P_l(x) = \frac{1}{2^l l!} \frac{d^l}{dx^l} (x^2-1)^l. \tag{2.14}$$

Likewise, for nonnegative integers $p \geq q$, the associated Laguerre polynomials $L_p^q(\rho)$ are polynomials of degree $p - q$, defined by

$$L_p^q(\rho) = \frac{d^q}{d\rho^q} L_p(\rho), \quad L_p(\rho) = e^\rho \frac{d^p}{d\rho^p} (\rho^p e^{-\rho}). \tag{2.15}$$

Since each individual wavefunction of Eqs. (2.11–2.13) is properly normalized by its own square-root factor, so is the total atomic wavefunction

---

[1] The "quantization" of the angular motion arises from the boundary conditions $\Phi(\varphi + 2\pi) = \Phi(\varphi)$, and finite $\Theta(0)$ and $\Theta(\pi)$ on the solutions of Eqs. (2.5) and (2.6): these conditions are granted only by integer values of the quantum numbers m_l and $l \geq |m_l|$ [16]. Likewise, the discrete energies arise from the condition that the solution $R(r)$ behaves regularly for $r \to 0$ and $r \to \infty$.

$$\psi_{n\,l\,m_l}(r, \theta, \varphi) = R_{n\,l}(r)\, \Theta_{l\,m_l}(\theta)\, \Phi_{m_l}(\varphi) \tag{2.16}$$

representing the atomic energy eigenstate $|n, l, m_l\rangle$. Explicitly, the *orthonormality* relation reads:

$$\begin{aligned}
\langle n, l, m_l | n', l', m_l' \rangle &= \int r^2 dr \, \sin\theta d\theta \, d\varphi \, \psi^*_{n\,l\,m_l}(r, \theta, \varphi) \, \psi_{n'\,l'\,m_l'}(r, \theta, \varphi) \\
&= \delta_{nn'} \, \delta_{ll'} \, \delta_{m_l m_l'}.
\end{aligned} \tag{2.17}$$

In addition to all these *bound states*, a continuum of *unbound states* at all positive energies represents the electron moving far away from the nucleus, leaving the neutral atom *ionized* (H$\to$H$^+$), or the 1-electron ion further ionized (He$^+$$\to$He^{2+}).

2.1.1 The Energy Spectrum

The energy eigenvalues Eq. (2.10) of the nonrelativistic one-electron atom are functions of the principal quantum number n only, and exhibit the characteristic structure sketched in Fig. 2.1. In particular, the *ground state* is $|n, l, m_l\rangle = |1, 0, 0\rangle$. For hydrogen ($Z = 1$), its energy $\mathcal{E}_1 = -0.5\,E_{Ha}\,\mu/m_e = -13.5983$ eV. Due to the reduced-mass correction $\mu/m_e = 0.999456$, this energy is slightly less negative than $-1/2\,E_{Ha}$, where $1/2\,E_{Ha} \simeq 13.6057$ eV is named Rydberg energy.

Above this lowest-energy state we find a sequence of energy levels. The lowest excited ($n = 2$) level is associated to $|2, 0, 0\rangle, |2, 1, -1\rangle, |2, 1, 0\rangle$, and $|2, 1, 1\rangle$: it is 4-

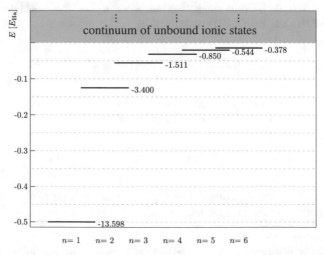

Fig. 2.1 The 6 lowest-energy levels $\mathcal{E}_n$ of atomic hydrogen, as resulting from the nonrelativistic Schrödinger theory, Eq. (2.10). Energies in eV are indicated next to each level. The energy zero marks the onset of a continuum of unbound states

Fig. 2.2 The observed emission line spectrum of atomic hydrogen: **a** across a broad IR-UV range; **b** a closeup of the visible (Balmer series) plus the overlapping near-infrared series; **c** a further closeup of the Balmer series, with its four $H\alpha$ – $H\delta$ lines in the visible range, plus a sequence of near-UV lines

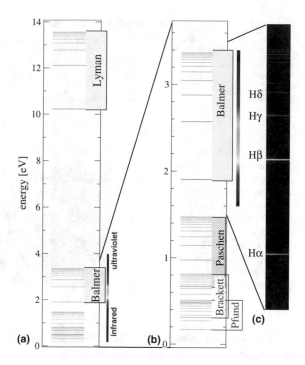

fold degenerate. Its energy is $\mathcal{E}_2 = -\frac{1}{2}(Z/2)^2 E_{\text{Ha}} \frac{\mu}{m_e} = -\frac{1}{8} Z^2 E_{\text{Ha}} \frac{\mu}{m_e}$. In hydrogen this level is $\mathcal{E}_2 - \mathcal{E}_1 = \frac{3}{8} E_{\text{Ha}} \frac{\mu}{m_e} \simeq 10.1987$ eV above the ground state. Successive $\mathcal{E}_3, \mathcal{E}_4 \ldots$ levels exhibit an increasing *degeneracy*, because of the multiple values $l = 0, \ldots n - 1$ compatible with $n > 1$, and the values $m_l = 0, \pm 1, \ldots \pm l$ compatible with each l. The m_l-degeneracy ($2l + 1$ states) is a general feature of central potentials, representing the possibility for the orbital angular momentum to point in any direction in 3D space without affecting the energy of the atom. In contrast, the extra l-degeneracy (n values of l, generating n^2 states in total) is peculiar of the Coulombic potential energy $U(r) \propto -r^{-1}$: no such "accidental" degeneracy occurs for a different radial dependence of $U(r)$, e.g. as appropriate for many-electron atoms, Sect. 2.2.5, and in diatomic molecules, Sect. 3.3.1.

Transitions between any two energy levels are observed, Fig. 2.2. Historically, the close agreement of the H-atom Schrödinger spectrum of Eq. (2.10) with accurate spectroscopic observations marked an early triumph of QM. The transitions $n_i \rightarrow n_f$ group in *series*. Each series of transitions is characterized by the same *lower*-energy level, i.e. the same final level n_f in emission spectra. Each series of hydrogen is observed in a characteristic spectral region and is named after the scientist who carried out its earliest investigation, see Fig. 2.2 and Table 2.1. Remarkably, the Lyman and Balmer series do not overlap with any other series, since the energy distance between their lower level ($n = 1$ or 2) and the next one ($n + 1$) exceeds the entire range of bound-state energies from level $n + 1$ to the ionization threshold.

Table 2.1   The 4 lowest-energy series of spectral lines of atomic hydrogen

name	lower n	lowest energy [eV]	max energy [eV]	spectral region
Lyman series	1	10.2	13.6	UV
Balmer series	2	1.89	3.40	Visible-UV
Paschen series	3	0.66	1.51	IR
Brackett series	4	0.31	0.85	IR

Fig. 2.3 A high-resolution line spectrum of the Balmer Hα emission of an isotope mixture. ^{1}H emits the lower-energy peak near 1.8887 eV; deuterium ^{2}H emits the higher-energy features near 1.8892 eV. The finer doublet structure is due to relativistic effects analyzed in Sect. 2.1.7 below

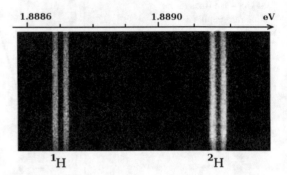

The prefactor $\mu/m_e = (1 + m_e/M)^{-1}$ is responsible for a weak dependence of the spectrum on the nuclear mass M. Accordingly, mixtures of different *isotopes*, e.g. regular hydrogen ^{1}H and twice as heavy *deuterium* ^{2}H (sometimes indicated by D), exhibit a duplication of the spectral lines, due to a tiny relative energy difference, here $\simeq 0.03\%$. Figure 2.3 shows this duplication for the $n = 3 \to 2$ "Hα" line.

Importantly, note the Z^2 dependence of the eigenvalues $\mathcal{E}_n$, Eq. (2.10). This dependence makes one half of the lines of one half of the series of the He$^+$ ion (and one third of those of Li^{2+}, …) almost coincident with the lines of hydrogen, as illustrated in Fig. 2.4. Small deviations are due to different reduced-mass corrections μ/m_e, and to relativistic effects discussed in Sect. 2.1.7 below.

2.1.2   The Angular Wavefunction

The angular solutions (2.11) and (2.12) combine to form the *spherical harmonics* $Y_{lm_l}(\theta, \varphi) = \Theta_{lm_l}(\theta)\, \Phi_{m_l}(\varphi)$, which are the normalized eigenfunctions[2] of the angular motion of a freely rotating quantum-mechanical particle. Y_{lm_l} contains complete information about an important observable: the *orbital angular momentum*. In detail, $\hbar^2 \times$ [the angular part of $-\nabla^2$ occurring at the left side of Eq. (2.6)] represents the squared orbital angular momentum $|\mathbf{L}|^2$ of the rotating two-body system. $\hbar^2 \lambda = \hbar^2 l(l + 1)$ are the eigenvalues of $|\mathbf{L}|^2$. Likewise, $\hbar m_l$ are the eigenvalues of

---

[2] We stick to the standard convention of physicists for the phase and normalization of $Y_{lm_l}(\theta, \varphi)$.

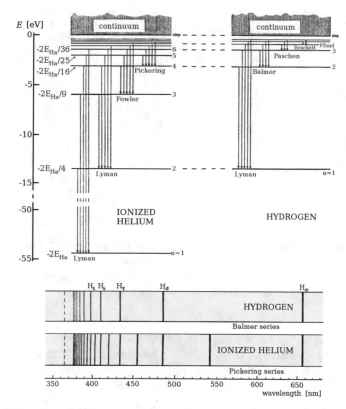

Fig. 2.4 Comparison of the emission spectra of one-electron atoms/ions with different nuclear charge: H ($Z = 1$) and He$^+$ ($Z = 2$). The odd-n series of ionized helium find no correspondence in the spectrum of hydrogen; in contrast, one half of the spectral lines of the even-n series almost coincide with the lines of hydrogen

the angular-momentum component L_z represented by $-i\hbar\frac{\partial}{\partial\varphi}$, thus clarifying the physical meaning of Eq. (2.5). Y_{lm_l} contains only statistical information about the L_x and L_y components, since these other components do not commute with L_z. Note however that the choice of the $\hat{\mathbf{z}}$ direction in 3D space (related to the choice of the spherical coordinate system) is arbitrary: due to spherical symmetry, any alternative choice would lead to the same observable results.

The spherical harmonics carry complete information about the angular distribution of the vector $\mathbf{r} = r\,(\sin\theta\,\cos\varphi,\ \sin\theta\,\sin\varphi,\ \cos\theta)$ joining the nucleus to the electron. In a state $|l, m_l\rangle$ whose squared angular momentum is fixed by l and whose z-projection is given by m_l, the angular probability distribution equals $|\langle\theta,\varphi|l,m_l\rangle|^2 d\Omega \equiv |Y_{lm_l}(\theta,\varphi)|^2 d\Omega$, where $d\Omega = \sin\theta\,d\theta\,d\varphi$ is the infinitesimal solid angle. Equation (2.11) implies that $|Y_{lm_l}|^2$ is independent of the azimuthal angle φ, but only depends on θ. For a few values of l and m_l, the polar plots of Fig. 2.5 illustrate this θ dependence of $|Y_{lm_l}|^2$. It is apparent that $l - |m_l|$ counts the number

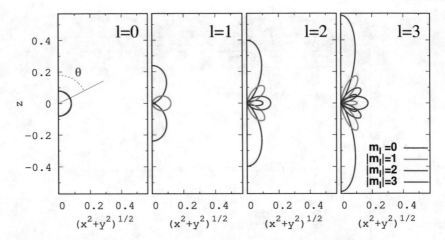

Fig. 2.5 Polar plots of the lowest-l angular probability distributions: the radial distance from the origin equals $\left|Y_{lm_l}(\theta, \varphi)\right|^2$, as a function of the angle θ from the positive $\hat{\mathbf{z}}$ axis, varying from 0° (upward) to 180° (downward), as sketched. Because $\left|Y_{lm_l}(\theta, \varphi)\right|$ is independent of φ, these probability distributions are axially symmetric around the $\hat{\mathbf{z}}$ axis

of zeros (*nodes*) of Y_{lm_l}, i.e. the values of the polar angle θ where the Legendre function $P_l^{m_l}(\cos\theta)$ vanishes and changes sign. A large number of nodes indicates large angular-momentum components perpendicular to $\hat{\mathbf{z}}$. Several textbooks and web sites provide useful alternative visualizations of the Y_{lm_l} functions.

Observations: (i) The increase of $|Y_{l0}(0, \varphi)|^2$ with l (quite evident in Fig. 2.5) does not contradict normalization (2.17), because of the $\sin\theta$ integration factor. (ii) If the r^l factor taken from the radial wavefunction (2.13) is grouped together with $Y_{lm_l}(\theta, \varphi)$, one can express $r^l\, Y_{lm_l}(\theta, \varphi)$ in Cartesian components (r_x, r_y, r_z), obtaining a homogeneous polynomial of degree l. For example,

$$r\, Y_{10}(\theta, \varphi) = \sqrt{\frac{3}{4\pi}}\, r_z, \qquad r\, Y_{1\pm1}(\theta, \varphi) = \sqrt{\frac{3}{8\pi}}\, (\mp r_x - i\, r_y). \qquad (2.18)$$

(iii) Observation (ii) implies that the parity of $Y_{lm_l}(\theta, \varphi)$ (i.e. its character under the reflection $\mathbf{r} \to -\mathbf{r}$) is the same as that of its degree l, i.e. $(-1)^l$. (iv) Y_{l0} spherical harmonics are real functions. (v) $m_l \neq 0$ spherical harmonics can be combined in pairs to construct real wavefunctions, e.g.

$$\psi_{p_x} = \frac{Y_{1-1} - Y_{11}}{\sqrt{2}}, \qquad \psi_{p_y} = i\,\frac{Y_{11} + Y_{1-1}}{\sqrt{2}}. \qquad (2.19)$$

(vi) It is easy and useful to retain the expression for the simplest spherical harmonic function, a polynomial of degree 0, i.e. a constant: $Y_{00}(\theta, \varphi) = (4\pi)^{-1/2}$.

Important notation: Spectroscopists have adopted a letter code for the value of the orbital angular momentum, namely: s, p, d, f, g, h..., for $l = 0, 1, 2, 3, 4, 5...$ respectively. This coding is a quite standard and widely-adopted notation.

2.1.3   The Radial Wavefunction

The radial wavefunction $R_{nl}(r)$, Eq. (2.13), is structured as the product of (i) a normalization factor, (ii) a power r^l (mentioned above in relation to Y_{lm_l}), (iii) a polynomial of degree $n - l - 1$ in r (with a nonzero r^0 term), and (iv) the exponential of $-Zr/(an)$. The power term is responsible for the $R_{nl}(r) \propto r^l$ behavior at small r. The Laguerre polynomial $L_{n+l}^{2l+1}(\rho)$ vanishes at as many different $\rho > 0$ points as its degree $(n - l - 1)$: each of these zeroes produces a *radial node*, i.e. a spherical shell of radius r where $R_{nl}(r)$, and thus the overall wavefunction, vanishes and changes sign. The exponential decay dominates at large r, where $R_{nl}(r) \propto r^{n-1} \exp(-Zr/na)$. Figure 2.6 illustrates these features for the squared moduli of the lowest-n radial eigenfunctions.

The squared wavefunction $|R(r)|^2$ of Fig. 2.6 and the *radial probability distribution* $P(r) = r^2|R(r)|^2$ of Fig. 2.7 convey different information about the quantum state. $P(r)\, dr$ yields the probability that the nucleus-electron distance is within dr of r, regardless of the direction where **r** points. The r^2 weighting factor is precisely the spherical-coordinates Jacobian, proportional to the surface of the sphere of radius r, or equivalently the volume of the spherical shell "between r and $r + dr$". In Fig. 2.7, note that for increasing n, $P(r)$ peaks at larger and larger distance from the origin. In contrast, the probability that the electron is found at a specific position **r** relative to the nucleus is not $P(r)$, but is rather given by

$$P_{3D}(\mathbf{r})\, d^3\mathbf{r} = |\psi(\mathbf{r})|^2\, d^3\mathbf{r} = |\psi_{nlm_l}(r, \theta, \varphi)|^2\, d^3\mathbf{r} = |R_{nl}(r)|^2\, |Y_{lm_l}(\theta, \varphi)|^2\, d^3\mathbf{r},$$

$$(2.20)$$

where the spherical coordinates are those representing that point **r**. Equation (2.20) indicates that $|\psi(r, \theta, \varphi)|^2$ gives the actual 3D probability distribution in space. Thus, when we move following a radial line (specified by fixing θ and φ), the probability density changes as $|R(r)|^2$, multiplied by a constant $|Y_{lm_l}(\theta, \varphi)|^2$. Note that all s eigenfunctions have a nonzero value at the origin, $R_{n0}(0) = 2[Z/(an)]^{3/2}$. Moreover, $r = 0$ is a cusp-type absolute *maximum* of $|R_{n0}(r)|^2$, see Fig. 2.6. It should be no surprise that the most likely location for s electrons coincides with the nucleus, $\mathbf{r} = \mathbf{0}$, the spot where the potential energy $U(\mathbf{r})$ is the most attractive. This fact is hidden by the vanishing of $P(r)$ for $r \to 0$, a consequence of the r^2 weight in $P(r)$.

For $l > 0$, even the probability density $|R_{nl}(r)|^2$ vanishes at the origin, because the repulsive "effective potential" centrifugal term $\propto \lambda/r^2 = l(l+1)/r^2$ in Eq. (2.7) diverges there. The vanishing of $|R_{nl}(r)|^2$ reflects the impossibility of a point particle carrying nonzero angular momentum to reach the origin of a central potential. The wave-mechanical reason is the following: for $l > 0$, $\mathbf{r} = \mathbf{0}$ is a common point of one or several nodes of the angular wavefunction; for this reason ψ would be multiple-valued unless it vanishes there.

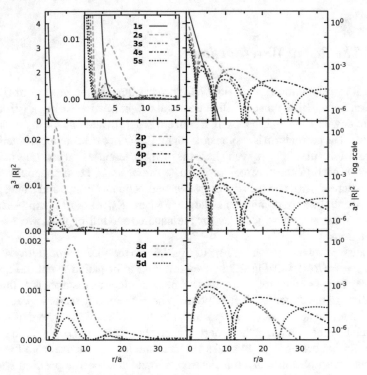

Fig. 2.6 Hydrogen ($Z = 1$) s p and d squared radial eigenfunctions $|R_{nl}(r)|^2$, for $n = 1$ to 5. Due to squaring, the $(n - l - 1)$ radial nodes, where $R_{nl}(r)$ vanishes and changes sign, appear as tangencies to the horizontal axis of the linear plots (*left*), and as downward kinks in the log-scale plots (*right*). The inset of the first panel zooms in the small-probability region

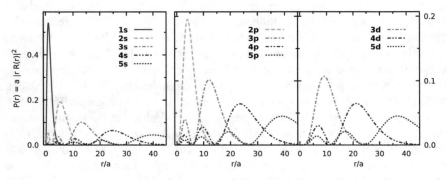

Fig. 2.7 Hydrogen ($Z = 1$) s p and d radial probability-distribution functions $P(r) = |r\, R_{nl}(r)|^2$.

For increasing nuclear charge Z, $|R_{nl}(r)|^2$ and thus $P(r)$ move in closer and closer to the origin[3]: the mean electron-nucleus separation decreases as $\propto Z^{-1}$, and this fact combined with $V_{ne} \propto Z/r$ explains the $\propto Z^2$ dependence of the eigenenergies (2.10). The simplest radial wavefunction, the one of the ground state, exemplifies well this Z dependence:

$$R_{10}(r) = \sqrt{\frac{k^3}{2}} e^{-kr/2} = 2\left(\frac{Z}{a}\right)^{3/2} \exp\left(-\frac{Zr}{a}\right). \qquad (2.21)$$

Accordingly, the ground state of He^+ has the same overall shape but half the size of that of H.

Notation: the hydrogenic kets/eigenfunctions $|n, l, m_l\rangle$ of Eq. (2.16) are often shorthanded as $n[l]$, where n is the principal quantum number and $[l]$ is the relevant letter s, p, d, ... for that value of l. For example, 4p refers to any of $\psi_{41-1}, \psi_{410}, \psi_{411}$. This notation is both incomplete and ambiguous, because (i) information about m_l is lacking, and (ii) the same 4p symbol implies different radial dependences $R_{41}(r)$ for nuclei with different charge and/or mass.

2.1.4 Orbital Angular Momentum and Magnetic Dipole Moment|

The angular momentum of an orbiting charged particle such as an electron is associated to a magnetic dipole moment. This is illustrated (Fig. 2.8) for a classical point particle with mass m and charge q rotating along a circular orbit of radius r at speed v. Its angular momentum $\mathbf{L} = \mathbf{r} \times \mathbf{p} = m\, r\, v\, \hat{\mathbf{n}}$, where $\hat{\mathbf{n}}$ is the unit vector perpendicular to its trajectory. As the rotation period is $\tau = 2\pi r/v$, the current along the loop $I = q/\tau = q\, v/(2\pi r)$. The magnetic moment of a ring current equals the product of the current times the loop area:

$$\boldsymbol{\mu} = I\, \pi r^2 \hat{\mathbf{n}} = \frac{qv}{2\pi r}\pi r^2 \hat{\mathbf{n}} = \frac{q}{2}\, v\, r\, \hat{\mathbf{n}} = \frac{q}{2m}\mathbf{L}. \qquad (2.22)$$

This equality can be shown to hold for an arbitrary shape of the periodic orbit.

Relation (2.22) holds even in QM, as an operatorial relation. For an electron (charge $q = -q_e$), where the angular momentum is quantized in units of $\hbar$, it is convenient to write Eq. (2.22) as

---

[3]The scaling laws are $R_{nl}^{[Z]}(r) = Z^{3/2} R_{nl}^{[1]}(rZ)$, and $P^{[Z]}(r) = Z\, P^{[1]}(rZ)$.

Fig. 2.8 The relation
between the mechanical
angular momentum **L** and
the magnetic moment $\boldsymbol{\mu}$
generated by an electron of
charge $-q_e$ orbiting
circularly. The curved lines
represent the magnetic
induction field **B** produced
by the circulating current

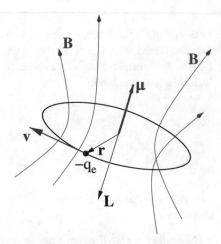

$$\boldsymbol{\mu} = -\frac{q_e}{2m_e}\mathbf{L} = -\frac{\hbar q_e}{2m_e}\frac{\mathbf{L}}{\hbar} = -g_l\,\mu_\mathrm{B}\,\frac{\mathbf{L}}{\hbar}, \qquad (2.23)$$

where the *Bohr magneton* $\mu_\mathrm{B} = \hbar q_e/(2m_e) \simeq 9.27401 \times 10^{-24}$ J T^{-1} (alias A m^2) is the natural scale of atomic magnetic moments. $g_l = 1$ is the orbital g-factor,[4] introduced for uniformity of notation with those situations, discussed below, with nontrivial proportionality factors $g \neq 1$ between $\mu_\mathrm{B}\,\mathbf{L}/\hbar$ and $\boldsymbol{\mu}$. Such g-factors arise because, while angular momenta are universal (i.e. simple algebraic multiples of $\hbar$), magnetic moments can be non-universal multiples of μ_B.

The atomic angular momenta can be detected by letting the associated magnetic moments interact with a magnetic field. If the field **B** is uniform, it induces a precession of $\boldsymbol{\mu}$ around the direction of **B** with a frequency (the Larmor frequency) $\omega = q_e B/(2m_e)$, routinely detected in microwave resonance experiments. If the field is nonuniform instead, a net force acts on the atom, as we discuss in the next section.

2.1.5 The Stern-Gerlach Experiment

The interaction energy of a magnetic moment with a magnetic field is

$$H_{\mathrm{magn}} = -\boldsymbol{\mu} \cdot \mathbf{B}. \qquad (2.24)$$

---

[4] The counter-current produced by the orbiting motion of the positively-charged nucleus around the common center of mass, decreases g_l to $1 - m_e/M$. We will ignore this small correction.

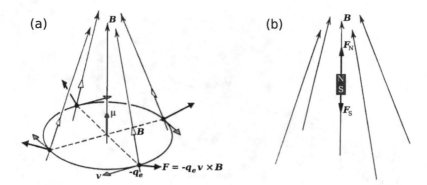

Fig. 2.9 The origin of the force that a nonuniform magnetic field produces on a magnetic dipole. **a** If the dipole is seen as a circulating current, the net force originates from a force component consistently pointing in the direction of increasing **B**. **b** If the dipole is viewed as a pair magnetic monopoles, a net force arises from the unbalance between the forces on the individual monopoles

Unless some external mechanism alters the angle between μ and **B**, this energy is conserved in time. A force arises on the magnetic dipole when the field **B** is nonconstant in space:

$$\mathbf{F} = -\nabla(-\mu \cdot \mathbf{B}) = \nabla(\mu \cdot \mathbf{B}). \tag{2.25}$$

In particular, in a magnetic field with a dominant B_z component, the z force component F_z is proportional to the derivative of B_z along the same direction:

$$F_z \simeq \mu_z \nabla_z B_z = \mu_z \frac{\partial B_z}{\partial z}. \tag{2.26}$$

Fig. 2.9 illustrates the "microscopical" origin of this force. The observation that a nonuniform magnetic field produces a force proportional to a magnetic-moment component is at the basis of the Stern-Gerlach experiment.

As illustrated in Fig. 2.10, a collimated beam of neutral atoms at thermal speeds is emitted from an oven into a vacuum chamber: there it traverses a region where a strongly inhomogeneous magnetic field produces a force F_z which deflects individual atoms proportionally to their magnetic-moment z component. The atoms are eventually collected at a suitable detector. Basically, the Stern-Gerlach apparatus is an instrument for measuring the component of the atomic magnetic moment in the field-gradient direction. This makes it one of the key tools in QM, as detailed in the initial sections of Ref. [17]. The original (1922) experiment was carried out using Ag atoms, but a similar pattern of deflections is observed using atomic H.

The main outcome of the Stern-Gerlach experiment is that the z component of μ is not distributed continuously as one would expect for a component of a classical vector pointing randomly in space, but rather peaked at discrete values. Figure 2.10c shows the observed clustering of the deflected H atoms in two lumps.

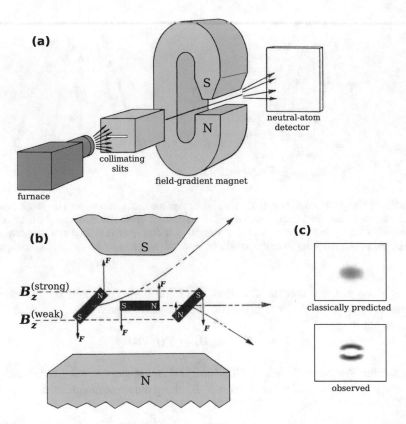

Fig. 2.10 **a** In the Stern-Gerlach experiment, a collimated beam of atoms emitted from an oven travels a region of inhomogeneous magnetic field generated by a magnet with asymmetric core expansions: the outcoming atoms are detected at a collector plate. **b** In an inhomogeneous magnetic field, a magnet experiences a net force which depends on its orientation. **c** The deflection pattern recorded on the detecting plate in a Stern-Gerlach measurement of the z component of the magnetic dipole moment of Ag atoms (the outcome would be the same for H atoms). Contrary to the classical prediction of an even distribution of randomly-oriented magnetic moments, two discrete components are observed, due to the quantization of an angular-momentum component

According to QM, the $\hat{\mathbf{z}}$-component L_z of angular momentum (and thus μ_z of magnetic moment) should indeed exhibit discrete eigenvalues. However, (i) the number of eigenvalues of L_z must be an *odd* integer $2l + 1$—see Eq. (2.9), and (ii) the ground state of hydrogen has $l = 0$, thus H should show no magnetic moment at all, and one undeflected lump should be observed, rather than two deflected lumps. The Stern-Gerlach evidence hints at some extra degree of freedom which must play a role in the one-electron atom.

2.1.6 Electron Spin

The outcome of the Stern-Gerlach experiment, the multiplet fine structure of the spectral lines (e.g the fine doublets of Fig. 2.3), and the Zeeman splitting of the spectral lines (see Sect. 2.1.10 below) are three pieces of evidence pointing at the existence of an extra degree of freedom of the electron, beside its position in space. W. Pauli introduced a nonclassical internal degree of freedom, later named *spin*, with properties similar to orbital angular momentum. Even though this picture is imprecise, the electron spin may be viewed as the intrinsic angular momentum of the rotation of the electron around itself. When spin is measured along a given direction, say $\hat{z}$, one detects eigenvalues of S_z, $\hbar m_s$, where the quantum number m_s takes $(2s + 1)$ values $m_s = -s, \ldots s$, like the orbital L_z takes the $(2l + 1)$ values $m_l = -l, \ldots l$. Since a Stern-Gerlach deflector splits an H beam into *two* lumps, $2 = 2s + 1$ components are postulated, requiring that the intrinsic angular momentum of the electron should be $s = 1/2$. In turn, this quantum number s is associated to a squared spin angular-momentum operator $|S|^2$ whose eigenvalue is $1/2 (1/2 + 1)\hbar^2 = 3/4 \hbar^2$.

A full wavefunction, that carries complete information on all degrees of freedom of the electron, is slightly more complicated than $R(r) Y_{lm_l}(\theta, \varphi)$: an extra spin dependence must be inserted. Assuming, as apparent from the nonrelativistic Hamiltonian (1.1), that spin and orbital motions do not interact, the full eigenfunction of a one-electron atom with spin pointing either up ($\uparrow$, i.e. $m_s = 1/2$) or down ($\downarrow$, i.e. $m_s = -1/2$) along a fixed orientation is written

$$\psi_{n l m_l m_s}(r, \theta, \varphi, \sigma) = R_{n l}(r) Y_{l m_l}(\theta, \varphi) \chi_{m_s}(\sigma). \tag{2.27}$$

Here σ is the variable for the spin degree of freedom, which spans values $\pm 1/2$, in checking if the electron spin points up or down in the $\hat{z}$ direction. The quantum number $m_s = \pm 1/2$ indicates which way the spin of this specific state is actually pointing relative to the reference direction. These basis spin functions are therefore simply $\chi_{m_s}(\sigma) = \langle \sigma | m_s \rangle = \delta_{m_s \sigma}$.

Less trivial spin wavefunctions arise when the spin points in some direction other than $\hat{z}$ (non-S_z eigenstates). A Stern-Gerlach apparatus can be employed to purify a *spin polarized* beam of atoms with spins pointing in some oblique direction. One such beam can then be analyzed by a second apparatus to measure the spin component σ along the reference $\hat{z}$ direction. For the oblique-spin state, the (now nontrivial) spin wavefunction $\chi(\sigma)$ bears the standard significance of a wavefunction in QM: $|\chi(\uparrow)|^2$ is the probability that, when measuring S_z, one observes $+1/2\hbar$, and likewise $|\chi(\downarrow)|^2$ is the probability to observe $-1/2\hbar$. When a $\hat{z}$-Stern-Gerlach measurement is carried out, the spin z component is necessarily found pointing either up or down, therefore the total probability $\sum_\sigma |\chi(\sigma)|^2 = |\chi(\uparrow)|^2 + |\chi(\downarrow)|^2 = 1$.

A remarkable feature of spin is that the separation of the $\uparrow$ and $\downarrow$ sub-beams in a Stern-Gerlach apparatus is consistent with a g-factor $g_s \simeq 2$, quite distinct from the orbital $g_l = 1$. The precise value $g_s = 2.00232$, fixing the electron intrinsic magnetic moment projection to $\pm 1/2 g_s \mu_B$, is measured extremely accurately by electron spin resonance (ESR) experiments, where the electron spin interacts with a uniform magnetic field which splits the energies of the $|\uparrow\rangle$ and $|\downarrow\rangle$ states.

At the present basic level of understanding, the electron spin is just an extra quantum number which, in the absence of magnetic fields, only provides an extra twofold degeneracy to all the levels of the one-electron atom: the total degeneracy of the n-th level is $2n^2$, rather than n^2. In fact, spin will affect the atomic levels when magnetic effects are included, as discussed in the following sections.

2.1.7   Fine Structure

Tiny (<0.1 meV) splittings, see Fig. 2.3, in the spectrum of H hint at some small correction, likely due to the relativistic effects neglected in the original Hamiltonian (1.1). We come to investigate these corrections in detail.

2.1.7.1   Spin-Orbit Coupling

Consider first the action of the magnetic field experienced by the electron spin due to its own orbital motion. This is a subtle relativistic effect, due to the Lorentz transformation of the nuclear electric field into the frame of reference of the electron. Call $\mathbf{v}$ the electron velocity in the nuclear rest frame. In the electron frame of reference, the nucleus is seen to move with velocity $-\mathbf{v}$, and thus carries a "current element" $-Zq_e\mathbf{v}$. According to the Biot-Savart law of electromagnetism, this moving charge generates a magnetic field

$$\mathbf{B}(\mathbf{r}) = -\frac{1}{4\pi\epsilon_0 c^2}\frac{\mathbf{r}\times(-Zq_e\mathbf{v})}{|\mathbf{r}|^3} = \frac{\mathbf{E}(\mathbf{r})\times\mathbf{v}}{c^2} \tag{2.28}$$

at the point where the electron sits. In this formula $\mathbf{r}$ is the vector reaching from the nucleus to the electron, Eq. (2.1). Equation (2.28) identifies this magnetic field as a relativistic effect [of order $(v/c)^2$], and expresses it in terms of the electric field generated by the nucleus at the electron location

$$\mathbf{E}(\mathbf{r}) = \frac{Zq_e}{4\pi\epsilon_0}\frac{\mathbf{r}}{|\mathbf{r}|^3}. \tag{2.29}$$

In Eq. (2.28) we recognize the orbital angular-momentum operator:

$$\mathbf{B}(\mathbf{r}) = \frac{Zq_e}{4\pi\epsilon_0 c^2}\frac{\mathbf{r}\times\mathbf{v}}{|\mathbf{r}|^3} = \frac{Zq_e}{4\pi\epsilon_0 c^2 m_e}\frac{\mathbf{L}}{|\mathbf{r}|^3}. \tag{2.30}$$

This magnetic field $\mathbf{B}(\mathbf{r})$ acts on the electron as it moves through space. By analogy with Eq. (2.24), the interaction energy of the electron-spin magnetic moment $\boldsymbol{\mu}_s$

with $\mathbf{B}$ should equal $-\boldsymbol{\mu}_s \cdot \mathbf{B}(\mathbf{r})$. However, this energy must actually be reduced by a factor $1/2$ (first recognized by L.H. Thomas) due to the electron frame of reference being accelerated [1]. The correct magnetic interaction energy operator is therefore:

$$H_{\text{s-o}} = -\frac{1}{2}\boldsymbol{\mu}_s \cdot \mathbf{B}(\mathbf{r}) = \frac{1}{2} g_s \mu_B \frac{\mathbf{S}}{\hbar} \cdot \left(\frac{Z q_e}{4\pi\epsilon_0 c^2 m_e} \frac{\mathbf{L}}{|\mathbf{r}|^3}\right) = \frac{Z e^2}{2 m_e^2 c^2} \frac{1}{r^3} \mathbf{S} \cdot \mathbf{L}.$$
(2.31)

This operator, named *spin-orbit interaction*, exhibits nonzero but tiny diagonal and off-diagonal elements $\langle n, l, m_l, m_s | H_{\text{s-o}} | n', l, m_l', m_s' \rangle$ connecting states with equal or different n, m_l, m_s. States with $n \neq n'$ have vastly different nonrelativistic energies (2.10): the n-off-diagonal spin-orbit couplings perturb these energies negligibly (see Appendix C.5.2 and C.9), and we safely ignore them. Considering only the n-diagonal matrix elements of $H_{\text{s-o}}$, we rewrite Eq. (2.31) as

$$H_{\text{s-o}} \simeq \frac{Z e^2}{2 m_e^2 c^2} \sum_{n,l} |n, l\rangle \langle n, l | r^{-3} | n, l \rangle \langle n, l | \, \mathbf{S} \cdot \mathbf{L}.$$
(2.32)

The radial average of r^{-3} can be evaluated for $l \geq 0$ hydrogenic wavefunctions R_{nl}, obtaining

$$\langle n, l | r^{-3} | n, l \rangle = \int_0^\infty r^{-3} [R_{nl}(r)]^2 r^2 \, dr = \left(\frac{Z}{a}\right)^3 \frac{2}{n^3 l(l+1)(2l+1)}.$$
(2.33)

The spin-orbit Hamiltonian is thus conveniently rewritten as

$$H_{\text{s-o}} = \sum_n \sum_{l=1}^{n-1} \xi_{nl} |n, l\rangle \langle n, l | \frac{\mathbf{S} \cdot \mathbf{L}}{\hbar^2},$$
(2.34)

where the projectors $|n, l\rangle\langle n, l|$ select the radial wavefunction of the $n[l]$ orbital, and the spin-orbit energy prefactor is

$$\xi_{nl} = \frac{Z e^2 \hbar^2}{2 m_e^2 c^2} \left(\frac{Z}{a}\right)^3 \frac{2}{n^3 l(l+1)(2l+1)} = Z^4 \alpha^2 E_{\text{Ha}} \frac{(\mu/m_e)^3}{n^3 l(l+1)(2l+1)}.$$
(2.35)

The last equality exploits the expression for the mass-rescaled atomic length scale $a = a_0 m_e/\mu$, the definition (1.9) of the Hartree energy, and the expression $\alpha^2 = E_{\text{Ha}}/(m_e c^2)$ for the fine-structure constant. In this form, it is apparent that the typical spin-orbit energy scale ξ_{nl}

- is positive, and therefore $H_{\text{s-o}}$ favors *antiparallel* alignment of $\mathbf{L}$ and $\mathbf{S}$;
- is a leading $\alpha^2 \simeq (v/c)^2$ relativistic correction;
- is $\alpha^2 \simeq 5.3 \times 10^{-5}$ times smaller than the typical orbital energies;

- grows as Z^4, reflecting the increase in nuclear field intensity $\propto Z$ and the reduction $\propto Z^{-1}$ of the average electron-nucleus distance so that $\langle n, l | r^{-3} | n, l \rangle \propto Z^3$;
- decreases as n^{-3}, reflecting the increase $\propto n$ of the average electron-nucleus distance, and the r^{-3} dependence of the interaction energy (2.31);
- decreases roughly as l^{-3}, due to the r^l suppression of the radial wavefunction close to the origin, which is where the spin-orbit interaction (2.31) dominates.

The energy scale of ξ_{nl} amounts to $\alpha^2 E_{Ha} (\mu/m_e)^3 \simeq 2.318 \times 10^{-22} \, \text{J} = 1.447 \, \text{meV}$ for hydrogen. Note however that the lowest-energy subshell for which spin-orbit is relevant, namely 2p, has $n^3 l(l+1)(2l+1) = 48$, thus $\xi_{2p} = 0.0301$ meV only. For all higher levels ξ_{nl} is even smaller.

The spin-orbit interaction (2.34) is small because it is proportional to α^2, i.e. second-order in α. We treat this small perturbation with the methods of *first-order* (in α^2) perturbation theory for degenerate unperturbed states. To do this, consider the remaining operatorial part in Eq. (2.34): $\mathbf{S} \cdot \mathbf{L}/\hbar^2$. In the basis $|l, s, m_l, m_s\rangle$ where spin and orbital motion are uncoupled, the operator $\mathbf{S} \cdot \mathbf{L}$ has plenty of nonzero off-diagonal matrix elements. In computing the first-order-corrected eigenvalues and eigenstates with the methods of Appendix C.9, one needs to pre-diagonalize $\mathbf{S} \cdot \mathbf{L}$, within each (0-th order degenerate) space at fixed l and s.

We couple spin and orbital angular momenta with the methods of Appendix C.8.1: the $|l, s, j, m_j\rangle$ coupled basis introduced in Eq. (C.73) does the trick: $\mathbf{S} \cdot \mathbf{L}$ is diagonal in this coupled basis. To prove this result, take the square of $(\mathbf{L} + \mathbf{S}) = \mathbf{J}$ and invert it as follows:

$$\mathbf{S} \cdot \mathbf{L} = \frac{|\mathbf{J}|^2 - |\mathbf{S}|^2 - |\mathbf{L}|^2}{2}. \tag{2.36}$$

All operators at the right hand side are of course diagonal in the $|l, s, j, m_j\rangle$ basis, and so is $\mathbf{S} \cdot \mathbf{L}/\hbar^2$. Following Eq. (2.36), the expression for its matrix elements

$$\langle l, s, j, m_j | \frac{\mathbf{S} \cdot \mathbf{L}}{\hbar^2} | l, s, j', m_j' \rangle = \frac{j(j+1) - s(s+1) - l(l+1)}{2} \delta_{jj'} \delta_{m_j m_j'}. \tag{2.37}$$

In summary: in the coupled basis $|l, s, j, m_j\rangle$, the spin-orbit interaction is diagonal and its corrections to the eigenvalues are given by ξ_{nl}, Eq. (2.35), multiplied by the diagonal expression in Eq. (2.37).

Notation: states of the coupled basis are commonly indicated with the *term symbol* $^{2s+1}[l]_j$, where $[l]$ is the capital letter S, P, D... appropriate for that value of $l = 0, 1, 2 \ldots$ Information about n is encoded elsewhere, e.g. in the $n[l]$ notation, and information about m_j is usually omitted. For example, 3d $^2D_{3/2}$ stands for any of the four $|n = 3, l = 2, j = 3/2, m_j\rangle$ kets, with $m_j = -3/2, -1/2, 1/2,$ or $3/2$.

As an example of spin-orbit-split states, for a p level of a one-electron atom, the $\mathbf{S} \cdot \mathbf{L}/\hbar^2$ operator has two possible eigenvalues

$$\langle 1, 1/2, j, m_j | \frac{\mathbf{S} \cdot \mathbf{L}}{\hbar^2} | 1, 1/2, j, m_j \rangle = \begin{cases} -1 & \text{for } j = 1/2 \\ +1/2 & \text{for } j = 3/2 \end{cases}.$$

Accordingly, the spin-orbit coupling splits the $3 \times 2 = 6$ orbital$\times$spin states of a p level into a doublet $^2P_{1/2}$ plus a quartet $^2P_{3/2}$, separated by an energy $3/2\,\xi_{n1}$. For the 2p level of hydrogen this splitting amounts to 45.2 μeV. Likewise, the spin-orbit coupling splits all non-s levels into two finely-separated levels corresponding to $j = l \pm 1/2$. This mechanism provides an explanation for the observed twofold-split fine structure of the spectral lines of 1-electron atoms, Fig. 2.3.

2.1.7.2 The Relativistic Kinetic Correction

We need to consider a second relativistic correction of the same order $(v/c)^2$ as spin-orbit. This energy contribution accounts for the *leading correction to the kinetic-energy* expression $p^2/(2\mu)$. We expand the exact relativistic kinetic energy as

$$T_r = \sqrt{\mu^2 c^4 + p^2 c^2} - \mu c^2 = \mu c^2 \left(1 + \frac{1}{2}\frac{p^2}{\mu^2 c^2} - \frac{1}{8}\frac{p^4}{\mu^4 c^4} + \cdots - 1 \right)$$

$$= \frac{p^2}{2\mu} - \frac{p^4}{8\mu^3 c^2} + \cdots \tag{2.38}$$

Like for $H_{s\text{-}o}$, to treat the weak perturbation $-p^4/(8\mu^3 c^2)$ at first order, we just need the n-diagonal matrix elements of this operator. Although p^4 looks like a formidable differential operator, the trick $p^4 = (p^2)^2 = [2\mu(H_{\text{Coul}} - V_{ne})]^2$ allows us to rewrite the diagonal matrix elements of p^4 in terms of simple radial integrals of r^{-1} and r^{-2}. The final result is

$$\langle n, l| - \frac{p^4}{8\mu^3 c^2} |n, l\rangle = -\frac{Z^4 \alpha^2}{n^3} E_{\text{Ha}} \left(\frac{\mu}{m_e}\right)^3 \left(\frac{1}{2l+1} - \frac{3}{8n}\right), \tag{2.39}$$

where we omit either m_l, m_s or j, m_j, which are irrelevant for such radial integrals.

By combining the spin-orbit and kinetic correction

$$H_{\text{rel}} = H_{s\text{-}o} - \frac{p^4}{8\mu^3 c^2}, \tag{2.40}$$

we obtain the diagonal matrix elements of the total relativistic correction to order α^2:

$$\langle n, l, j|H_{\text{rel}}|n, l, j\rangle = \frac{Z^4 \alpha^2}{n^3} E_{\text{Ha}} \left(\frac{\mu}{m_e}\right)^3$$

$$\times \left[\frac{j(j+1) - s(s+1) - l(l+1)}{2l(l+1)(2l+1)} - \frac{1}{2l+1} + \frac{3}{8n}\right]$$

$$= -\frac{Z^4 \alpha^2}{n^3} E_{\text{Ha}} \left(\frac{\mu}{m_e}\right)^3 \left(\frac{1}{2j+1} - \frac{3}{8n}\right), \tag{2.41}$$

where the final simplification is based on spin being $s = 1/2$, thus $j = l \pm 1/2$. A separate derivation shows that this Eq. (2.41) (unlike all intermediate steps) holds for s states too.

Expression (2.41) can be combined with the nonrelativistic eigenvalues (2.10) to obtain the following expression for the energy eigenvalues, correct to order α^2:

$$\langle n, l, j | H_{\text{tot}} + H_{\text{rel}} | n, l, j \rangle = -\frac{E_{\text{Ha}}}{2} \frac{\mu}{m_e} \frac{Z^2}{n^2} \left[1 + \left(Z\alpha \frac{\mu}{m_e} \right)^2 \frac{1}{n} \left(\frac{2}{2j+1} - \frac{3}{4n} \right) \right]. \quad (2.42)$$

This remarkable relation yields a quantitative prediction for the spectrum that can be directly compared to experiment: all n-levels should be split as j takes different values, but not, for given n and j, when l takes different values, e.g. $^2S_{1/2}$ and $^2P_{1/2}$ should remain degenerate. This extra l-degeneracy is retained even in the solutions of Dirac's equation, which is exact to all orders in α, not just α^2 as Eq. (2.42).

2.1.7.3   The Lamb Shift

As the l-degeneracy is rather surprising, the possibility of an energy splitting between states with equal n and j, but different l, was investigated closely, both theoretically and experimentally. Eventually, quantum fluctuations of the electromagnetic field and the finite nuclear size lift this degeneracy, introducing tiny splittings named *Lamb shift*. Fig. 2.11b reports the expected spectral fine structure of the Balmer Hα line, including both the relativistic corrections and the Lamb shift.

Due mainly to Doppler broadening (see Sect. 1.2), the spectral lines are usually not resolved sharply enough for these tiny energy differences. To circumvent Doppler broadening and acquire the high-resolution spectrum of Fig. 2.11a, the authors of Ref. [18] devised a trick based on double resonance. An intense tunable monochromatic light beam is split into a strong interruptible "pump" beam plus a second weak "probe" beam, Fig. 2.11c. When the light frequency matches a resonant transition, absorption takes place and the probe beam is attenuated, as in a regular absorption spectroscopy experiment. This absorption is strongly reduced if the pump beam happens to "saturate" the transition in the sample, with the mechanism discussed quantitatively in Sect. 4.4 below. The spectrum of Fig. 2.11a records the probe-beam absorption *difference* between time intervals when the pump beam is on and when it is off. All atoms with a sizable translational velocity component in the beam direction are Doppler shifted in opposite directions relative to the two beams, which are almost antiparallel: these atoms do not contribute to the difference signal. This double-resonance condition with identical pump and probe frequencies selects a subset of atoms, namely those with practically null instantaneous translational velocity component, thus negligible Doppler shift. This selection gets rid of the Doppler broadening, allowing the Lamb shift to be detected, Fig. 2.11a.

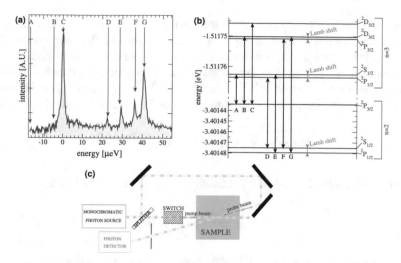

Fig. 2.11 **a** A direct spectroscopic observation of the Lamb shift in the Hα line. *Arrows* point at the predicted dipole-allowed transition energies. **b** A level scheme of the $n = 2$ and $n = 3$ levels involved in the Balmer Hα line, including the relativistic corrections and the Lamb shift [18]. Peaks labeled A−C involve the $n = 2$ $^2P_{3/2}$ state. Peaks labeled D−G involve the $n = 2$ $j = 1/2$ states. E.g., the separation between the B and G peaks measures the spin-orbit splitting in the 2p level: it agrees with the theoretical prediction of $\simeq 45$ μeV. The splitting between (D, E) and (F, G) is due to the spin-orbit coupling in the $n = 3$ multiplet. Instead, the splittings between (D and E) and between (F and G) is due to the Lamb shift, which is thus measured in the order of 5 μeV. Without the Lamb shift, the four D−G peaks would collapse to 2 peaks only. **c** A scheme of the double-resonance experiment [18] used to record the data of panel (**a**)

2.1.8 Nuclear Spin and Hyperfine Structure

Like electrons, nuclei have a "spin" angular momentum $\mathbf{I}$, and for many nuclei it does not vanish. For example, the proton has spin $I = 1/2$. Unsurprisingly, the nuclear magnetic moment $\boldsymbol{\mu}_n$ is proportional to its angular momentum:

$$\boldsymbol{\mu}_n = g_n \mu_N \frac{\mathbf{I}}{\hbar}.$$

The typical scale of nuclear magnetic moments is provided by the nuclear magneton μ_N, defined by analogy with the Bohr magneton, but with the proton mass replacing that of the electron:

$$\mu_N = \frac{q_e \hbar}{2m_p} = \mu_B \frac{m_e}{m_p}. \tag{2.43}$$

The nuclear g-factor g_n is a number of order unity, whose value depends on the inner nuclear structure. For example, the proton has $g_n \simeq 5.58569$.

At the same distance, the magnetic field generated by a nuclear spin is much weaker than that of an electron spin, because the former is suppressed by the ratio

$m_e/m_p \simeq 1/1836$. Through this field, the nuclear and electron magnetic moments interact. Due to the r^l term in Eq. (2.13), the electron hardly ever comes close to the nucleus when in $l > 0$ states: here we ignore the resulting extremely weak interaction. Focusing then on s orbitals, the only nonzero electronic magnetic moment is associated to the electron spin **S**. Like spin-orbit, the interaction Hamiltonian is proportional to the simplest scalar combination of the involved vector quantities:

$$H_{\text{SI}} = -C \mu_n \cdot \mu_e = C \, g_n g_s \, \mu_B^2 \, \frac{m_e}{m_p} \, \frac{\mathbf{I} \cdot \mathbf{S}}{\hbar^2}.$$

The coupling factor C is the relevant radial matrix element, which equals

$$C = \frac{2}{3} \frac{1}{4\pi\epsilon_0 c^2} \, |R_{n0}(0)|^2.$$

Substituting $R_{n0}(0) = 2[Z/(an)]^{3/2}$, the characteristic coupling energy

$$\xi_N = C g_n g_s \mu_B^2 \frac{m_e}{m_p} = \frac{2}{3} \, g_n g_s \, \frac{Z^3}{n^3 m_e c^2} \, E_{\text{Ha}}^2 \, \frac{m_e}{m_p}$$

$$= \frac{4}{3} \, g_n \, \frac{Z^3 \alpha^2}{n^3} \, \frac{m_e}{m_p} \, E_{\text{Ha}} \simeq g_n \frac{Z^3}{n^3} \times 1.05 \, \mu\text{eV}. \qquad (2.44)$$

In the second line, we also substituted α^2 for $E_{\text{Ha}}/(m_e c^2)$. For the ground state of ^{1}H, Eq. (2.44) yields $\xi_N \simeq 5.88$ μeV.

Following the general rules given in Appendix C.8.1, the electron and nuclear spins couple to a grand total angular momentum $\mathbf{F} = \mathbf{I} + \mathbf{S}$. By analogy with Eq. (2.36), we obtain

$$\mathbf{I} \cdot \mathbf{S} = \frac{|\mathbf{F}|^2 - |\mathbf{I}|^2 - |\mathbf{S}|^2}{2},$$

so that the expectation value of $\mathbf{I} \cdot \mathbf{S}/\hbar^2$ equals $[f(f+1) - i(i+1) - s(s+1)]/2$ in the coupled basis, the one where $|\mathbf{F}|^2$ and $|F_z|$, rather than I_z and S_z, are diagonal. As $s = 1/2$, two coupled states $f = i \pm 1/2$ occur, with an energy separation of $\xi_N (i + 1/2)$.

The nucleus of ^{1}H, the proton, has spin $i = 1/2$, so that $\hbar^{-2} \langle \mathbf{I} \cdot \mathbf{S} \rangle = -3/4$ and $1/4$ for $f = 0$ and 1, respectively. The separation between these two hyperfine-split states equals therefore $\xi_N \simeq 5.88$ μeV: in the electromagnetic spectrum this photon energy corresponds to a wavelength $2\pi\hbar c \, \xi_N^{-1} \simeq 21$ cm, and a frequency $\xi_N (2\pi\hbar)^{-1} \simeq 1.42$ GHz. This transition, in the radio-frequency range at 1420405751.8 Hz, was discovered in 1951 in astrophysical spectra, has been adopted as a frequency standard, and is now used to map the distribution of interstellar atomic ^{1}H.

2.1.9 *Electronic Transitions, Selection Rules*

Not all conceivable transitions are equally easy to observe. Experimentally, certain transitions proceed at a fast rate, while others occur immensely more slowly. This fact can be explained by a QM analysis of the interaction of the mechanical system with the electromagnetic field. The probability per unit time that an object (e.g. an atom) decays radiatively from an initial state $|i\rangle$ to a final state $|f\rangle$ is given by

$$\gamma_{if} = \frac{1}{3\pi\epsilon_0\hbar^4c^3}\,\mathcal{E}_{if}^3\,|\langle f|\mathbf{d}|i\rangle|^2, \tag{2.45}$$

where $\mathcal{E}_{if} = \hbar\omega_{if} = E_i - E_f$, and $\mathbf{d}$ is the operator describing its coupling to the radiation field. In the approximation that the radiating object is much smaller than the radiation wavelength (see Fig. 1.2), this operator is the electric-dipole operator $\mathbf{d} = -q_e\mathbf{r}$. All transitions for which the matrix element $\langle f|\mathbf{d}|i\rangle$ vanishes are "forbidden" in the *electric-dipole approximation*: this means that they occur at much lower rates, associated to higher-order terms in the multipolar expansion.

The matrix elements of the dipole operator of the one-electron wavefunction are:

$$\langle n_f, l_f, m_{lf}|\mathbf{d}|n_i, l_i, m_{li}\rangle = \int \psi^*_{n_f l_f m_{lf}}(\mathbf{r})\,\mathbf{d}\,\psi_{n_i l_i m_{li}}(\mathbf{r})\,d^3r. \tag{2.46}$$

This integration is carried out conveniently in spherical coordinates. We express the dipole operator as $\mathbf{d} = -q_e r\,(\sin\theta\cos\varphi,\ \sin\theta\sin\varphi,\ \cos\theta)$. By inverting Eq. (2.18), we observe that

$$r_x = r\sqrt{\frac{2\pi}{3}}\,\left[Y_{1-1}(\theta,\varphi) - Y_{11}(\theta,\varphi)\right]$$

$$r_y = r\,i\sqrt{\frac{2\pi}{3}}\,\left[Y_{1-1}(\theta,\varphi) + Y_{11}(\theta,\varphi)\right]$$

$$r_z = r\sqrt{\frac{4\pi}{3}}\,Y_{10}(\theta,\varphi).$$

As a result, the squared dipole matrix element is proportional to

$$|\langle f|\mathbf{r}|i\rangle|^2 = |\langle f|r_x|i\rangle|^2 + |\langle f|r_y|i\rangle|^2 + |\langle f|r_z|i\rangle|^2$$

$$= |\langle f|r|i\rangle|^2\,\frac{2\pi}{3}\left(|\langle f|Y_{1-1} - Y_{11}|i\rangle|^2 + |\langle f|Y_{1-1} + Y_{11}|i\rangle|^2 + 2|\langle f|Y_{10}|i\rangle|^2\right)$$

$$= |\langle f|r|i\rangle|^2\,\frac{4\pi}{3}\left(|\langle f|Y_{1-1}|i\rangle|^2 + |\langle f|Y_{11}|i\rangle|^2 + |\langle f|Y_{10}|i\rangle|^2\right),$$

with factored radial and angular matrix elements. Explicitly, in terms of the wave-functions

$$\left|\langle n_f, l_f, m_{l\,f}\,|\,\mathbf{d}\,|n_i, l_i, m_{l\,i}\rangle\right|^2 = q_e^2 \left|\int_0^\infty R_{n_f l_f}(r)\, r\, R_{n_i l_i}(r)\, r^2\, dr\right|^2 \times \qquad (2.47)$$

$$\frac{4\pi}{3} \sum_m \left|\int_0^\pi \sin\theta\, d\theta \int_0^{2\pi} d\varphi\, Y^*_{l_f\, m_{l\,f}}(\theta, \varphi) Y_{1\,m}(\theta, \varphi) Y_{l_i\, m_{l\,i}}(\theta, \varphi)\right|^2 .$$

The radial matrix element of r can be computed analytically between any initial $R_{n_i l_i}(r)$ and any final $R_{n_f l_f}(r)$: nonzero values are obtained for arbitrary n_i, l_i, n_f, l_f. For increasing difference $|n_i - n_f|$, the absolute value of this radial integral decreases rapidly toward zero, because $R_{n_i l_i}(r)$ and $R_{n_f l_f}(r)$ take sizable values at increasingly distant regions of the r line, so that the product $R_{n_i l_i}(r)\, R_{n_f l_f}(r)$ is small everywhere.

The angular factor yields much stricter results. The integration in the second row of Eq. (2.47) represents the angular overlap of a state with angular momentum $l = l_f$ with the product $Y_{1\,m}\, Y_{l_i\, m_{l\,i}}$. By applying the rule (C.72) of angular-momentum coupling, this product of objects carrying angular momenta $l = 1$ and $l = l_i$ can be decomposed[5] into the following values of the total coupled angular momentum:

$$l_{\text{tot}} = |l_i - 1|,\ l_i,\ l_i + 1. \qquad (2.48)$$

Final states characterized by angular momentum l_f *not* equaling one of the l_{tot} values in Eq. (2.48) are guaranteed to make the angular integral vanish. Moreover, the angular integral vanishes also when $l_f = l_i$. The reason is that the parity of $Y_{1\,m}$, i.e. the parity of $\mathbf{r}$, is $(-1)^1 = -1$, while the parity of $Y^*_{l_i\, m_{l\,f}}\, Y_{l_i\, m_{l\,i}}$ is $(-1)^{2l_i} = 1$. As a result, the parity of the integrated product function $Y^*_{l_i\, m_{l\,f}}\, Y_{1\,m}\, Y_{l_i\, m_{l\,i}}$ is odd, and therefore its integration over the entire solid angle vanishes.

In conclusion, in the dipole approximation, a nonzero matrix element can occur only for transitions involving states with l increasing or decreasing by exactly unity. The *allowed transitions* have

$$\Delta l = l_f - l_i = \pm 1. \qquad (2.49)$$

This equality represents the *electric-dipole selection rule* regarding l.

The fact that the dipole operator is associated to an orbital $l = 1$ implies also that its component $m = -1, 0, 1$. Accordingly, the only values of $m_{l\,f}$ for which the angular integral is nonzero, are obtained by adding m to $m_{l\,i}$. From this observation we formulate the m_l-selection rule:

---

[5]The angular integral within the absolute value in Eq. (2.47) can be shown to be proportional to the Clebsch-Gordan coefficient $C^{l_f m_{l\,f}}_{l_i\, m_{l\,i}\ 1\,m}$, see Eq. (C.73).

$$\Delta m_l = m_{l\,f} - m_{l\,i} = 0, \pm 1. \tag{2.50}$$

Until this point, we ignored spin, because the dipole operator is purely spatial: it does nothing to spin, thus it acts as the identity of spin space. In the uncoupled basis $|n, l, m_l, m_s\rangle$, **d** only affects the spatial degrees of freedom. As a result,

$$\Delta s = s_f - s_i = 0, \tag{2.51}$$
$$\Delta m_s = m_{s\,f} - m_{s\,i} = 0. \tag{2.52}$$

For a one-electron atom where $s \equiv 1/2$ anyway, the first spin selection rule (2.51) is trivial, but it will become relevant for many-electron atoms. The second rule (2.52) informs us that spin does not flip in electric-dipole allowed transitions.

By analyzing the composition of the coupled-basis states $|n, l, j, m_j\rangle$ in terms of the $|n, l, m_l, m_s\rangle$ uncoupled basis, see Eq. (C.73), one obtains the following selection rules for the coupled states:

$$\Delta j = j_f - j_i = 0, \pm 1 \tag{2.53}$$

$$\Delta m_j = m_{j\,f} - m_{j\,i} = 0, \pm 1. \tag{2.54}$$

After determining which transitions are allowed in the dipole approximation, we make use of Eq. (2.45) to evaluate typical rates of allowed radiative atomic transitions. We note that (i) $\mathcal{E}_{if} \approx Z^2 E_{Ha}$ and (ii) the order of magnitude of $|\langle f|\mathbf{d}|i\rangle| \approx q_e a_0/Z$ (as long as n_f and n_i are not too distant). We can estimate

$$\gamma_{if} = \frac{\mathcal{E}_{if}^3 \, |\langle f|\mathbf{d}|i\rangle|^2}{3\pi\epsilon_0 \hbar^4 c^3} \simeq \frac{Z^4}{\epsilon_0 \hbar^3 c^3} E_{Ha}^2 \, \omega_{if} \, q_e^2 \frac{a_0^2}{Z^2} \simeq \frac{e^2 Z^2}{(\hbar c)^3} e^4 \omega_{if} = Z^2 \left(\frac{e^2}{\hbar c}\right)^3 \omega_{if}$$
$$= Z^2 \alpha^3 \omega_{if}, \tag{2.55}$$

where we have dropped factors of order unity, and used $E_{Ha} a_0 = e^2$ and $\alpha = e^2/(\hbar c)$. As $\alpha^3 \simeq 10^{-7}$ and in one-electron atoms $\omega_{if} = \mathcal{E}_{if}/\hbar \simeq Z^2 \, 10^{16}$ Hz, we expect typical radiative transition rates in the order $\gamma_{if} \simeq Z^4 \, 10^9 \, s^{-1}$, i.e. decay times $\gamma_{if}^{-1} \simeq Z^{-4}$ ns. The rapid energy dependence $\gamma_{if} \propto \mathcal{E}_{if}^3$ makes this decay rate faster for more energetic transitions, and slower for low-energy transitions. *Dipole-forbidden transitions* proceed at far slower rates, associated to weaker higher-order multipoles of the electromagnetic field (magnetic dipole, electric quadrupole...).

Other *nonradiative* transitions can also occur due to collisions with other atoms in the gaseous sample and/or with the vessel walls. Nonradiative mechanisms can dominate the decay of long-lived *metastable* states, namely those lacking fast dipole-allowed decay transitions.

2.1.10   Spectra in a Magnetic Field

We have now all conceptual tools needed for analyzing the atomic spectra in the condition where a maximum of information can be extracted from them: namely when the atomic sample is immersed in a uniform magnetic field. In this condition, the atom, depending on the component of its magnetic moment along the field direction, acquires a little extra energy which can then be detected by spectroscopy.

In the presence of both orbital and spin angular momenta, the total atomic magnetic moment is the vector sum

$$\boldsymbol{\mu} = \boldsymbol{\mu}_l + \boldsymbol{\mu}_s = -\mu_B \frac{g_l \mathbf{L} + g_s \mathbf{S}}{\hbar} \simeq -\mu_B \frac{\mathbf{L} + 2\mathbf{S}}{\hbar}, \qquad (2.56)$$

involving both spin and orbital magnetic moments with different coefficients. Following Eqs. (2.24) and (2.56), the coupling with the external magnetic field can be expressed as:

$$H_{\text{magn}} = -\mathbf{B} \cdot \boldsymbol{\mu} = \mu_B \, \mathbf{B} \cdot \frac{\mathbf{L} + 2\mathbf{S}}{\hbar} = \mu_B \, B_z \, \frac{L_z + 2S_z}{\hbar}, \qquad (2.57)$$

taking the $\hat{\mathbf{z}}$ axis aligned with $\mathbf{B}$. This operator is diagonal in the uncoupled $|l, s, m_l, m_s\rangle$ basis. We recall (Sect. 2.1.7) that $H_{\text{s-o}}$ is not diagonal in that basis, but rather in the coupled basis $|l, s, j, m_j\rangle$. Unfortunately, $H_{\text{s-o}}$ and H_{magn} cannot be diagonalized simultaneously, because they do not commute. To obtain the eigenenergies and eigenkets of $H_{\text{s-o}} + H_{\text{magn}}$ one must diagonalize it, within each $(2l + 1) \cdot (2s + 1)$-dimensional subspace at fixed n, l, and s. This diagonalization is not especially complicated, but it is perhaps more instructive to understand in detail the limiting cases where one of the characteristic energy scales, either $\mu_B \, |\mathbf{B}|$ or ξ, dominates.

The limit $\mu_B |\mathbf{B}| \gg \xi$, where the magnetic field energy far exceeds the spin-orbit coupling energy, is straightforward. How large $\mathbf{B}$ is needed to reach this limit depends on the considered atom and level: for hydrogen 2p, the strong-field limit is reached for $|\mathbf{B}| \gg 0.5$ T, while for He$^+$ 2p it takes a magnetic field exceeding $|\mathbf{B}| \gg 8$ T, due to the Z^4-dependence of the spin-orbit energy, Eq. (2.35). In this limit of very strong field, the coupled basis is ill suited, because the full rotational invariance of the atom is badly broken. The uncoupled basis $|l, s, m_l, m_s\rangle$ works fine instead: spin and orbital moments align relative to the field with separate energy contributions depending on their different g-factors. In this basis, the dominating interaction H_{magn} is diagonal, with the following eigenvalues:

$$E_{\text{magn}}(m_l, m_s) = \langle m_l, m_s | H_{\text{magn}} | m_l, m_s \rangle = \mu_B \, B_z (m_l + 2m_s) \qquad (2.58)$$

(*Paschen-Back limit*). Small $H_{\text{s-o}}$ corrections can then be added perturbatively.

In the opposite, more common, weak-field limit ($|\mathbf{B}| \ll \xi/\mu_B$), spherical symmetry is just weakly perturbed. The states $|l, s, j, m_j\rangle$ of the coupled basis are exact eigenstates of $H_{\text{s-o}}$ and approximate eigenstates of $H_{\text{s-o}} + H_{\text{magn}}$. Following the results of Appendix C.8.2, to first order in $\mu_B |\mathbf{B}|/\xi$, the energy contribution of H_{magn} is given by Eq. (C.75):

$$E_{\text{magn}}(m_j) \simeq \langle j, m_j | H_{\text{magn}} | j, m_j \rangle = \langle j, m_j | -\mu_z B_z | j, m_j \rangle = g_j \, \mu_B \, B_z m_j \qquad (2.59)$$

(*Zeeman limit*), where g_j is the Landé g-factor, Eq. (C.78).

In the intermediate-field regime $\mu_B |\mathbf{B}| \simeq \xi$ neither basis is appropriate and neither Eq. (2.58) nor Eq. (2.59) are accurate expressions for the energy levels. Figure 2.12 displays the exact pattern of splittings of the six ^{2}P states, as a function of the magnetic field strength, from the Zeeman (weak field) to the Paschen-Back (strong-field) limit. The initial slopes of the energy curves at $B \to 0$, divided by the relevant m_j, measure the values of the Landé g_j.

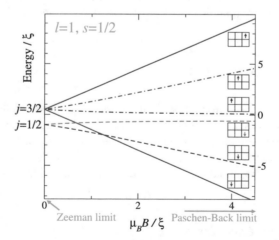

Fig. 2.12   Combined spin-orbit and magnetic splittings of a ^{2}P multiplet. Introducing the ratio $b = \mu_B B/\xi$, the expressions for the $m_j = \pm 3/2$ eigenenergies (*solid lines*) are trivially $(1/2 \pm 2b)\,\xi$. The eigenenergies of the four other levels are $1/4(-1 \pm 2b + d)\xi$ (*dot-dashed lines*) and $1/4(-1 \pm 2b - d)\xi$ (*dashed lines*), where $d = \sqrt{9 + 4b(1+b)}$. The box notation identifies the uncoupled-basis main component at strong field

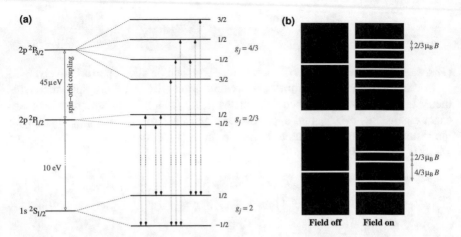

Fig. 2.13 **a** A scheme of the Zeeman-split $1s \longleftrightarrow 2p$ lowest Lyman line of hydrogen. The magnetic energy amounts to $m_j \, g_j \, \mu_B B$, with the indicated Landé g-factors g_j. **b** The resulting line spectrum and splittings

The experimentally observed spectra confirm the theory outlined here. For H, provided that a sufficiently strong magnetic field is applied, the Paschen-Back effect is straightforward to observe as a triplication of all lines, corresponding to $\Delta m_l = 0, \pm 1$. At very high spectral resolution, even the weak-field Zeeman splitting of the H lines shown in the scheme of Fig. 2.13 can be detected. The irregular spacing of the lines in this pattern qualifies this Zeeman effect as "anomalous".

2.2 Many-Electron Atoms

The novelty of many-electron atoms is the electron-electron repulsion, Eq. (1.6), which makes the associated Schrödinger equation impossible to solve exactly. The other fundamental ingredient that plays a central role in the QM of many-electron systems is the identity of electrons. This identity affects the very structure of the Hilbert space of states, and therefore it needs to be discussed first.

2.2.1 Identical Particles

The concept of indistinguishable particles is central to all the physics of matter wherever $N > 1$ electrons are involved. Recall that in classical mechanics each particle is labeled by its own position and momentum: one can track the trajectories of individual classical particles along their motion, and thus tell particles i and j apart at any time, regardless of they being different or identical. In QM, all particles explore simultaneously all points of space, with certain probability amplitudes. As a result, identical particles are indistinguishable at the deepest level. There is no way, even as a matter of principle, to ever tell, e.g., two electrons apart.

QM implements perfect indistinguishableness through symmetry: any many-identical-particles ket has a definite symmetry "character" (i.e. eigenvalue) of the permutation operator Π_{ij} swapping any pair ij of identical particles. As this permutation symmetry is a discrete symmetry which, when applied twice, leads back to the original state, the eigenvalues of Π_{ij} can only be $+1$ or -1.

The particles for whose swap the overall system ket $|a\rangle$ is *symmetric* are called *bosons*. In other words, for bosons the eigenvalue of Π_{ij} is $+1$, i.e. $\Pi_{ij}|a\rangle = |a\rangle$.

The particles for whose swap the overall ket $|a\rangle$ is *antisymmetric* are called *fermions*. For fermions the eigenvalue of Π_{ij} is -1, i.e. $\Pi_{ij}|a\rangle = -|a\rangle$.

All elementary particles of "matter" (electrons, protons, neutrons[6]...) are examples of fermions, all with spin $1/2$. Elementary bosons are carriers of "interactions". For example, photons (spin 1) are the quanta of the electromagnetic field, see Sect. 4.3.2.2. A simple general rule connects the spin of a particle to the permutational symmetry of bunches identical particles of such kind: *integer-spin particles are bosons, half-odd-integer-spin particles are fermions*.

A collection of bosons and fermions lumped together (e.g. an atom or a nucleus) can behave as a single point-like particle. This occurs when the internal dynamics is associated to high excitation energy, so that the lump remains in its ground state, possibly degenerate according to the projection of the lump's total angular momentum. When several such *identical* lumps move around together, it is important to know the eigenvalue for the permutation of two such *identical* lumps (*composite bosons/fermions*). This is simply answered by counting the number of (-1)'s generated by the permutations of pairs of identical fermions. For example, hydrogen atoms in the same hyperfine state are identical bosons, since a -1 is generated by the permutation of the two protons and a second -1 is generated by the permutation of the two electrons, in total $(-1) \cdot (-1) = +1$. Likewise, ^{238}U atoms (92 protons + 146 neutrons + 92 electrons) are bosons, while ^{3}He helium isotopes (2 protons + 1 neutron + 2 electrons) are fermions. Identical nuclei are bosons or fermions, too, according to whether they contain an even or odd number of nucleons (protons plus neutrons); for example deuteron nuclei D$^+$ (1 proton + 1 neutron bound in a $i = 1$ nuclear-spin state) are bosons. Due to the angular-momentum coupling rule (C.72), all composite particles fulfill the general integer/half-odd-integer spin rule.

The permutational antisymmetry affects fundamentally the dynamics of many electrons and, specifically, the structure of many-electron atoms. As discussed in greater detail below, antisymmetry prevents electrons to occupy identical quantum states. Antisymmetry makes electrons avoid each other in different and often more effective ways than the direct electron-electron repulsion V_{ee}. Without antisymmetry all electrons would occupy the same 1s shell in the atomic ground state. If that

---

[6]Contrary to electrons, protons and neutrons are not really elementary. They are composite fermionic particles, and precisely bound states of triplets of elementary spin-$1/2$ quarks.

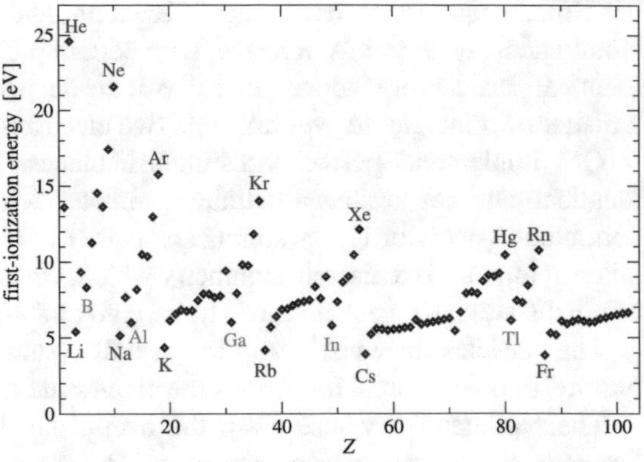

Fig. 2.14 The first-ionization energy of the atoms as a function of their atomic number Z. This quantity expresses the minimum energy required to remove one electron from the neutral atom ($N = Z$) in its ground state. The final state is the positive ion ($N = Z - 1$) (in its ground state too) plus a free electron

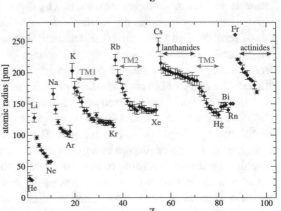

Fig. 2.15 Empirical atomic radii as a function of Z. The error bars represent the standard deviation. The labels TM1 TM2 and TM3 mark the ranges of the three series of transition metals, corresponding to the progressive filling of subshells 3d 4d and 5d, respectively (Data from Ref. [19])

happened (as it could if electrons were distinguishable particles—or bosons—rather then identical fermions), then the *first ionization energy* of atoms should increase roughly as Z^2 and their size should decrease roughly as Z^{-1}. In stark contrast, relatively mild non-monotonic Z dependences of both these quantities are observed, see Figs. 2.14 and 2.15: overall, the first ionization energy and the atomic size exhibit weak respectively decreasing and increasing trends with Z.

2.2.2 The Independent-Particles Approximation

The explicit solution of the Schrödinger equation associated to Hamiltonian (1.1) requires the determination of a N-electron wavefunction. For increasing N, very rapidly this becomes a formidable task, because the N-electron wavefunction describing the correlated motion of all the N electrons depends on all position and spin coordinates of the N electrons at the same time. The amount of information carried by a general N-electron wavefunction grows exponentially with N: there is no way to store (let alone compute!) the full wavefunction for the ground state of many interacting electrons.

Most available approximate methods exploit the observation that a basis of the Hilbert space of N-particle states can be built as the product of single-particle basis states. Consider the complete set of orthonormal states $\{|\alpha\rangle\}$ for a single particle. Here α is a full set of quantum numbers, e.g. n, l, m_l, m_s in the example of an electron in an atom, and takes all allowed values. Then the set of all tensor products

$$|\alpha_1, \alpha_2, \ldots \alpha_N\rangle = |\alpha_1\rangle \otimes |\alpha_2\rangle \otimes \ldots |\alpha_N\rangle \qquad (2.60)$$

realizes a basis for N distinguishable particles when all possible choices of the single-particle quantum numbers $\alpha_1, \alpha_2, \ldots \alpha_N$ are explored. For indistinguishable particles, the correct permutational symmetry is imposed to the product state (2.60) by taking properly symmetrized linear combination of these tensor products:

$$
\begin{aligned}
|\alpha_1, \alpha_2, \ldots \alpha_N\rangle^{S/A} &= \frac{1}{\sqrt{N_\Pi}} \sum_\Pi (\pm 1)^{\{\Pi\}} \Pi |\alpha_1, \alpha_2, \ldots \alpha_N\rangle \\
&= \frac{1}{\sqrt{N_\Pi}} \sum_\Pi (\pm 1)^{\{\Pi\}} |\alpha_{\Pi_1}, \alpha_{\Pi_2}, \ldots \alpha_{\Pi_N}\rangle.
\end{aligned} \qquad (2.61)
$$

Here Π indicates a generic permutation of the N states α_j, and the sum extends over all N_Π permutations; for example, $N_\Pi = N!$ if the states α_j happen to be all different. $\{\Pi\}$ in the exponent indicates the parity of the permutation Π, i.e. the number of pair swaps composing Π.

A fully symmetrized basis state $|\alpha_1, \alpha_2, \ldots \alpha_N\rangle^S$ realizes the correct permutational symmetry of N bosons. For bosons, with positive $(+1)^{\{\Pi\}} = +1$, no restriction applies to the quantum numbers α_j: any number of them may coincide.

An antisymmetric combination $|\alpha_1, \alpha_2, \ldots \alpha_N\rangle^A$ involving nontrivial $(-1)^{\{\Pi\}}$ signs can play the role of basis state for N fermions. Due precisely to these signs, *for fermions, all quantum numbers must necessarily be different*. If two were equal, say $\alpha_i = \alpha_j$, in the sum (2.61), the kets $|\alpha_{\Pi_1}, \ldots \alpha_{\Pi_i}, \ldots, \alpha_{\Pi_j}, \ldots \alpha_{\Pi_N}\rangle$ and $|\alpha_{\Pi_1}, \ldots \alpha_{\Pi_j}, \ldots, \alpha_{\Pi_i}, \ldots \alpha_{\Pi_N}\rangle$ would be equal, but with opposite parity phase factor $(-1)^{\{\Pi\}}$: these terms would then all cancel in pairs in the sum, so that the total ket $|\alpha_1, \ldots \alpha_i, \ldots \alpha_j, \ldots \alpha_N\rangle^A$ would vanish. As a result, the product basis kets for N fermions are characterized by N *different* single-fermion quantum numbers: this property expresses the *Pauli exclusion principle*, according to which *two identical fermions cannot occupy the same quantum state*.

The wavefunction representation of a N-fermion basis ket is obtained starting from a general eigenket of position and spin for the j-th fermion, shorthanded to $|w_j\rangle = |\mathbf{r}_j, \sigma_j\rangle$. The corresponding N-fermions product converts to a bra as $\langle w_1, w_2, \ldots w_N|$, and must then be properly antisymmetrized. The wavefunction associated to Eq. (2.61) is then

$$\Psi_{\alpha_1,\dots\alpha_N}(w_1,\dots w_N) = {}^A\langle w_1, w_2, \dots w_N | \alpha_1, \alpha_2, \dots \alpha_N \rangle^A$$

$$= \frac{1}{N!} \sum_{\Pi'} (-1)^{\{\Pi'\}} \sum_{\Pi} (-1)^{\{\Pi\}} \langle w_{\Pi'_1}, \dots w_{\Pi'_N} | \alpha_{\Pi_1}, \dots \alpha_{\Pi_N} \rangle$$

$$= \frac{1}{N!} \sum_{\Pi'} (-1)^{\{\Pi'\}} \sum_{\Pi} (-1)^{\{\Pi\}} \psi_{\alpha_{\Pi_1}}(w_{\Pi'_1}) \dots \psi_{\alpha_{\Pi_N}}(w_{\Pi'_N})$$

$$= \sum_{\Pi} (-1)^{\{\Pi\}} \psi_{\alpha_{\Pi_1}}(w_1) \dots \psi_{\alpha_{\Pi_N}}(w_N), \tag{2.62}$$

where the sum over the permutations Π' of the variables w_j generates $N!$ identical copies of the same wavefunction, and is therefore eliminated. The sum in the last line is the determinant of a matrix whose ij element is $\psi_{\alpha_i}(w_j)$:

$$\Psi_{\alpha_1,\dots\alpha_N}(w_1,\dots w_N) = \frac{1}{\sqrt{N!}} \begin{vmatrix} \psi_{\alpha_1}(w_1) & \psi_{\alpha_1}(w_2) & \cdots & \psi_{\alpha_1}(w_N) \\ \psi_{\alpha_2}(w_1) & \psi_{\alpha_2}(w_2) & \cdots & \psi_{\alpha_2}(w_N) \\ \vdots & \vdots & \ddots & \vdots \\ \psi_{\alpha_N}(w_1) & \psi_{\alpha_N}(w_2) & \cdots & \psi_{\alpha_N}(w_N) \end{vmatrix}. \tag{2.63}$$

This basis wavefunction is called a *Slater determinant*.[7]

An (anti)symmetrized product ket (2.61) contains an amount of information increasing just linearly, rather than exponentially, with the number N of identical particles. This makes it substantially simpler to deal with than a general boson/fermion ket. Despite their simplicity, the kets of Eq. (2.61) provide a basis of the Hilbert space with the proper permutational symmetry: any physical N-boson/N-fermion ket $|a^{B/F}\rangle$, e.g. an exact energy eigenstate of N interacting identical particles, can be expressed as a linear combination of the factorized basis kets:

$$|a^{B/F}\rangle = \sum_{\alpha_1,\alpha_2,\dots\alpha_N} c^a_{\alpha_1,\alpha_2,\dots\alpha_N} |\alpha_1, \alpha_2, \dots \alpha_N\rangle^{S/A}, \tag{2.64}$$

where $c^a_{\alpha_1,\alpha_2,\dots\alpha_N}$ are the complex coefficients defining the linear combination. The number of these coefficients grows exponentially with N: the complexity of the correlated state $|a^{B/F}\rangle$ is now encoded in its expansion coefficients $c^a_{\alpha_1,\alpha_2,\dots\alpha_N}$.

Several approximate methods of solution of the Schrödinger problem for many-electron systems replace the (lowest) exact eigenstate with *one* (as smart as possible) basis state $|\alpha_1, \alpha_2, \dots \alpha_N\rangle^A$ constructed with single-particle eigenstates $|\alpha_1\rangle, |\alpha_2\rangle, \dots$ of some appropriate single-electron Hamiltonian. This scheme is named *independent-particles approximation*. For example, the simplest scheme of this kind is realized when the electron-electron Coulomb interaction V_{ee} of Eq. (1.6) is neglected entirely.

---

[7]In Eq. (2.63) we introduce an extra normalization factor $(N!)^{-1/2}$. This standard factor allows one to carry out integration (e.g. for wavefunction normalization) over an unrestricted domain of all w_j variables, instead of the appropriate "hyper-triangle" $w_1 > w_2 > \cdots > w_N$.

2.2.3 Independent Electrons in Atoms

For atoms, if there was no V_{ee}, the Schrödinger problem for the N electrons would factorize exactly: each electron would move independently of the others in the field V_{ne} of the nucleus (carrying charge Zq_e). Neglecting relativistic effects, the single-electron eigenstates are represented by hydrogenic wavefunctions defined by Eq. (2.27) and Eqs. (2.11–2.13).[8] Thus, in this atomic context the label α_i of a single-electron state stands for a complete set of quantum numbers $n_i, l_i, m_{l\,i}, m_{s\,i}$.

In this approximate scheme, we choose a list of N *different* α_i, and combine them to construct a N-electron ket $|\alpha_1, \alpha_2, \ldots \alpha_N\rangle^A$, as in Eq. (2.61). For example, a possible $N = 4$-electron state is

$$|1, 0, 0, \uparrow,\ 3, 1, 1, \uparrow,\ 3, 1, 0, \uparrow,\ 3, 1, -1, \downarrow\rangle^A. \qquad (2.65)$$

The standard spectroscopic notation $1s^1 3p^3$ for this state lists the occupied single-particle orbitals, with the corresponding numbers of electrons as exponents: all information about the m_l's and m_s's is omitted. Instead, the box notation $1s\,\boxed{\uparrow}\ 3p\,\boxed{\uparrow|\uparrow|\uparrow}$ collects all these details.

The binding energy E^{bind} of an atomic state is defined as the work needed to decompose the atom from that given bound state to an isolated nucleus plus the N individual electrons at rest at infinite reciprocal distance. The total atomic energy is the opposite of the binding energy: $E^{\text{tot}} = -E^{\text{bind}}$.

For non-interacting electrons E^{tot} is obtained simply by summing their (negative) single-electron energies. The total energy of the example $1s^1 3p^3$ state in Eq. (2.65) is $E^{\text{tot}}_{1s3p^3} = \mathcal{E}_1 + 3\mathcal{E}_3 = -\frac{1}{2}\left(\frac{1}{1^2} + 3 \times \frac{1}{3^2}\right) Z^2\, E_{\text{Ha}} = -\frac{2}{3} Z^2\, E_{\text{Ha}}$. This $1s^1 3p^3$ state is not the one with the lowest possible energy. The ground state can be obtained by minimizing the energy of each single-particle orbital, *without violating the Pauli principle*. This minimization is realized through the building-up principle, or *aufbau* rule, consisting in filling the orbitals with the lowest available energy before occupying higher levels. With $N = 4$ electrons, each of $1s\,\boxed{\uparrow|\downarrow}\,2s\,\boxed{\uparrow|\downarrow}$, or $1s\,\boxed{\uparrow}\,2s\,\boxed{\uparrow}\,2p\,\boxed{\uparrow}$ plus 11 similar states, or $1s\,\boxed{\uparrow|\downarrow}\,2p\,\boxed{\uparrow|\uparrow}$ plus 14 similar states, for total of 1 ($1s^2 2s^2 2p^0$) +12 ($1s^2 2s2p$) +15 ($1s^2 2s^0 2p^2$) = 28 individual states, yields the lowest energy

$$E^{\text{tot}}_{1s^2 2s^2} = 2\mathcal{E}_1 + 2\mathcal{E}_2 = -\frac{1}{2}\left(2 \times \frac{1}{1^2} + 2 \times \frac{1}{2^2}\right) Z^2\, E_{\text{Ha}} = -\frac{5}{4} Z^2\, E_{\text{Ha}}. \qquad (2.66)$$

Real electrons do interact with each other. In a neutral atom ($N = Z$), the electron-electron repulsion V_{ne} is of the same order of magnitude (but opposite sign) as the attraction V_{ne} to the nucleus. This means that the extreme simplification of a complete neglect of V_{ee} is a terrible approximation, doomed to yield unphysical predictions.

---

[8] In many-electron atoms, the displacements of the nucleus allowed by the finiteness of its mass introduce correlations among the electronic motions, which add to those of Coulombic origin. Here we neglect entirely these tiny effects, assuming a practically infinite nuclear mass. When applying one-electron-atom formulas to many-electron atoms, we shall then take $\mu \equiv m_e$, i.e. $a \equiv a_0$.

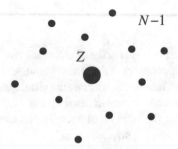

Fig. 2.16 When one electron moves far away from an atom/ion, the nucleus of charge Zq_e and the remaining $(N-1)$ electrons attract that electron as an effective point charge $(Z-N+1)\,q_e$

Let us discuss a few. Since the maximum occupation $2n^2$ of the n-th hydrogenic level grows rapidly with n, the minimum energy required to remove an electron from a neutral atom (the first ionization energy) would increase with Z only marginally more slowly than Z^2, at variance with experiment (Fig. 2.14). Moreover, in this model any atom (regardless of Z) would be able to accept any number N of electrons, always forming bound states. Experimentally, however, only certain atoms can form negatively-charged ionic bound states, but never with more than 1 extra charge, i.e. $N \leq Z+1$.

The main reason for the failures of this model is illustrated in Fig. 2.16: in reality while an electron moves far from the atom, it does not feel the bare nuclear attraction $-Ze^2/r$ but rather, due to electron-electron repulsion and according to the divergence law of electromagnetism, the substantially weaker combined effect of the nuclear charge and that of the other $N-1$ electrons, $V(r) \simeq (-Z+N-1)\,e^2/r$. This phenomenon, named *screening*, reduces quite substantially the electron binding strength, and consequently the ionization energy of the atom. We need to include this screening for a fair description of the atomic states and energies.

The next Sect. 2.2.4 sketches a simple yet instructive approach to account for V_{ee} approximately: perturbation theory, applied to the two-electron atom ($N=2$), e.g. He ($Z=N=2$). Alas, this perturbative method fails for $N \geq 3$. The following Sect. 2.2.5 sketches a much smarter independent-electron method for the inclusion of screening, that produces accurate results for any N.

2.2.4   The 2-Electron Atom

The simplest polyelectronic atoms have $N=2$ electrons: He, Li$^+$, Be^{2+}.... Here the independent-electron wavefunction reads

$$\Psi_{\alpha_1,\alpha_2}(w_1, w_2) = \frac{1}{\sqrt{2}} \begin{vmatrix} \psi_{\alpha_1}(w_1) & \psi_{\alpha_1}(w_2) \\ \psi_{\alpha_2}(w_1) & \psi_{\alpha_2}(w_2) \end{vmatrix} = \frac{\psi_{\alpha_1}(w_1)\psi_{\alpha_2}(w_2) - \psi_{\alpha_1}(w_2)\psi_{\alpha_2}(w_1)}{\sqrt{2}}. \quad (2.67)$$

In states where all orbital quantum numbers coincide ($n_1 = n_2, l_1 = l_2, m_{l1} = m_{l2}$), the orbital part can be factored out of the Slater determinant. Only spin remains in the antisymmetric combination:

$$\Psi_{n,l,m_l,\uparrow,\ n,l,m_l,\downarrow}(w_1, w_2) = \psi_{n,l,m_l}(\mathbf{r}_1)\,\psi_{n,l,m_l}(\mathbf{r}_2)\,\frac{1}{\sqrt{2}}\begin{vmatrix} \chi_\uparrow(\sigma_1) & \chi_\uparrow(\sigma_2) \\ \chi_\downarrow(\sigma_1) & \chi_\downarrow(\sigma_2) \end{vmatrix}. \tag{2.68}$$

Here the Slater determinant is simply $\chi_\uparrow(\sigma_1)\chi_\downarrow(\sigma_2) - \chi_\uparrow(\sigma_2)\chi_\downarrow(\sigma_1)$. This combination of the two spin wavefunctions is an eigenstate of the square modulus of the total electron spin $\mathbf{S} = \mathbf{s}_1 + \mathbf{s}_2$, with null eigenvalue $|\mathbf{S}|^2 \to S(S+1)\hbar^2 = 0$. Eigenstates of $|\mathbf{S}|^2$, like the one of Eq. (2.68), are useful because the matrix elements of the (hitherto neglected) Coulomb repulsion between states of different S vanish since V_{ee} is an orbital operator, which does not act on spin. The other $S = 0$ states (spin singlets), those involving two different sets of orbital quantum numbers, are:

$$\Psi^{S=0}_{n_1,l_1,m_{l1},\ n_2,l_2,m_{l2}}(w_1, w_2) = \tag{2.69}$$
$$\frac{\psi_{n_1,l_1,m_{l1}}(\mathbf{r}_1)\,\psi_{n_2,l_2,m_{l2}}(\mathbf{r}_2) + \psi_{n_1,l_1,m_{l1}}(\mathbf{r}_2)\,\psi_{n_2,l_2,m_{l2}}(\mathbf{r}_1)}{\sqrt{2}}\,\frac{1}{\sqrt{2}}\begin{vmatrix} \chi_\uparrow(\sigma_1) & \chi_\uparrow(\sigma_2) \\ \chi_\downarrow(\sigma_1) & \chi_\downarrow(\sigma_2) \end{vmatrix}.$$

These singlet states are characterized by an orbital part of the wavefunction which is symmetric under electron exchange Π_{12}, with the spin part taking care alone of the required antisymmetry. Note that the states of Eq. (2.69) and those of Eq. (2.68) share the advantage of being $S = 0$ eigenstates of the total spin $|\mathbf{S}|^2$; instead, generic Slater determinants of the type of Eq. (2.67) are not necessarily $|\mathbf{S}|^2$ eigenstates.

The other value of S allowed by the composition rule (C.72) is $S = 1$. The spin part of the wavefunctions of the $S = 1$ spin-triplet states is any of:

$$\begin{aligned} \chi^{S=1,\ M_S=1}(\sigma_1, \sigma_2) &= \chi_\uparrow(\sigma_1)\,\chi_\uparrow(\sigma_2) \\ \chi^{S=1,\ M_S=0}(\sigma_1, \sigma_2) &= \frac{1}{\sqrt{2}}[\chi_\uparrow(\sigma_1)\,\chi_\downarrow(\sigma_2) + \chi_\uparrow(\sigma_2)\,\chi_\downarrow(\sigma_1)]. \\ \chi^{S=1,\ M_S=-1}(\sigma_1, \sigma_2) &= \chi_\downarrow(\sigma_1)\,\chi_\downarrow(\sigma_2) \end{aligned} \tag{2.70}$$

These three spin states are symmetric under electron exchange Π_{12}. Therefore the orbital part has to take care of antisymmetry:

$$\Psi^{S=1,M_S}_{n_1,l_1,m_{l1},\ n_2,l_2,m_{l2}}(w_1, w_2) = \tag{2.71}$$
$$\frac{\psi_{n_1,l_1,m_{l1}}(\mathbf{r}_1)\,\psi_{n_2,l_2,m_{l2}}(\mathbf{r}_2) - \psi_{n_1,l_1,m_{l1}}(\mathbf{r}_2)\,\psi_{n_2,l_2,m_{l2}}(\mathbf{r}_1)}{\sqrt{2}}\,\chi^{S=1,\ M_S}(\sigma_1, \sigma_2).$$

Table 2.2 The characters of the fixed-total-spin-S two-electron basis states

	$S = 0$ (spin-singlet) states	$S = 1$ (spin-triplet) states
Orbital quantum number	Any	$(n_1, l_1, m_{l1}) \neq (n_2, l_2, m_{l2})$
Orbital wavefunction	Symmetric	Antisymmetric
Spin quantum numbers	$\uparrow$ and $\downarrow$ (different)	Any
Spin wavefunction	Antisymmetric	Symmetric

In these spin-triplet states, at least one of the orbital quantum numbers for the two electrons needs to be different: $(n_1, l_1, m_{l1}) \neq (n_2, l_2, m_{l2})$, otherwise the wavefunction vanishes. Table 2.2 summarizes the basic properties of the spin singlet ($S = 0$) and triplet ($S = 1$) basis states $|n_1, l_1, m_{l1}, n_2, l_2, m_{l2}, S, M_S\rangle$.

Following the notation of Appendix C.9, we use these simultaneous eigenstates of $H_0 \equiv T_e + V_{ne}$ and of $|\mathbf{S}|^2$ as convenient 0-th order states for a perturbation theory in $V = V_{ee}$, assumed "small". The "unperturbed" energies are

$$E_{n_1,n_2}^{\text{tot}(0)} = -\frac{1}{2}\left(\frac{1}{n_1^2} + \frac{1}{n_2^2}\right) Z^2 E_{\text{Ha}} = -2 E_{\text{Ha}}\left(\frac{1}{n_1^2} + \frac{1}{n_2^2}\right), \qquad (2.72)$$

like in the example of Eq. (2.66). Here the last expression refers to He ($Z = 2$). The ground state $n_1 = n_2 = 1$, $l_1 = m_{l1} = l_2 = m_{l2} = 0$ (in spectroscopic notation 1s^2) is necessarily a spin singlet. Its unperturbed energy $E_{1,1}^{\text{tot}(0)} = -4 E_{\text{Ha}} \simeq -109$ eV.

As a next step, following standard perturbation theory (Appendix C.9) the effect of V_{ee} is accounted for by evaluating its diagonal matrix elements over the 0-th order eigenkets. Equation (C.85) provides the first-order (additive) correction $E^{\text{tot}(1)} = \langle V_{ee}\rangle$ to the eigenenergies:

$$E_{n_1,l_1,m_{l1},\, n_2,l_2,m_{l2},\, S,M_S}^{\text{tot}(1)} = \qquad (2.73)$$
$$\langle n_1, l_1, m_{l1}, n_2, l_2, m_{l2}, S, M_S | V_{ee} | n_1, l_1, m_{l1}, n_2, l_2, m_{l2}, S, M_S\rangle.$$

The detailed calculation of these 6-dimensional Coulomb integrals is a rather intricate mathematical exercise. Here we discuss its main qualitative outcomes. Since the electron-electron repulsion is positive, this correction is *always positive*. The largest of these $E^{\text{tot}(1)}$ corrections occurs for the most localized wavefunction, the one where the two electrons stay nearest to each other, both in the most compact orbital: the ground state 1s^2. The average inter-electron distance is approximately a_0, thus the Coulomb integral $E_{1,0,0,\, 1,0,0,\, 0,0}^{\text{tot}(1)}$ is of order $\sim e^2/a_0 = E_{\text{Ha}}$. The precise value obtained from integration [5] is $5/4\, E_{\text{Ha}} \simeq 34$ eV. This correction brings the estimated ground-state energy of He to $E_{1,1}^{\text{tot}(0)} + E_{1,0,0,\, 1,0,0,\, 0,0}^{\text{tot}(1)} = -74.8$ eV, in fair agreement with the experimental value -79.00 eV (minus the sum of the first and second ionization energies of He).

For the spectroscopically most relevant states, those with one electron sitting in 1s and the second one in an excited state $n_2[l_2]$, the first-order correction accounts for several experimental observations:

- All Coulomb integrals are smaller than the one for the ground state, and tend to *decrease for increasing n_2*: the Coulomb correction become less and less important as the electrons move farther apart from each other.
- The Coulomb integrals at given n_2 depend weakly on l_2: they usually *increase for increasing l_2*. By inspecting the hydrogenic radial distributions Fig. 2.7, one notes that indeed the excited electron, on average, sits slightly closer to 1s when l_2 is larger. This is an important novelty, since it breaks the 1-electron-atom l-degeneracy of the shells, putting $\mathcal{E}_{ns} < \mathcal{E}_{np} < \mathcal{E}_{nd} < \cdots$, in accordance to experimental finding, Fig. 2.17.
- The Coulomb integrals depend on S, clearly not through the spin wavefunction which has nothing to do with the purely spatial operator V_{ee}, but through the different electron-electron correlation in the Π_{12}-symmetric ($S = 0$) or Π_{12}-antisymmetric ($S = 1$) spatial wavefunction. In particular, Eq. (2.71) shows that the triplet wavefunction $\Psi_{n_1,l_1,m_{l1}, n_2,l_2,m_{l2}}^{S=1,M_S}(w_1, w_2)$ vanishes for $\mathbf{r}_1 \to \mathbf{r}_2$. On the contrary, the singlet wavefunction $\Psi_{n_1,l_1,m_{l1}, n_2,l_2,m_{l2}}^{S=0}(w_1, w_2)$ is finite at $\mathbf{r}_1 = \mathbf{r}_2$. Therefore, on average, two electrons in a spin-triplet state avoid each other more effectively than in the spin-singlet state with the same orbital quantum numbers.[9] Indeed, Coulomb integrals are systematically smaller for $S = 1$ than for $S = 0$ states, as an explicit evaluation of the integral (2.73) shows. This result accounts for the experimental observation that each triplet level lies lower than the corresponding singlet (Fig. 2.17). This type of Coulomb splitting between states which differ uniquely in their total spin, here $S = 0$ or 1, is known as *exchange splitting*.

The perturbative approach presented here is useful mostly as a conceptual tool, to understand qualitative trends, and general concepts such as those listed above. Perturbation theory is relatively successful for the 2-electron atom, but for $N > 2$ electrons the repulsion that a given electron experiences from the other $N - 1$ electrons is comparable to the attraction generated by the nucleus, and any attempt to treat it as a small perturbation fails. A better approximate approach, based on a mean-field self-consistent evaluation of the electron-electron repulsion, yields fair quantitative accuracy for any N, and is regularly used to date. The reliability of this and similar self-consistent-field methods have made them standard tools for understanding experiments and making predictions of atomic properties of matter from first principles.

---

[9]This indicates that the fixed-spin states include a degree of correlation of the electronic motion induced by the symmetry properties of the spatial wavefunction. Adopting a basis of H_0 eigenkets where the perturbation V_{ee} is diagonal within each unperturbed degenerate space is required to make Eq. (C.85) applicable. The same strategy is adopted in Appendix C.8.1 and Sect. 2.1.7, choosing the $|l, s, j, m_j\rangle$ basis to have $H_{\text{s-o}}$ diagonal within the degenerate multiplets.

Fig. 2.17 Energy levels (*bold horizontal lines*) of atomic He, with several electric-dipole transitions (*oblique lines*). While one electron remains in the $n_1 = 1$ single-electron state, the marked quantum numbers n_2 and $[l_2]$ refer to the second (excited) electron. Note that each triplet state (*right side*) sits at a systematically lower energy than the corresponding singlet (*left side*). The energy zero is the ground state $1s^2$ (note the broken vertical scale). The *horizontal dashed line* at 24.59 eV marks the first-ionization threshold (Experimental energies from Ref. [20])

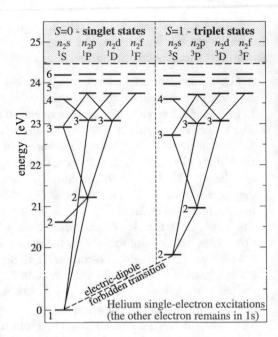

2.2.5 The Hartree-Fock Method

The idea is attempting to approximate in the best possible way the exact N-electron solution of the Schrödinger equation with an antisymmetrized product (a Slater determinant) of one-electron states, called *atomic orbitals*. This best Slater determinant is obtained following the variational principle, see Appendix C.5: the average energy $E^{\text{var}}[a] = \langle a|H_{\text{tot}}|a\rangle$ of any state $|a\rangle$ is greater than or equal to that of the ground state. The lower $E^{\text{var}}[a]$ is, the better $|a\rangle$ approaches the ground state. For a generic Slater determinant, the total energy that we need to minimize is:

$$E^{\text{var}}[\psi_{\alpha_1}, \ldots \psi_{\alpha_N}] = \langle \alpha_1, \ldots \alpha_N|^A H_{\text{tot}}|\alpha_1, \ldots \alpha_N\rangle^A \tag{2.74}$$

$$= \int dw_1, \ldots dw_N \, \Psi^*_{\alpha_1, \ldots \alpha_N}(w_1, \ldots w_N) H_{\text{tot}} \Psi_{\alpha_1, \ldots \alpha_N}(w_1, \ldots w_N)$$

$$= \sum_i^N \langle \alpha_i|H_1|\alpha_i\rangle +$$

$$+ \frac{1}{2} \sum_{i,j}^N \Big[\int dw \, dw' \, |\psi_{\alpha_i}(w)|^2 \, v_{ee}(w, w') \, |\psi_{\alpha_j}(w')|^2$$

$$- \int dw \, dw' \, \psi^*_{\alpha_i}(w)\psi^*_{\alpha_j}(w') \, v_{ee}(w, w') \, \psi_{\alpha_j}(w)\psi_{\alpha_i}(w') \Big]$$

under arbitrary variations of the N single-particle wavefunctions ψ_{α_i} composing the Slater determinant. We only require the ψ_{α_i} to remain mutually orthonormal $\int dw\, \psi_{\alpha_i}^*(w)\psi_{\alpha_j}(w) = \delta_{ij}$. In Eq. (2.74), the "one-particle term" $H_1(w) = \left[-\frac{\hbar^2}{2m_e}\nabla_{\mathbf{r}}^2 - \frac{Ze^2}{|\mathbf{r}|} \right] \otimes \mathbb{1}_{\text{spin}}$ describes the individual motion of each electron in the field of the nucleus; $v_{ee}(w, w') = \frac{e^2}{|\mathbf{r}-\mathbf{r}'|} \otimes \mathbb{1}_{\text{spin}}$ represents the electron-electron Coulomb repulsive energy of two electrons at locations $\mathbf{r}$ and $\mathbf{r}'$.

The mathematical problem of finding a minimum of E^{var} is that of minimizing a *functional*, i.e. a function whose independent variable is a set of functions. This constrained minimization problem is formally solved if the atomic orbitals ψ_{α_i} satisfy the nonlinear integro-differential equation called *Hartree-Fock (HF) equation*:

$$
\overbrace{H_1(w)\psi_\alpha(w)}^{1} + \overbrace{\int \sum_\beta |\psi_\beta(w')|^2\, v_{ee}(w, w')\, dw'\, \psi_\alpha(w)}^{2} + \tag{2.75}
$$

$$
- \overbrace{\int \sum_\beta \psi_\beta^*(w')\, v_{ee}(w, w')\, \psi_\beta(w)\, \psi_\alpha(w')\, dw'}^{3} = \epsilon_\alpha \psi_\alpha(w).
$$

Each numbered term in Eq. (2.75) derives from a corresponding term in the total energy (2.74). If one pretends that all ψ_β functions are given *known* functions (rather than the unknown functions they really are), then Eq. (2.75) becomes *linear* in ψ_α. This equation for ψ_α has a Schrödinger-like form. Term 1 contains the kinetic energy plus the Coulomb attraction of the nucleus. Term 2 represents the Coulomb repulsion that an electron at position&spin w experiences due to the average charge distribution of all electrons, because $-q_e \sum_\beta |\psi_\beta(w')|^2$ represents the electric-charge-density distribution of all N electrons at position&spin w'. Term 3 is a nonclassical nonlocal exchange term which, among other effects, removes the unphysical repulsion of each electron with itself introduced by term 2, because the $\beta = \alpha$ contributions in the sums of terms 2 and 3 cancel.[10] The HF equation realizes a natural way to deal with the electron-electron repulsion as accurately as possible at the mean-field level.

Terms 2 and 3 of Eq. (2.75) depend explicitly on the (unknown) wavefunctions ψ_β. A standard strategy for the solution of the HF equation is based on repeatedly pretending that all ψ_β in Eq. (2.75) are known, following the scheme of Fig. 2.18: start from some arbitrary initial set of N orthonormal one-electron wavefunctions, put them in place of all ψ_β's in Eq. (2.75), thus generating a first approximation for the effective potential energy acting on the single electrons; solve (usually numerically) the linear equation for ψ_α; from the list of solutions, take the N eigenfunctions with lowest *single-particle eigenenergy* ϵ_β (*aufbau* rule); re-insert them into Eq. (2.75) in place of the ψ_β's thus generating a better approximation for the effective potential energy; iterate this procedure as long as needed. After several iterations ($\approx$10–100,

---

[10] Note that in the β sum of term 3, only $m_{s\,\beta} = m_{s\,\alpha}$ terms contribute, as v_{ee} is purely orbital and does not modify spin. As a consequence, this negative exchange term favors high-spin states.

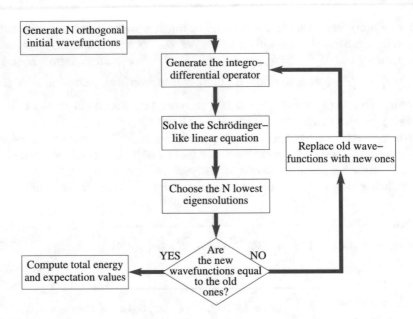

Fig. 2.18   An idealized scheme for the self-consistent resolution of the HF equation (2.75)

depending on the starting ψ_β), *self-consistency* is usually reached, i.e. the wavefunctions do not change appreciably from one iteration to the next. With the converged wavefunctions one can compute several observable quantities, for example the total HF energy, by means of Eq. (2.74).

The sum of the nuclear potential plus the repulsion of the charge distribution of the other electrons represents the *self-consistent potential* energy V_{HF} driving the motion of each electron. Until now, we refrained from making assumptions about the symmetry of $V_{HF}(w)$: this is a safe approach in a general context. When applying HF to atoms one usually makes a crucial simplifying assumption: that the electron charge distribution and therefore the self-consistent V_{HF} are spherically symmetric functions, like the attraction to the nucleus V_{ne}. This assumption allows us to separate variables in Eq. (2.75), like in the Schrödinger problem for the one-electron atom, and write each HF single-particle solution ψ_α as the product of a radial wavefunction $R_{nl}(r)$ × a spherical harmonic Y_{lm_l} × a spinor $\chi_{m_s}(\sigma)$, as in Eq. (2.27). In a spherical HF atom, one-electron wavefunctions are labeled by hydrogen-like quantum numbers $\alpha = (n, l, m_l, m_s)$. Here the quantum numbers l, m_l, m_s label exactly the same angular and spin dependence as in Eq. (2.27) for the one-electron atom. In contrast, the radial wavefunctions $R_{nl}(r)$ differ from those of Eq. (2.13), and are usually determined numerically. Despite their differences, the HF radial functions and those of the one-electron atom share the following properties: (i) the number of radial nodes $(n - l - 1)$ defines n, and (ii) near the nucleus $R_{nl}(r) \propto r^l$.

Fig. 2.19 A sketch of $r \times$ the one-electron Hartree-Fock effective potential energy for atomic oxygen, $N = Z = 8$ (*solid line*). This quantity interpolates between the hydrogen $-e^2$ value (*dashed*) at large distance, and the value $-8\,e^2$ generated by the bare oxygen nucleus in the context of a O^{7+} ion (*dotted*) at small distance

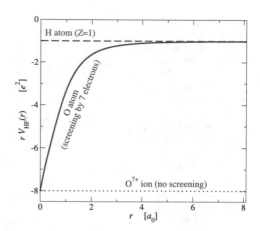

The N-electron ground state is built by filling the single-electron levels for increasing energy ϵ_α, starting from $\alpha = 1s, 2s, 2p\ldots$ upward. As expected, the spherically symmetric self-consistent potential $V_{HF}(r)$ acting on each electron is close to $\simeq (N - 1 - Z)\, e^2/r$ for large r and to $\simeq -Ze^2/r$ for small r, see Fig. 2.19. Because the potential has not a simple Coulomb shape, the single-electron levels ϵ_α do *not* coincide with those of the one-electron atom, Eq. (2.10). Importantly these HF levels depend on l, not only on n: $\epsilon_\alpha = \epsilon_{nl}$. Indeed, an ns orbital sits lower in energy than np, because ns has a larger probability near the nucleus (where the effective HF potential is more strongly attractive) than np. Thus the HF method accounts quite naturally for the observed l-*ordering* $ns, np, nd \ldots$ of the single-electron levels observed in the atomic spectra (illustrated for He in Fig. 2.17). Also, the faster-than-Coulombic raise of the effective potential of many-electron atoms induces a n-dependence of ϵ_{nl} which is more rapid than the $-n^{-2}$ dependence of the one-electron atom.

Figure 2.20 reports the radial distribution of the one-electron states composing the ground-state HF wavefunction of argon. The typical radii of the shells vary substantially with n, from $\approx a_0/Z$ of $1s$, to a few times a_0 for the *valence* (=outermost occupied) shell, here $n = 3$. Three peaks, associated to the $n = 1, 2$, and 3 filled shells, emerge prominently in the total probability distribution $P(r)$, Fig. 2.20b.

The independent-electron self-consistent spherical-field HF model has become far more than just an approximation to the actual atomic state: it provides the common language of atomic physics. The electron occupations of the single-electron orbitals composing the Slater determinant which best overlaps the actual correlated many-electron atomic eigenstate are adopted routinely as a label of that eigenstate. For example, the electronic ground-state of Ar ($N = Z = 18$) is usually labeled with its HF *configuration* $1s^2 2s^2 2p^6 3s^2 3p^6$.

Fig. 2.20 The radial distributions associated to the HF wavefunction of a many-electron atom: Ar, $N = Z = 18$. **a** The radial probability distribution for the individual filled single-electron states. Note that the characteristic radius of the innermost shell $n = 1$ is $\approx a_0/Z$, while the outermost filled shell ($n = 3$) is slightly wider than a_0. **b** The total radial probability distribution $P(r)$ and effective integrated charge $Z_{\mathrm{eff}}(r)$ generating the effective one-electron HF potential acting on each electron (Data from Ref. [1])

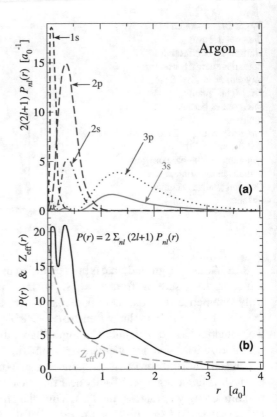

2.2.6 Electronic Structure Across the Periodic Table

It is instructive to examine the electronic configurations of atoms across the periodic table, within the HF scheme. The HF $1s^2$ ground configuration of He has total energy $E_{1s^2}^{\mathrm{tot}} = -77.8$ eV, ~ 1.2 eV above the measured ground-state energy; the radial dependence of the one-electron wavefunction is of course nonhydrogenic. In lithium, a third electron adds into 2s, to form a $1s^2 2s$ configuration. Its HF ground-state total energy is $-7.4328\,E_{\mathrm{Ha}}$ [5], to be compared with the experimental (1st + 2st + 3rd) ionization energy $7.4755\,E_{\mathrm{Ha}}$, with a deviation $\simeq 1$ eV. The first ionization energy can be evaluated by $E_{\mathrm{gs}}^{\mathrm{tot}}(N = Z - 1) - E_{\mathrm{gs}}^{\mathrm{tot}}(N = Z)$, requiring two HF calculations. For Li, this method yields a ionization energy $\simeq 5.34$ eV, in good accord with the experimental value 5.39 eV. This energy is much smaller than that of He (24.59 eV), because binding for the 2s electron is much weaker than for 1s, even with smaller Z. Beryllium has a $1s^2 2s^2$ ground state. For these atoms with $N \leq 4$ electrons, involving only s orbitals, the spherical assumption is exact.

Starting from boron, electrons occupy progressively a degenerate p subshell: as the p orbitals are not spherically symmetric, the spherical assumption for the self-consistent field becomes questionable. The 2p subshell is completely filled as neon,

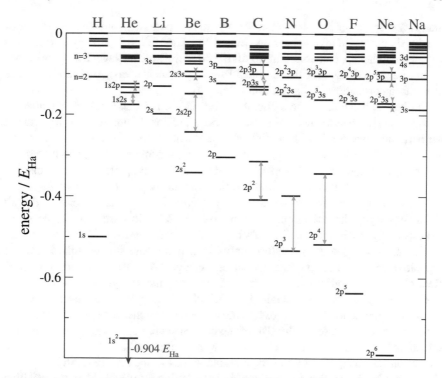

Fig. 2.21 An energy diagram comparing the observed ground state and low-energy excitations of the $1 \leq Z \leq 11$ atoms. The zero of the energy scale (top of the figure) is set at the ground energy of each monocation. The energy spread of the atomic multiplets for fixed configuration, highlighted by the *vertical arrows*, is due to Coulomb exchange and correlation effects, see Sect. 2.2.9.3

the next noble gas, is reached for $N = Z = 10$. Again Ne is a spherically symmetric atom, because $\sum_{m_l} |Y_{l\,m_l}(\theta, \varphi)|^2$ is independent of θ and φ. The ionization energy of Ne is again very large, but not as much as that of He (see Fig. 2.21).

The next atom, Na, involves one valence electron in the 3s subshell, much higher in energy than the deep 2p level. Again the ionization energy has a dip (Fig. 2.14), due to the new $n = 3$ shell being only weakly bound as it starts to fill up. As $Z = N$ further increases, the filling of the $n = 3$ shell proceeds fairly smoothly, with 3s and 3p becoming more and more strongly bound until the next noble gas Ar is reached.[11]

For Z this large, the l-dependence of the HF levels ϵ_{nl} has become so strong that the HF self-consistent field places the 3d subshell above 4s! Indeed experiment shows that the ground state of potassium ($N = Z = 19$) is $1s^2 2s^2 2p^6 3s^2 3p^6 4s$ rather

---

[11]For argon, $Z = 18$, the HF approximation obtains a total energy of $-526.817\,E_{\text{Ha}}$, which is $0.791\,E_{\text{Ha}} = 21.5$ eV in excess of the experimental energy [21]. The absolute error is rather large, which indicates that the neglect of dynamical correlations in the electronic motion is a serious drawback of HF. However, the relative error is $\sim 0.15\%$ only, and the excess energy per electron amounts to approximately 1 eV, indicating that the HF mean-field approximation captures the vast majority of the electron-electron repulsion.

than $1s^2 2s^2 2p^6 3s^2 3p^6 3d$. Many atomic properties of potassium are similar to those of previously discussed alkali elements, Li and Na. After filling 4s at $Z = 20$ (Ca), and before involving 4p, electrons start to fill the 3d shell. Note however a few inversions (as in Cr and Cu) indicating that 4s and 3d are very close in energy, and subtler effects of electron correlation play a relevant role.

Further shell intersections associated to a sizable l-dependence of the subshell energies ϵ_{nl} manifest themselves as 4d, 4f, 5d, and 5f are being filled, as reported in the periodic table. Similar properties of all elements with a given number of s or p electrons in the valence shell suggest the overall arrangement of the periodic table in groups. Many "low-energy" properties of atoms with incomplete d shells (transition metals) and f shells (lanthanides and actinides) are fairly uniform across each series.

The size of the atoms (a not especially well-defined property) computed with HF agrees well with the empirical trends of Fig. 2.15. In particular, noble gases are especially small and alkali atoms especially large relative to other atoms with similar Z; overall, the size of neutral atoms tends to increase slowly with Z. Cations (=positive ions) can be produced with any charge $(Z - N)q_e$. A cation with N electrons often exhibits the same formal ground electronic configuration as the neutral atom $(Z = N)$: this holds especially for small charging, $N \lesssim Z$. However all single-electron wavefunctions shrink closer to the nucleus in cations than in the neutral atom with the same N: as a result the size of the cations decreases as shells empty up and screening becomes less and less effective. Many but not all atoms can accept up to 1 electron (i.e. $N \leq Z + 1$) and form stable anions (=negative ions) in gas phase. The HF model signals that a certain ionic configuration is unstable by never reaching self consistence. The HF stability or lack thereof often agrees with experiment. For example for all halogens (F, Cl, Br, I, At), characterized by a relatively "deep" in energy np shell (Fig. 2.21) with 5 electrons, an extra electron easily joins this outer shell, forming a stable np^6 configuration. It takes some amount of work, called the *electron affinity* (defined as the first ionization energy of the negative ion), to remove that extra electron and return to the neutral atom in its ground state.

Beside ionic states, one can use HF even to compute (to some extent) excited states and excitation energies. For example, after computing the ground-state properties of Na in its configuration $1s^2 2s^2 2p^6 3s$ involving the lowest single-electron levels, one could run a new self-consistent calculation putting the 11-th (valence) electron in 3p rather than in 3s: configuration $1s^2 2s^2 2p^6 3p$. The self-consistent field turns out different, and the total energy is larger. The difference in total energy between these two calculations provides a fair estimation of the excitation energy, here of the 3s$\rightarrow$3p transition of Na, approximately 2 eV, see Fig. 2.24.

2.2.7 Fundamentals of Spectroscopy

To rationalize the spectra of many-electron atoms, first of all observe that the electric-dipole operator driving electromagnetic transitions is a one-electron operator. In concrete, **d** is the sum of the individual dipoles of the single electrons:

$$\mathbf{d} = \sum_i \mathbf{d}_i = -q_e \sum_i \mathbf{r}_i. \tag{2.76}$$

Consider the initial and final states, described in the HF approximation as two permutation-antisymmetrized states, Eq. (2.61). Between two such states, the matrix element of the dipole operator is[12]

$$\langle \beta_1, \ldots \beta_N |^A \sum_i \mathbf{d}_i | \alpha_1, \ldots \alpha_N \rangle^A =$$

$$= \frac{1}{N!} \sum_i \sum_{\Pi \Pi'} (-1)^{\{\Pi\}+\{\Pi'\}} \langle \beta_{\Pi_1} | \alpha_{\Pi'_1} \rangle \ldots \langle \beta_{\Pi_i} | \mathbf{d}_i | \alpha_{\Pi'_i} \rangle \ldots \langle \beta_{\Pi_N} | \alpha_{\Pi'_N} \rangle$$

$$\simeq \frac{1}{N!} \sum_i \sum_{\Pi \Pi'} (-1)^{\{\Pi\}+\{\Pi'\}} \delta_{\beta_{\Pi_1},\alpha_{\Pi'_1}} \ldots \delta_{\beta_{\Pi_{i-1}},\alpha_{\Pi'_{i-1}}} \langle \beta_{\Pi_i} | \mathbf{d}_i | \alpha_{\Pi'_i} \rangle \ldots \delta_{\beta_{\Pi_N},\alpha_{\Pi'_N}}$$

$$= \frac{1}{N!} \sum_i \sum_{\Pi} \langle \beta_{\Pi_i} | \mathbf{d}_i | \alpha_{\Pi_i} \rangle \prod_{j \neq i} \delta_{\beta_{\Pi_j},\alpha_{\Pi_j}} = \sum_i \langle \beta_i | \mathbf{d}_i | \alpha_i \rangle \prod_{j \neq i} \delta_{\beta_j,\alpha_j}.$$

This formula relates the N-electron electric-dipole matrix element to that of the involved single-electron states. The angular part of each single-electron wavefunction is a standard spherical harmonic Y_{lm}. Whenever the $\langle \beta_i | \mathbf{d}_i | \alpha_i \rangle$ factor vanishes due to violating the single-electron electric-dipole selection rules, the overall matrix element vanishes. Therefore the i-th electron executing the transition must respect the basic electric-dipole selection rule $\Delta l_i = \pm 1$ derived in Sect. 2.1.9 for the one-electron atom. The above sum of products expresses the following rule: dipole-allowed transitions occur only between N-electron states differing in *one electron* that *makes a dipole-allowed transition*, with all other $N - 1$ electrons remaining in their initial single-particle state. The i sum tells us that any of the N electrons in the initial states $|\alpha_i\rangle$ can make its own transition to any initially empty state $|\beta_i\rangle$, following the above rule.

Let us list a few examples of *allowed* transitions of beryllium: $1s^2 2s^2 \rightarrow 1s^2 2s2p$, $1s^2 2s2p \rightarrow 1s^2 2s4d$, and $1s^2 2s^2 \rightarrow 1s2s^2 3p$; and a few examples of *forbidden* transitions: $1s^2 2s^2 \nrightarrow 1s^2 2s3d$, $1s^2 2s^2 \nrightarrow 1s^2 2p^2$, and $1s^2 2s^2 \nrightarrow 1s2s2p3p$. In Sect. 2.2.10 we will discuss further rules, restricting dipole-allowed transitions depending on the total angular momenta obtained by coupling the spins and orbital angular momenta of the individual electrons following the schemes of Sect. 2.2.9.3.

---

[12]The first simplification is a result of all one-electron angular products vanishing unless each of $\beta_{\Pi_k} = \alpha_{\Pi'_k}$. This statement is rigorous under the simplifying assumption that the single-electron basis states composing the initial state $|\alpha_1, \ldots \alpha_N\rangle^A$ are eigenstates of the same effective Schrödinger problem as to those composing the final state $|\beta_1, \ldots \beta_N\rangle^A$. The elimination of the Π' summation is due to the $N - 1$ Kronecker deltas forcing the Π' permutation to be identical to Π. The final deletion of the sum over Π is due to this sum leading to $N!$ copies of the same matrix elements.

2.2.8   Core Levels and Spectra

According to the HF theory, the screening of the deep (strongly bound) one-electron *core* states is scarce, so that their energy is highly negative, and changes with Z essentially as $\propto -Z^2$. Indeed, in the independent-electron language, it should be possible to excite electrons out of the core shells 1s, 2s,... Consider, for example, a configuration such as $1s^1 2s^2 2p^6 3s^2$ of sodium, with an inner 1s electron promoted to the outer 3s shell. To generate this state it takes a large excitation energy (in the order of 1 keV), to the extent that one might suspect that such a highly unbound state has no right to exist. Indeed, an atom in this excited state has plenty of electric-dipole-allowed transitions to get rid of big chunks of this excitation energy, to configurations at a much lower energy, e.g. $1s^2 2s^2 2p^5 3s^2$. According to Eq. (2.45), the decay transition rate is very large as it grows with the third power of the energy associated to the transition, which dominates over the reduction in dipole matrix element due to the small size of the initial shell, to an overall $\sim Z^4$ dependence—see Eq. (2.55). As a result, the natural broadening due to the short lifetime of *core excitations* is often huge, exceeding $\hbar\gamma \approx 1$ eV. Despite such huge broadening, core-hole states are not just a theoretical prediction of the independent-electron model, but they are routinely observed in UV and X-ray spectroscopies.

Several experiments probing core-electron spectra with photons fit the conceptual scheme of either absorption, Fig. 1.3, or emission, Fig. 1.4. Absorption data (Fig. 2.22) exhibit a remarkable regularity of the spectra above ≈ 50 eV, and systematic changes in peak positions and intensities with the atomic number Z. A characteristic feature of X-ray absorption spectra is the asymmetry of the peaks, which exhibit a steep *edge* at the low-energy side and a slow decrease at the high-energy side. This edge occurs because, below the minimum excitation energy from the core level to the lowest empty level, no absorption can take place due to Pauli's

Fig. 2.22 The observed X-ray absorption coefficient of the elements in the third row of the periodic table, showing, for increasing Z, the regular displacement of the K edge (*right side*), and a gradual buildup and shift of the L edge (*left side*) to higher energy as Z increases

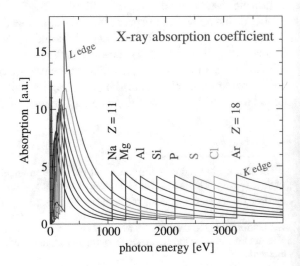

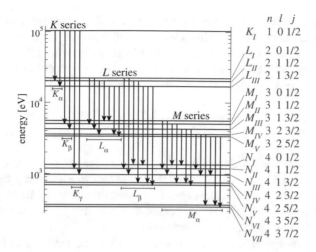

Fig. 2.23 The observed core-level structure of uranium. States are labeled by the quantum numbers $n\,l$ and j of the core hole. *Vertical arrows* identify the electric-dipole–allowed emission transitions. Inter-shell transitions are characterized by a "fine" structure involving large splittings related to l-dependent subshell energies, and to the spin-orbit interaction that splits core-hole states with different j

principle. Above this threshold, the core electron can be promoted to many empty bound and unbound states of the atom (whether in gas or condensed phase), leading to continuum absorption. Far above the edge, the slow intensity decay is due to the increase of the ejected-electron kinetic energy (which equals the difference between the absorbed photon energy and the energy of the atomic excited core state): the final state becomes increasingly orthogonal to the initial core level. As a result, the dipole matrix element (2.46) decreases steadily. For this same reason, an X-ray photon hitting an atom is more likely to extract a core electron than a weakly-bound outer-shell electron which would be ejected with a far higher kinetic energy.

Core emission spectra exhibit the same regularity as absorption spectra. Initial core excitations are usually prepared by collisions with high-energy electrons. The subsequent emission involves transitions only from levels for which enough energy is made available by the primary excitation. For example, if 2 keV electrons are used to excite the sample, X-ray emission involving the 1s shell is observed for all $Z \leq 14$ (Si), but not for P and higher-Z atoms (see Fig. 2.22).

Yet another uncalled-for traditional notation haunts core states and X-ray spectra: a *hole* (=missing electron) in shell $n = 1, 2, 3, 4, \ldots$ is labeled $K, L, M, N, \ldots$ The substructures related to states with different l and j acquire a Roman counting subscript (e.g. L_{III} for $2p\,^2P_{3/2}$), see Fig. 2.23. Dipole-allowed X-ray emission transitions are organized in series according to the initial shell, with a Greek-letter subscript for the final shell. For example, the transition $K \rightarrow L$ (in other words the decay $1s2s^22p^6\ldots \rightarrow 1s^22s^22p^5\ldots$) produces the K_α emission lines, $K \rightarrow M$ generates K_β, and $L \rightarrow M$ generates L_α, see Fig. 2.23.

In the early 20th century, when the structure and classification of atoms were still being understood, H. G.-J. Moseley acquired and compared characteristic X-ray emission spectra of numerous elements. Moseley discovered that the reciprocal wavelength (or equivalently, energy) of K_α is roughly proportional to Z^2. This inverse wavelength fits an approximate phenomenological law:

$$\frac{1}{\lambda_{K_\alpha}} \approx C\,(Z - a)^2. \tag{2.77}$$

In this phenomenological dependence, the proportionality constant C is close to $3/8\,E_{\mathrm{Ha}}/(2\pi\hbar c) \simeq 8.23 \times 10^6$ m^{-1}, and the parameter a, accounting for screening, is in the order of unity. The discovery of this regularity allowed Moseley to identify the actual value of Z of each atomic species, thus correcting a few mistakes in early versions of the periodic table of the elements.

The fair accuracy of Moseley's fit suggests that one could estimate the core-electron energy levels (within, say, 20%) without going through a full self-consistent HF calculation. Indeed, the energy of a core level is close to that of an hydrogenic state in an effective $-Z_{\mathrm{eff}}\,e^2/r$ Coulomb potential, where the value of Z_{eff} is the average of the effective potential, e.g. the solid curve of Fig. 2.19, weighted by the radial distribution $P(r)$ of the single-electron wavefunction of that level. In this scheme, the energy of a core shell can be estimated by means of Eq. (2.10), replacing Z with $Z_{\mathrm{eff}} \simeq Z - 2$ for the K shell, $Z_{\mathrm{eff}} \simeq Z - 10$ for the L shell, and in general $Z_{\mathrm{eff}} \simeq Z-$ (number of electrons in inner shells up to and including the target shell).

Nowadays X-ray spectroscopies are used routinely in research and applications, e.g. as position-sensitive analytic tools, as local probes of the near chemical environment of different atomic species. Numerous other applications of X-ray spectroscopies are reported in the scientific literature and at the web sites of X-ray photon facilities [22].

2.2.9 Optical Spectra

The systematic regularity of the core spectra in the hard-UV—X-rays region is entirely lost when the valence electrons are involved, in the IR—visible—soft-UV spectra. Neighboring atoms in the periodic table often exhibit entirely different spectra. A few trends and patterns can be recognized, though.

We first need to identify the relevant quantum numbers of many-electron atoms. Beside the shell occupations $1s^2 2s^2 \ldots$, we must assess the global quantum numbers of the atom. An atom is invariant under overall rotations around its nucleus. As a result the *total-angular-momentum* $|\mathbf{J}|^2$ and J_z operators commute with H_{tot}, and therefore can be made diagonal in the basis of energy eigenstates. If, as occurs for small Z, the coupling of spin and orbital motions is weak, also the total orbital L^2 and total spin S^2 operators are diagonal. In addition to the shell occupations, the total quantum numbers L, S, J, and M_J, are sometimes sufficient to characterize the atomic states. These quantum numbers (except M_J) are collected to form a term symbol $^{2S+1}[L]_J$. For example $^3P_2$ stands for $S = 1$, $L = 1$, $J = 2$.

2.2.9.1 Alkali Atoms

The alkali atoms have $Z - 1$ electrons filling a number of *close* (=completely filled) subshells which are energetically well separated ($\sim$20 eV) from the outermost half-occupied n_0s level. The latter is very shallow, with a binding energy ranging from 5.4 eV in Li to 3.9 eV in Cs. The electrons in the inner shells form an essentially "frozen" spherically symmetric core (null spin and orbital angular momentum), which provides an effective potential (Fig. 2.19) for the motion of the outer (valence) electron. Like for hydrogen, the ground-state label is $^2S_{1/2}$, because the total angular momenta of an alkali atom coincide with those of its valence electron.

This outer electron is the protagonist of all excitations in the optical spectra of the alkali elements. As shown in Fig. 2.24, these spectra resemble the spectrum of hydrogen. The two main differences are the inaccessibility of $n < n_0$ levels to the valence electron (due to Pauli's principle), and the sizable energy gaps between states characterized by the same n but different l. In emission or absorption spectra, dipole-allowed optical transitions appear as oblique lines in such level schemes, namely s $\leftrightarrow$ p, p $\leftrightarrow$ d, d $\leftrightarrow$ f,...like in Fig. 2.17. A characteristically bright transition in the visible (Li and Na) or near IR (K, Rb, and Cs) spectrum originates from

Fig. 2.24 Comparison of the observed $n \le 6$ energy levels of hydrogen, lithium, and sodium. In the vertical scale, the reference energy is set at the ground state of each atom's monocation, like in Fig. 2.21. Note the significant dependence of energy not only on n but also on l. The 2s $\leftrightarrow$ 2p transition of Li (*red lines* at 670.776 and 670.791 nm) and the 3s$\leftrightarrow$3p transition of Na (*yellow "D-lines"* at 588.995 and 589.592 nm) are characteristically bright optical transitions in the visible range. All other oblique $\Delta l = \pm 1$ transitions are electric-dipole allowed too. (Data from Ref. [20])

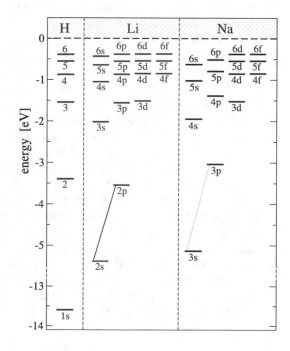

Table 2.3 Comparison of the spin-orbit splittings and effective nuclear charge $Z_{\text{eff s-o}}$ of the lowest excited p level of hydrogen and of the alkali atoms

Element	H	Li	Na	K	Rb	Cs
Z	1	3	11	19	37	55
Single-electron excited level n_0p	2p	2p	3p	4p	5p	6p
Spin-orbit splitting $\frac{3}{2}\xi_{n_0p}$ [meV]	0.045	0.042	2.1	7.2	29.5	68.7
$Z_{\text{eff s-o}}$	1	0.98	3.5	6.0	10.0	14.2

the ~ 2 eV n_0s $- n_0$p separation, with no analogue in hydrogen spectroscopy. This n_0s $\leftrightarrow n_0$p transition is especially intense due to the large electric-dipole matrix element involving strongly overlapping and fairly extended wavefunctions.

The spin-orbit coupling affects all non-s states. The natural generalization of Eqs. (2.32) and (2.35) to a generic radial potential yields the following microscopic estimate of the spin-orbit energy prefactor ξ_{nl} for a given subshell:

$$\xi_{nl} = \frac{\hbar^2}{2\,m_e^2\,c^2}\,\langle n, l | \frac{1}{r}\,\frac{dV_{\text{eff}}(r)}{dr} | n, l \rangle = Z_{\text{eff s-o}}^4\,\alpha^2 E_{\text{Ha}}\frac{1}{n^3 l(l+1)(2l+1)}\,, \quad (2.78)$$

where $|n, l\rangle$ represents the HF radial wavefunction of that subshell. $Z_{\text{eff s-o}}$ is implicitly defined by this equation. A guess of $Z_{\text{eff s-o}}$ provides a rough estimate of ξ_{nl}. Due to the strong localization of the mean field $r^{-1}\,dV_{\text{eff}}(r)/dr$ near the origin, the effective charge $Z_{\text{eff s-o}}$ is usually larger than the quantity Z_{eff} introduced above for estimating the energy level: $Z_{\text{eff}} < Z_{\text{eff s-o}} < Z$. Table 2.3 reports the observed spin-orbit splittings of the lowest excited p state of the alkali atoms. The spin-orbit level splittings determine a fine structure of all optical transitions. These splittings are usually larger than those of hydrogen: for example the 3p splitting originates a well-resolved doublet structure in the optical spectrum of sodium vapors: the characteristic yellow 3s–3p lines at wavelengths 588.99 and 589.59 nm. Remarkably, the spin-orbit splitting of 2p is smaller in Li than in H: the reason is that most of the Li 2p wavefunction lies well outside the compact 1s screening shell.

The fine-structure splittings of higher excited non-s states are smaller than those of the n_0p states, see Fig. 2.25. In the Li spectrum all doublets are split with the smaller J at lower energy, as expected by $\xi_{nl} > 0$, Eq. (2.78). However, the spectra of Na and K display an inverted splitting of the $l \geq 2$ doublets, due to nontrivial correlation effects (neglected in the mean-field HF approximation) involving the valence and core electrons [6] prevailing over the weak spin-orbit coupling.

Within the HF model, Eq. (2.78) can be used to estimate the spin-orbit energy ξ_{nl} for any shell of any atom, not just for the excited shells of alkali atoms. For example, the large effective $Z_{\text{eff s-o}}$ accounts for the colossal spin-orbit splittings (tens or hundreds eV) of the core shells of heavy atoms, observed in X-ray spectra (e.g. the L_{II}–L_{III} splitting shown in Fig. 2.23).

Fig. 2.25 The fine structure of a few low-energy excited levels of sodium. All splittings are highly magnified to make them visible. The regular $P_{1/2}$–$P_{3/2}$ splitting of the p states is governed by the spin-orbit interaction. Note the inverted splitting of the $D_{5/2}$–$D_{3/2}$ doublets. The reference energy is the same as in Fig. 2.24

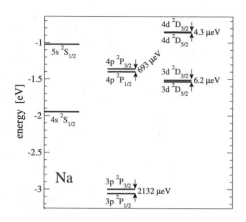

2.2.9.2 Atoms with Elementary Ground States

For a number of atoms, the electron occupations of the individual subshells are sufficient to determine uniquely all global symmetry properties of the atomic ground state, in particular its total angular momentum J (thus its degeneracy).

The ground-state symmetry of noble gases, alkaline earth and, in general, all atoms in close-shell configurations, including Zn, Cd, Hg, Yb, and No is trivial, as all orbital and spin angular momenta vanish: the ground states of all these atoms qualify as nondegenerate spherically symmetric $^1S_0$.

Likewise, the ground state of alkali elements is simply $^2S_{1/2}$, with a twofold degeneracy associated entirely to spin, with no orbital degeneracy.

B, Sc, and other atoms in groups 13 and 3, with a single electron in a degenerate p or d shell and all inner shells complete, are only marginally more complicated. Here the total spin and orbital angular momenta coincide with those of the unpaired electron. Spin-orbit coupling splits the $J = L \pm 1/2$ states, putting $J = L - 1/2$ lower. Accordingly, the ground-state term symbol of B is $^2P_{1/2}$ and that of Sc is $^2D_{3/2}$.

A final relatively simple class is that of the halogen atoms (group 17, p^5 configuration), characterized by one missing electron (a *hole*) in an otherwise full shell. This hole carries the same spin and orbital angular momentum ($S = 1/2$ and $L = 1$) as one electron in that shell. However, the sign of the effective spin-orbit interaction for the hole is reversed.[13] As a consequence, the halogen atoms have $^2P_{3/2}$ term symbol. Similarly, the term symbol of the 4f^{13} ground state of Tm is $^2F_{7/2}$.

---

[13]For N electrons in a degenerate shell, the spin-orbit operator is $H_{s\text{-}o} = \xi \sum_{i=1}^{N} \mathbf{l}_i \cdot \mathbf{s}_i$. When the shell is full but for one electron, i.e. it contains $N = 2(2l + 1) - 1 \equiv d - 1$ electrons, one can rewrite the previous expression as: $H_{s\text{-}o} = \xi \sum_{i=1}^{d} \mathbf{l}_i \cdot \mathbf{s}_i - \xi \mathbf{l}_d \cdot \mathbf{s}_d$. Observe that the first term vanishes, since in a completely full shell each product $\mathbf{l}_i \cdot \mathbf{s}_i$ has another product $\mathbf{l}_j \cdot \mathbf{s}_j$ which cancels the former one. Note also that $\mathbf{l}_d = \sum_i^d \mathbf{l}_i - \sum_i^N \mathbf{l}_i = 0 - \mathbf{L} = -\mathbf{L}$ and, likewise, $\mathbf{s}_d = -\mathbf{S}$. As a result, $H_{s\text{-}o} = -\xi \mathbf{L} \cdot \mathbf{S}$. In words, the effective spin-orbit interaction of a "missing" electron in an otherwise full shell (e.g. the p^5 configuration of a halogen atom) is the same as that of one electron, but with reversed coupling $\xi_{\text{eff}} = -\xi < 0$.

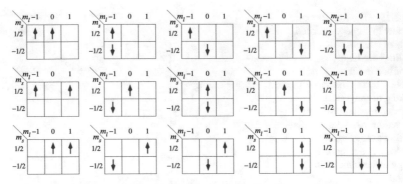

Fig. 2.26 The $6 \times 5/2 = 15$ Slater-determinant states of a p^2 configuration

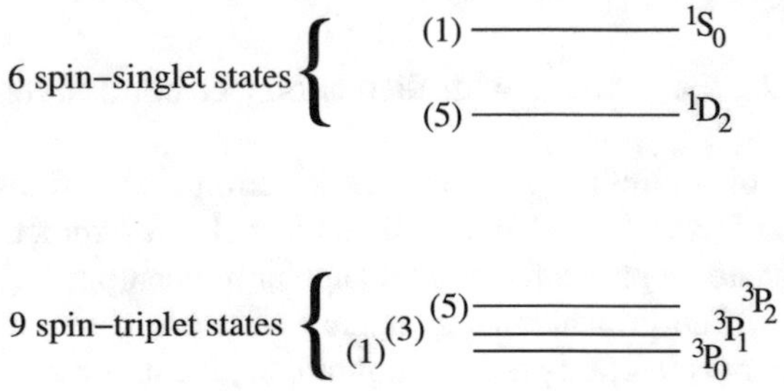

Fig. 2.27 The observed ordering of the p^2 multiplets of carbon and silicon, which follow Hund's rules. Labels adopt the LS coupling scheme. The appropriate $(2J + 1)$ degeneracies are indicated near the levels

6 spin–singlet states $\left\{ \begin{array}{l} (1) \quad\rule{2cm}{0.4pt}\quad {}^1S_0 \\[1em] (5) \quad\rule{2cm}{0.4pt}\quad {}^1D_2 \end{array} \right.$

9 spin–triplet states $\left\{ \begin{array}{l} (5)\rule{1cm}{0.4pt}\ {}^3P_2 \\ (3)\rule{1cm}{0.4pt}\ {}^3P_1 \\ (1)\rule{1cm}{0.4pt}\ {}^3P_0 \end{array} \right.$

2.2.9.3 Atoms with Incomplete Degenerate Shells

Things get intricate when several electrons occupy a degenerate shell, but are too few to fill it completely. The HF independent-electron ground state usually predicts incorrect degeneracies. For example, the two 2p electrons of carbon can settle in any of the 15 Slater-determinant states sketched in Fig. 2.26. HF puts all the six $M_S = \pm 1$ states of the leftmost and rightmost column at the same energy, and the remaining nine $M_S = 0$ states in the three central columns at a higher energy. This splitting is due to the exchange term in Eq. (2.75). However, the actual ground state of C is *not* 6-fold degenerate! The $M_S = \pm 1$ are the high-$|S_z|$ components of a ^3P spin and orbital triplet, consisting of $3 \times 3 = 9$ states. The three missing $M_S = 0$ states can be constructed as suitable linear combinations of the nine $M_S = 0$ Slater determinant, combined to build S^2 and L^2 eigenstates. These combinations involve several Slater determinants, and therefore the HF method cannot be applied to these combinations without specific adaptations. The 6 remaining $M_S = 0$ combinations are excited states, identified in Fig. 2.27 as $^1D_2$ and $^1S_0$. The splittings between these states is due to mechanisms discussed in the following.

In general, N electrons can occupy $d = 2(2l + 1)$ degenerate spin-orbitals in $\binom{d}{N} = d!/N!/(d - N)!$ different ways, corresponding to physically different (orthonormal) quantum states. The degeneracy of these states is partly, but generally incorrectly, lifted by the HF method. In an actual atom, the splitting of these

Table 2.4 All LS (Russell-Saunders) multiplet term symbols into which Coulomb exchange and correlation split a $[l]^N$ configuration, for $l = 0$–3 and all possible fillings N; when several states with the same $^{2S+1}[L]$ occur in that configuration, the number of occurrences is noted below the letter $[L]$ denoting the total orbital angular momentum. After Ref. [5]

$[l]^N$	$S = 0$	$1/2$	1	$3/2$	2	$5/2$	3	$7/2$
s		^{2}S						
p, p^5		^{2}P						
p^2, p^4	^{1}S D		^{3}P					
p^3		^{2}P D		^{4}S				
d, d^9		^{2}D						
d^2, d^8	^{1}S D G		^{3}P F					
d^3, d^7		^{2}P D F G H		^{4}P F				
		2						
d^4, d^6	^{1}S D F G I		^{3}P D F G H		^{5}D			
	2 2 2		2 2					
d^5		^{2}S P D F G H I		^{4}P D F G	^{6}S			
		3 2 2						
f, f^{13}		^{2}F						
f^2, f^{12}	^{1}S D G I		^{3}P F H					
f^3, f^{11}		^{2}P D F G H I K L		^{4}S D F G I				
		2 2 2 2						
f^4, f^{10}	^{1}S D F G H I K L N		^{3}P D F G H I K L M	^{5}S D F G I				
	2 4 4 2 3 2		3 2 4 3 4 2 2					
f^5, f^9		^{2}P D F G H I K L M N O		^{4}S P D F G H I K L M	^{6}P F H			
		4 5 7 6 7 5 5 3 2		2 3 4 4 3 3 2				
f^6, f^8	^{1}S P D F G H I K L M N Q		^{3}P D F G H I K L M N O		^{5}S P D F G H I K L		^{7}F	
	4 6 4 8 4 7 3 4 2 2		6 5 9 7 9 6 6 3 3		3 2 3 2 2			
f^7		^{2}S P D F G H I K L M N O Q		^{4}S P D F G H I K L M N	^{6}P D F G H I		^{8}S
		2 5 7 10 10 9 9 7 5 4 2		2 2 6 5 7 5 5 3 3				

multiplets is governed by (i) *Coulomb* exchange (mostly accounted for by HF) and correlation (i.e. the effects of the *residual electron-electron interaction* ignored at the HF level, inducing correlated electronic motion), and (ii) the *spin-orbit interaction*.

In the outer atomic shells and for $Z \lesssim 30$, the spin-orbit energy is a comparatively small ($\xi_{nl} \ll 1$ eV) and thus initially negligible interaction. The Coulomb repulsion induces much larger splittings, in the order of one to a few eV, as in the He example, Sect. 2.2.4. The eigenvalues of the total $|\mathbf{S}|^2$ and $|\mathbf{L}|^2$ operators provide a proper (although not always complete) labeling of the states in the multiplet manifold for given subshell occupation. Table 2.4 reports a complete list of the "multiplet" states labeled by their total spin and orbital angular momentum.

Coulomb exchange and correlation act on the degenerate HF states very much like the full Coulomb repulsion does in He (Sect. 2.2.4). First of all they split states with different total spin S: low-spin states sit higher in energy because the poorly correlated motion of the electrons allows them to come, on average, nearer to each other. The ground state will therefore have the highest possible spin: this result agrees with the empirical *Hund's first rule*.

In degenerate configurations there can occur several states with the same S, but different total orbital angular momentum L. Electrons avoid each other more efficiently when they rotate all together coordinately, in states with higher total L. Accordingly, the state with the largest possible L among those with given S sits lowest in energy. This is known as *Hund's second rule*.

Finally, once the total L and S are determined, the hitherto neglected spin-orbit interaction couples them together to a total angular momentum J. The allowed values of J are given by the usual angular-momentum rule (C.72), e.g. $^2P \rightarrow {}^2P_{1/2}, {}^2P_{3/2}$. The energy ordering of the J states is decided by the sign of the *effective spin-orbit* parameter for that partly filled shell. While the true spin-orbit parameter is necessarily positive, see Eq. (2.78), the effective parameter driving the coupling of total $\mathbf{L}$ with total $\mathbf{S}$ can be positive or negative, as discussed above for the halogens. In practice, the sign of the effective spin-orbit coupling reverses when more than $2l + 1$ electrons occupy a shell with room for $2(2l + 1)$ electrons. Accordingly, *Hund's third rule* predicts that the lowest-energy state has $J = |L - S|$ for a less-than-half-filled shell, but $J = L + S$ when the shell is more than half filled.

Coming back to the p^2 example, spin-orbit and Coulomb couplings rearrange the 15 states of Fig. 2.26 into 9 spin-orbit split triplet levels $^3P_0, {}^3P_1, {}^3P_2$, plus 6 singlet states $^1D_2, {}^1S_0$. The observed ordering of these levels, Figs. 2.27 and 2.28a, follows Hund's rules. No other term is compatible with Pauli's principle, see Table 2.4.

Similar Hund-rules level ordering is usually observed also in configurations involving several incomplete shells, as occur in atomic excited states—see Fig. 2.28b. Coulomb couplings scatter the multiplets generated by each configuration over energy ranges, shown as double arrows in Fig. 2.21 for a few low-Z atoms. The low-energy multiplets extend over several eV, while more excited multiplets scatter less, because the Coulomb repulsion is smaller for more extended wavefunctions. Note also that Hund's rules are phenomenological results, not exact laws of nature. Their predictions are quite reliable for ground levels, but less reliable for excited states. For example, for the $3d^2 4s^2$ configuration of Ti, Hund's rules predict an ordering $^3F < {}^3P < {}^1G < {}^1D < {}^1S$, while in reality ^1D lies below ^3P.

This coupling order, first all $\mathbf{s}_i$ together to a total $\mathbf{S}$ and all $\mathbf{l}_i$ together to a total $\mathbf{L}$ followed by spin-orbit coupling of $\mathbf{S}$ and $\mathbf{L}$ together, is called *LS* or *Russell-Saunders coupling*. It provides a satisfactory basis for the multiplet levels of low-Z atoms, where Coulomb exchange and correlation dominate over spin orbit. For increasing Z, the spin-orbit interaction grows rapidly, while the electron-electron repulsion remains in the few-eV range, and even weakens due to the valence orbitals spreading out. For very large $Z \geq 50$ spin-orbit dominates: $H_{\text{s-o}}$ must be accounted for before Coulomb terms. The spin-orbit interaction couples the spin and orbital moment of each electron to an individual $j_i = l_i \pm 1/2$. These individual-electron total angular momenta are then coupled to a total J by weaker Coulomb terms. This alternative coupling scheme for the angular momenta is called *jj coupling*, and provides another basis for the many-electron states. While the LS basis is nearly diagonal for small

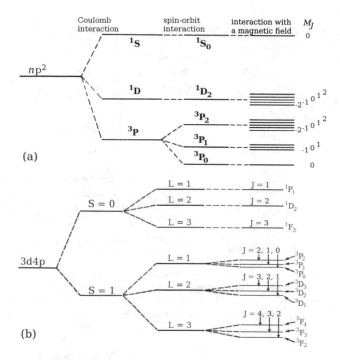

Fig. 2.28 The conceptual sequence of splittings of LS multiplets caused by interactions of decreasing strength. **a** The $6 \times (6-1)/2! = 15$ states of two "equivalent" electrons in a np^2 configuration, including Coulomb and spin-orbit splittings plus those induced by a weak external magnetic field (Zeeman limit). **b** The $6 \times 10 = 60$ states of two "nonequivalent" (different n and/or l) electrons in a 3d4p configuration, occurring, e.g., in the spectrum of excited titanium

Z, the jj basis is almost diagonal for large Z (see Fig. 2.29). For intermediate Z, both bases are nondiagonal: to construct the proper atomic eigenstates as linear combinations of the states of either basis, a matrix including both the Coulomb and the spin-orbit terms must be diagonalized.

2.2.9.4 Many-Electron Atoms in Magnetic Fields

When a uniform magnetic field acts on a many-electron atom, two very different behaviors are observed depending on whether the atom carries a magnetic moment or not. Atoms with total angular momentum $J = 0$ carry no permanent magnetic dipole available to align with the field: the field induces a small magnetic moment $\sim \mu_B (\mu_B B)/\Delta$, where Δ is the energy gap between the ground state and the lowest $J = 1$ excitation. We will ignore such tiny effects.

In contrast, atoms with a $J \neq 0$ ground state do carry a magnetic moment μ, whose component along the magnetic field $\mu_z = -g_J \mu_B J_z/\hbar$. For LS coupling, the appropriate Landé g-factor g_J is determined by Eq. (C.78), with the total angular

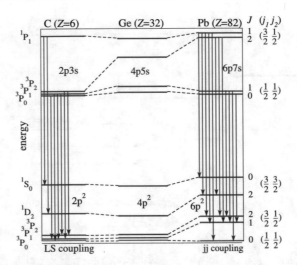

Fig. 2.29 A level correlation diagram, illustrating, for increasing Z, the effect of the increasing relative magnitude of the spin-orbit coupling over the Coulomb exchange and correlation terms: atomic spectra evolve from the pure LS coupling of carbon, to the intermediate coupling of germanium, with Coulomb and spin-orbit of similar magnitude, to the jj coupling of lead, where the spin-orbit interaction dominates

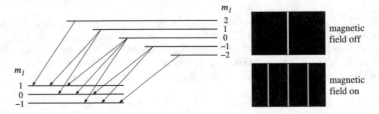

Fig. 2.30 A regular Zeeman spectrum, with its interpretation. It occurs in transitions between $S = 0$ states, which have the same initial and final Landé g-factors. For example, it is observed in the $2s3d\,^1D_2 \to 2s2p\,^1P_1$ emission of Be

momenta J, L, and S in place of the single-electron j, l, and s. As discussed in Sect. 2.1.10, this total magnetic moment, derived by the coupling of orbital and spin contributions, is relevant in the limit of weak external magnetic field (Zeeman limit). In practical experiments this is the relevant limit for many-electron atoms, due to the Z^4 increase of the spin-orbit energy ξ_{nl}, usually exceeding the maximum field strength accessible in the lab (in the order of 10 T). The opposite strong-field (Paschen-Back) limit can occasionally be realized, since electrons in highly-excited states are weakly affected by the field of the nucleus, and thus have a small ξ_{nl}.

The simplest splitting pattern (*regular Zeeman splitting*) consists of three equally spaced lines, corresponding to $\Delta M_J = -1$, 0, and 1. An example is shown in Fig. 2.30. This pattern occurs when the initial and final g-factors are equal, as occurs when $S = 0$, or when $L = 0$, or when both S and L remain unchanged in the transi-

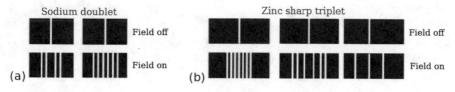

Fig. 2.31 Anomalous Zeeman splittings in the spectra of **a** Na and **b** Zn

tion. The simplest example of regular Zeeman splitting is observed in optical transitions between spin-singlet states ($S = 0$, so that $g = g_L = 1$). In most cases, the Zeeman spectrum involves more complicated patterns due to the different initial and final g-factors (*anomalous Zeeman splitting*, see Fig. 2.31).

To investigate the ground-state degeneracy and magnetic moment of many-electron atoms, the most straightforward technique is the Stern-Gerlach experiment (Sect. 2.1.5). The amount of deflection of the atoms in a field gradient measures the z component of the magnetic moment, and therefore the Landé g-factor g_J, according to Eq. (2.26):

$$\mathbf{F}_z = \mu_z \frac{\partial B_z}{\partial z} = -\mu_B \, g_J \, M_J \, \frac{\partial B_z}{\partial z}. \tag{2.79}$$

The number of sub-beams into which the inhomogeneous field splits the original beam measures directly the number of allowed M_J values, i.e. the ground-state degeneracy $2J + 1$.

2.2.10 Electric-Dipole Selection Rules

As discussed in Sect. 2.2.7, the main electric-dipole selection rule requires that a single electron jumps to another state satisfying

$$\Delta l = \pm 1 \quad \text{(for the one electron making the transition).} \tag{2.80}$$

A few extra electric-dipole selection rules for the total quantum numbers J, L, and S of many-electron atoms in LS coupling also apply. They are summarized below:

$$\text{Parity changes}, \tag{2.81}$$

$$\Delta S = 0, \tag{2.82}$$

$$\Delta L = 0, \pm 1, \tag{2.83}$$

$$\Delta J = 0, \pm 1 \quad \text{(no } 0 \to 0 \text{ transition)}, \tag{2.84}$$

$$\Delta M_J = 0, \pm 1 \quad \text{(no } 0 \to 0 \text{ transition if } \Delta J = 0). \tag{2.85}$$

As both S and L are good quantum numbers only in the limit of very small spin-orbit, in practice selection rules (2.82) and (2.83) are only approximate.

Figure 2.29 draws the allowed transitions in characteristic examples of LS coupling and jj coupling. In this latter scheme, specific dipole selection rules apply, which are described in advanced atomic-physics textbooks [6].

Final Remarks

The present Chapter summarizes the main concepts and experimental evidence in the field of atomic spectroscopy which, in the context of a general course on fundamentals, provide the language for the microscopic atomic structure, which lays the grounds of the physics of matter. Subtle conceptual points (e.g. the seniority scheme for the labeling of LS states when L, S and J are not sufficient), and modern spectroscopic techniques (e.g. Auger) are omitted altogether. For these and other more advanced subjects, including countless analytic chemical and astrophysical applications of atomic physics, we invite the reader to refer to specific textbooks [5–7].

Problems

A ⋆ marks advanced problems.

2.1 A beam of oxygen atoms emerges from an oven with approximately equal populations of its $1s^2 2s^2 2p^2$ triplet states: $^3P_0$, $^3P_1$, $^3P_2$. The average kinetic energy is $E_{\text{kin}} = 0.2$ eV. These atoms are sent through a $l = 0.3$ m–long Stern-Gerlach magnet, where a field gradient $\partial B / \partial z = 150$ T/m is present. Successively, the beam crosses a region of length $l' = 0.5$ m where the magnetic field is negligible. Compute the total number of beam components detected at the end of the SG apparatus and the distance between the most widely spaced components.

2.2 The two lowest electric-dipole-allowed optical transitions starting from the ground state of the Ne atom are observed at energies 16.8 and 19.8 eV. Determine the electronic configuration and the corresponding spectroscopic term of both final excited states involved in these transitions. Describe how these absorption lines change when a uniform magnetic field of 1 T acts on a gaseous Ne sample.

2.3 Construct the scheme of the core levels of Sn, indicating the excitation energies in eV. The K edge is observed at $\lambda_K = 42.5$ pm, and the wavelengths of the first two lines of the K series are: $K_\alpha = 51.7$ pm, $K_\beta = 43.7$ pm. Evaluate the minimum excitation energy needed to observe the L series emission lines after excitation. Evaluate the maximum kinetic energies of photoelectrons produced by the excitation of shells K, L, M of Sn induced by 32 keV X rays.

2.4 Two hydrogen atoms are excited in the quantum states described by the following wavefunctions:

$$(a) \quad \psi_{300} = \frac{1}{9\sqrt{3\pi}\, a_0^{3/2}} \left(3 - 2\frac{r}{a_0} + \frac{2}{9}\frac{r^2}{a_0^2} \right) e^{-r/3a_0}$$

and

$$(b) \quad \psi_{210} = \frac{1}{4\sqrt{2\pi}\, a_0^{3/2}} \frac{r}{a_0} e^{-r/2a_0} \cos\theta,$$

where a_0 is the Bohr radius. For each, evaluate the decay probability rate γ to the ground state

$$\psi_{100} = \frac{1}{\sqrt{\pi}\, a_0^{3/2}} e^{-r/a_0}$$

in the electric-dipole approximation.
[Recall that in this approximation $\gamma_{if} = (3\pi\epsilon_0)^{-1}\hbar^{-4}c^{-3}(\mathcal{E}_i - \mathcal{E}_f)^3 |\langle f|\mathbf{d}|i\rangle|^2$, where the Cartesian components of $\mathbf{d}$ are $-q_e r \sin\theta \cos\varphi$, $-q_e r \sin\theta \sin\varphi$, and $-q_e r \cos\theta$.]

2.5 The 7 valence electrons of the Fe^+ ion in its ground state assume the configuration $3d^6\,4s$. Evaluate the ground-state magnetic moment and the number of components into which the spin-orbit interaction splits the lowest-energy (according to Hund's first and second rules) degenerate configuration.

2.6 Determine the number of absorption lines that a gas-phase sample of atomic chlorine at 1000 K exhibits in the $3s^2 3p^5 (^2P) \rightarrow 3s^2 3p^4 4s (^2P)$ transition. Assume a significant thermal population of the excited state of the $3s^2 3p^5 (^2P)$ configuration, given that this excited state sits 109.4 meV above the ground state. If this sample is immersed in a static uniform magnetic field $|\mathbf{B}| = 1.5$ T, how many distinct sub-lines does each of the lines determined above split into?

2.7 The three components of the ground configuration $[Ar]3d^2 4s^2\,^3F$ of atomic titanium sit at energies 0 eV, 0.02109 eV, and 0.04797 eV. The absorption spectrum of Ti vapor at 1000 K involves lines related to transitions starting from the ground-state components to $3d^2 4s4p$ levels, which are organized in three groups of states with total orbital angular momentum $L = 2$, 3, and 4. Establish the quantum numbers of the final states of the electric-dipole allowed transitions in each of the three groups. Considering now uniquely the transitions starting from the ground state $^3F_2$, identify the most intense transition in each group.

2.8* Evaluate the magnitude of the magnetic moment $|\boldsymbol{\mu}|$ of atomic Sc, V, and Mn in their respective ground states $3d4s^2$, $3d^34s^2$, $3d^54s^2$. Beams of such atoms with equal kinetic energy are sent through one Stern-Gerlach apparatus. Identify for which of these three elements the least-deflected beam component is deflected the most.

2.9 A 30 keV X-ray beam hits a palladium target. Among the photoemitted electrons, several are measured at values of kinetic energy: 5650, 26,396, 26,670, and 26,827 eV. Given these data, evaluate the core-shell energies of Pd, assign the departing shell for the electrons of the smallest kinetic energy and evaluate the effective charge Z_{eff} (with three significant digits) relevant for the motion of an electron in that shell, consistently with the measured photoelectron energies.

2.10 Helium atoms in the excited state $1s2s(^3S_1)$ are generated and accumulated in a storage vessel for (on average) one hour. They leave it through a tiny orifice at an average speed 2000 m/s, and into a Stern-Gerlach apparatus characterized by a length 0.5 m and a field gradient $\partial B/\partial z = 9$ T/m. Given that the lifetime τ of the metastable $1s2s(^3S_1)$ state is 8000 s, evaluate the fraction of deflected atoms with a $M_J = 1$ component of total angular momentum over the total number of atoms leaving the storage vessel in a given time. Evaluate also the angular deflection of that component.

2.11 A 10 keV photon beam hits a cesium target. A part of the photoemitted electrons is collimated by suitably placed slits and sent into a transverse-field analyzer characterized by $|\mathbf{B}| = 0.1$ T. Electrons deflected with curvature radii 2.3 and 3.2 mm are observed. Based on these data, assign the core shells from which the electrons are emitted and evaluate their excitation energy.

2.12* In the standard formulation of the hydrogen-atom problem, the nucleus is taken as a point charge producing a potential energy $V_{\text{Coul}}(r) = -e^2/r$ for the electron at distance r. In fact the nucleus has a finite size. Modeling it as uniform positive charge in a sphere of radius $r_n = 0.9$ fm, the potential energy changes to $V_{\text{true}}(r) = -3e^2/(2r_n) + e^2r^2/(2r_n^3)$ inside the nucleus (no change outside, of course). Evaluate the effect of this modification on the hydrogen ground-state energy to first order in perturbation theory. [Hint: approximate $\exp(x) \simeq 1$ when $|x| \ll 1$.]

2.13 For the iron atom ($Z = 26$, configuration $[\text{Ar}]3d^64s^2$) the energy separation between the ground state and the lowest fine-structure excited level is 51.6 meV. Estimate the excitation energies of the 3 successive energy levels. Verify quantitatively whether a 3 T external magnetic field produces Zeeman or Paschen-Back splittings in case these atomic states were involved in a optical transition.

Chapter 3
Molecules

Molecules, particularly diatomic ones, are the simplest pieces of matter where several nuclei are bound together by their interaction with electrons. Their simplicity makes diatomic molecules the ideal playground to approach two central concepts of condensed-matter physics: the adiabatic separation of the electronic and nuclear motions, and chemical bonding.

3.1 The Adiabatic Separation

The full Hamiltonian (1.1) for a piece of matter, and likewise its eigenfunction Ψ, depend on the coordinates r of all electrons and R of all nuclei. This entangled dependence makes the resulting information content catastrophically and needlessly complicate. A disentanglement of the fast electron dynamics from the slow motion of the nuclei is a crucial conceptual step to make progress in understanding and interpreting the dynamical behavior of matter.

Since the electron mass is much smaller than the nuclear mass, the *motions of the nuclei and electrons* are likely to *occur over different* (and hopefully uncoupled) *time scales*. As masses appear at the denominators of the kinetic terms Eqs. (1.2) and (1.3), electrons move much faster than atomic nuclei. The fast electrons should be capable to follow the slow displacements of the nuclei with essentially no delay.

To implement this decoupling consider the following factorization of the total wavefunction:

$$\Psi(r, R) = \Phi(R)\,\psi_e(r, R). \tag{3.1}$$

Assume that the *electronic wavefunction* $\psi_e(r, R)$ is a solution $\psi_e^{(a)}(r, R)$ of the following *electronic equation*:

N. Manini, *Introduction to the Physics of Matter*, Undergraduate Lecture Notes in Physics, https://doi.org/10.1007/978-3-030-57243-3_3

$$[T_e + V_{ne}(r, R) + V_{ee}(r)]\,\psi_e^{(a)}(r, R) = E_e^{(a)}(R)\,\psi_e^{(a)}(r, R)\,, \qquad (3.2)$$

where (a) represents the set of quantum numbers characterizing a given N-electron eigenstate with energy $E_e^{(a)}$. The electronic eigenfunction $\psi_e^{(a)}(r; R)$ describes an electronic eigenstate compatible with a fixed geometrical configuration R of the ions: the R-dependence of the eigenfunction $\psi_e^{(a)}(r; R)$ and eigenenergy $E_e^{(a)}(R)$ is purely parametric [no ∇_R operators in Eq. (3.2)]. The factorization defined by Eqs. (3.1, 3.2), (known as the *adiabatic* or Born-Oppenheimer scheme, implies the assumption that, once an initial electronic eigenstate is selected, the atomic nuclei move slowly enough not to induce transitions to other electronic states. These transitions are prevented by the electronic energy gaps usually being much larger than the typical energies associated with the slow motion of the nuclei.

To make use of the *ansatz* (3.1), note first that $T_e \sim \nabla_r^2$ does not act on the R coordinates:

$$T_e\,[\Phi(R)\,\psi_e(r, R)] = \Phi(R)\,[T_e\psi_e(r, R)]\,. \qquad (3.3)$$

Then observe that the nuclear kinetic term acts on both factors:

$$\nabla_R^2[\Phi(R)\,\psi_e(r, R)] = \psi_e(r, R)\,\nabla_R^2\Phi(R) + 2\,[\nabla_R\psi_e(r, R)]\,\nabla_R\Phi(R) + \Phi(R)\,\nabla_R^2\psi_e(r, R)\,. \qquad (3.4)$$

We can now substitute this result into the Schrödinger equation $H_{tot}\Psi = E_{tot}\Psi$ with $\Psi(r, R) = \Phi(R)\,\psi_e(r, R)$. We obtain

$$-\sum_\alpha \frac{\hbar^2}{2M_\alpha}\left\{2\left[\nabla_{R_\alpha}\psi_e(r, R)\right]\nabla_{R_\alpha}\Phi(R) + \Phi(R)\nabla_{R_\alpha}^2\psi_e(r, R) + \psi_e(r, R)\nabla_{R_\alpha}^2\Phi(R)\right\}$$

$$+ \Phi(R)\,T_e\psi_e(r, R) + [V_{ne} + V_{ee} + V_{nn}]\,\Phi(R)\psi_e(r, R) = E_{tot}\,\Phi(R)\psi_e(r, R)\,, \qquad (3.5)$$

where we omit the obvious coordinate dependence of the potential-energy terms. We can rearrange Eq. (3.5) as follows:

$$-\sum_\alpha \frac{\hbar^2}{2M_\alpha}\left\{2\left[\nabla_{R\alpha}\psi_e(r, R)\right]\nabla_{R\alpha}\Phi(R) + \Phi(R)\,\nabla_{R\alpha}^2\psi_e(r, R)\right\} + \qquad (3.6)$$

$$+ \psi_e(r, R)\left[-\sum_\alpha \frac{\hbar^2\nabla_{R\alpha}^2}{2M_\alpha}\right]\Phi(R) + \Phi(R)\left[\underbrace{T_e + V_{ne} + V_{ee}} + V_{nn}\right]\psi_e(r, R) = \qquad (3.7)$$

$$= E_{tot}\,\Phi(R)\psi_e(r, R)\,,$$

in order to highlight the electronic Hamiltonian ($T_e + V_{ne} + V_{ee}$) of Eq. (3.2).

The two terms of line (3.6), involving derivatives ∇_R of the electronic wavefunction $\psi_e(r, R)$ are called *nonadiabatic terms*. These nonadiabatic terms (whose nature is nuclear kinetic) are much smaller (usually by a factor $\approx m_e/M_\alpha$) than the typical differences between electronic eigenenergies of Eq. (3.2), dominating Eq. (3.7). The *adiabatic approximation* consists precisely in the neglect of the terms of line (3.6).

We proceed now to substitute an eigensolution $\psi_e^{(a)}(r, R)$ of Eq. (3.2) for $\psi_e(r, R)$ in the remaining terms of Eq. (3.7):

$$\psi_e^{(a)}(r, R) \left[-\frac{\hbar^2}{2} \sum_\alpha \frac{\nabla_{R\alpha}^2}{M_\alpha} + E_e^{(a)}(R) + V_{nn}(R) \right] \Phi(R) = E_{\text{tot}} \, \psi_e^{(a)}(r, R) \, \Phi(R),$$

(3.8)

where the electronic wavefunction $\psi_e^{(a)}(r, R)$ is displaced to the left of the operator, to stress that the differential part only acts on the factor $\Phi(R)$ regarding the nuclei. Note that all three terms T_e, V_{ne} and V_{ee} are entirely (and in principle exactly) accounted for by the electronic eigenvalue $E_e^{(a)}(R)$, the inter-nuclear repulsion V_{nn} remains indicated explicitly, and only the nuclear kinetic term T_n is treated approximately, due to the neglect of the nonadiabatic corrections of line (3.6). The electronic wavefunction can now be dropped [formally by multiplying Eq. (3.8) by $\psi_e^{(a)*}(r, R)$ and integrating over all electronic coordinates r], to derive the equation for the adiabatic motion of the nuclei described by $\Phi(R)$:

$$\left[-\frac{\hbar^2}{2} \sum_\alpha \frac{\nabla_{R\alpha}^2}{M_\alpha} + E_e^{(a)}(R) + V_{nn}(R) \right] \Phi(R) = E_{\text{tot}} \, \Phi(R). \qquad (3.9)$$

The **electronic equation** (3.2) describes the motion of all electrons in the piece of matter. When the number of electrons is greater than one (as is usually the case), Eq. (3.2) involves at least the same technical difficulties as many-electron atoms discussed in Sect. 2.2, made worse by the lack of spherical symmetry. Equation (3.2) is often solved within an approximate quantum many-body method, e.g. the Hartree-Fock method sketched in Sect. 2.2.5, or the density-functional theory [23]. In the polyatomic context, theory faces the extra difficulty that the electronic problem depends explicitly on the position R of all nuclei, through V_{ne}. One should thus solve the electronic problem for very many geometric arrangements (classical *configurations* R) of the nuclei, to obtain a detailed knowledge of the parametric dependence of the electronic eigenfunction $\psi_e^{(a)}(r, R)$ and eigenvalue $E_e^{(a)}(R)$ on the $3N_n$ coordinate of the nuclei R. Within the adiabatic approximation, once the electronic eigenstate (a) is chosen, this state never mixes with other eigenstates (a'): the electronic state follows adiabatically the slow nuclear motion. In reality, the neglected nonadiabatic terms originate a small probability that the motion of the nuclei induces transitions to different electronic states.

Assuming that the electronic eigenvalue $E_e^{(a)}(R)$ is available as a function of R, the **Schrödinger equation for the nuclei** (3.9) describes the displacements of the nuclei as driven by a *total adiabatic potential energy*

$$V_{\mathrm{ad}}^{(a)}(R) = V_{nn}(R) + E_e^{(a)}(R), \qquad (3.10)$$

which is the sum of the Coulombic ion-ion repulsion $V_{nn}(R)$ plus the electronic eigenvalue $E_e^{(a)}(R)$. Both these energies change as a function of the $3N_n$ coordinates of the nuclei. $V_{nn}(R)$ is repulsive, and if it was alone it would of course drive an explosive Coulomb evaporation of any piece of matter. The second term $E_e^{(a)}(R)$, the "adiabatic electronic contribution", acts as a sort of "glue" which tends to keep the atoms close together.[1]

Of all the electronic eigenstates labeled by a, the *electronic ground state* $a =$ gs generates the especially important *lowest adiabatic potential energy* $V_{\mathrm{ad}}^{(\mathrm{gs})}(R)$, in short $V_{\mathrm{ad}}(R)$. Electronic excitations $a \neq$ gs generate different adiabatic potential energy surfaces.

In the adiabatic scheme, the total adiabatic potential $V_{\mathrm{ad}}(R)$ guides the displacements of the nuclei through Eq. (3.9). Since in common language the adjective "nuclear" recalls the internal dynamics of nuclei, in practice $V_{\mathrm{ad}}(R)$ is referred to as the potential energy governing the "*atomic*" or "*ionic*" motion.

As a function of the $3N_n$ ionic coordinates, $V_{\mathrm{ad}}(R)$ exhibits two general symmetries, both consequences of the symmetries of the original Hamiltonian (1.1) describing an isolated system:

- *Translational symmetry*: if all ions are displaced by an arbitrary (equal for all) translation $\mathbf{u}$, then V_{ad} remains unchanged. In formula: $V_{\mathrm{ad}}(\mathbf{R}_1 + \mathbf{u}, \mathbf{R}_2 + \mathbf{u}, \ldots, \mathbf{R}_{N_n} + \mathbf{u}) = V_{\mathrm{ad}}(\mathbf{R}_1, \mathbf{R}_2, \ldots, \mathbf{R}_{N_n})$.
- *Rotational symmetry*: if all ions are rotated by an arbitrary (equal for all) rotation A around a given arbitrary point in space, then V_{ad} also remains unchanged. In symbols: $V_{\mathrm{ad}}(A\mathbf{R}_1, A\mathbf{R}_2, \ldots, A\mathbf{R}_{N_n}) = V_{\mathrm{ad}}(\mathbf{R}_1, \mathbf{R}_2, \ldots, \mathbf{R}_{N_n})$.

These symmetries indicate that V_{ad} depends on the *relative positions* of the atoms. $V_{\mathrm{ad}}(R)$ often exhibits an absolute minimum, for the atoms placed at certain specific relative positions R_{M}: the *equilibrium configuration* or *equilibrium geometry*.

In the simplest example of a di-atom, a priori $V_{\mathrm{ad}}(R) = V_{\mathrm{ad}}(\mathbf{R}_1, \mathbf{R}_2)$ is a function of 6 coordinates, but due to the translational and rotational symmetries, it depends uniquely on the distance $R_{12} = |\mathbf{R}_1 - \mathbf{R}_2|$ between the two nuclei. For any two atoms picked at random from the periodic table, the potential energy as a function of R_{12} takes the qualitative shape shown in Fig. 3.1, with an equilibrium inter-ionic separation R_{M} at which $V_{\mathrm{ad}}(R_{12})$ is minimum. If the potential well is deep enough,

---

[1]Note that $E_e^{(a)}(R)$ is generally a complicated simultaneous function of all ionic coordinates $\mathbf{R}_1, \mathbf{R}_2, \ldots, \mathbf{R}_{N_n}$. More precisely, $E_e^{(a)}(R)$, and therefore $V_{\mathrm{ad}}^{(a)}(R)$, cannot usually be expressed as a simple sum of two-body contributions, as one can instead express, e.g. V_{nn}, as in Eq. (1.5).

Fig. 3.1 A sketch of the
adiabatic potential for a
diatom, as a function of the
inter-ionic separation R_{12}.
The energy zero is taken at
the sum of the total energies
of the two individual atoms
at large distance. *Horizontal
lines* represent possible
vibrational ground and
low-energy excited levels.
The actual number and
positions of these states
depends both on the shape of
$V_{ad}(R_{12})$ and on the diatom
reduced mass

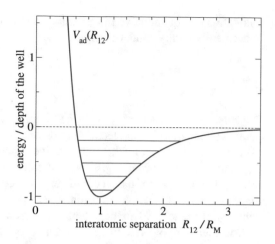

then at low temperature the ionic motion remains confined to a neighborhood of the
equilibrium position R_M: the ions execute small oscillations around R_M.

The ionic motion is sometimes treated within *classical* mechanics $[M_n \frac{d^2 R}{dt^2} = -\nabla_R V_{ad}(R)]$. This does not mean that the actual ionic motion is any classical, only
that under certain conditions (e.g. heavy ions), the classical limit can provide a fair
approximation to the actual quantum dynamics. The classical ionic dynamics often
yields useful insight into the quantum solution of Eq. (3.9). For example, the classical
problem of the normal modes describing *small independent harmonic oscillations*
around R_M is mapped to that of a set of *quantum* harmonic oscillators (only one
oscillator in the case of a diatomic molecule). In Sects. 3.3 and 5.3, we shall return
to this important problem in mechanics, and review its solutions.

In summary, the adiabatic scheme provides a fundamental separation of the coupled electron-ion dynamics into two conceptually and practically distinct problems:
the electronic equation (3.2) governs the motion of the fast electrons in the electric
field of the ions, imagined as instantaneously frozen; the Schrödinger equation (3.9)
for the slower motion of the ions is controlled by the adiabatic potential, which consists of the sum of the ion-ion Coulomb repulsion plus a "gluing" term provided by
the total energy of the electrons—Eq. (3.10).

3.2 Chemical and Non-chemical Bonding

The long-distance attractive and short-distance strongly repulsive behavior of the
adiabatic potential of a diatom, Fig. 3.1, is quite universal, and it generalizes even to
$N_n > 2$ atoms. It provides therefore the microscopic mechanism making collections
of atoms lump together, forming all kinds of bound states, including the molecular
gases, liquids, and solid objects of everyday experience. The short-distance repulsion
accounts for compressed matter resisting collapse to infinite density. To obtain both

qualitative and quantitative insight into the bonding nature of $V_{ad}(R)$, we consider conveniently simple model systems.

3.2.1 H_2^+

Our analysis starts naturally with the simplest di-atom: H_2^+, which is free of the complications of electron-electron repulsion. We construct a piece of evidence that the adiabatic energy of H_2^+ has a qualitative dependence of the inter-proton distance R_{12} of the kind sketched in Fig. 3.1: specifically, we verify that, as we reduce R_{12} from infinity, the decrease of the electronic eigenvalue $E_e^{(a)}(R_{12})$ exceeds the increase of the inter-nuclear Coulomb repulsion $V_{nn}(R_{12}) = e^2/R_{12}$, resulting in a net bonding behavior of the total adiabatic potential energy $V_{ad}^{(a)}(R_{12})$, Eq. (3.10).

Consider the potential energy V_{ne} acting on the electron of H_2^+. V_{ne} is the sum of the two attractive Coulomb terms $V_L + V_R$ produced by the left and right nuclei respectively. Eventually, we will displace the nuclei to explore the adiabatic potential energy $V_{ad}(R_{12})$, but for now we take the nuclei as stationary at a fixed positions $\pm R_{12}/2$ along the $\hat{z}$ axis. Figure 3.2 shows three "cuts" of $V_{ne}(\mathbf{r}, R_{12})$ as a function of the electron position $\mathbf{r}$. Observe that in the intermediate region, between the two nuclei, the total potential is roughly twice more negative than it would be if only one proton was present. This suggests that the electron moving in the field of both nuclei could take advantage of both attractions and lower its average potential energy by spending a significant fraction of its probability distribution in this intermediate extra-attractive region. We check if this mechanism generates bonding, i.e. if $V_{ad}(R_{12})$ decreases below its infinite-R_{12} value $V_{ad}(\infty) = E_{1s} = -1/2\,E_{Ha}$, namely the ground-state energy of one isolated hydrogen atom (no reduced-mass correction μ/m_e, because the nuclei are kept at fixed positions!).

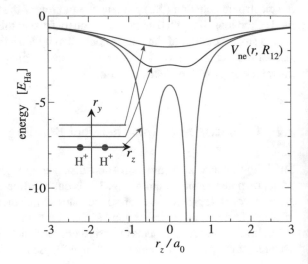

Fig. 3.2 The attractive potential energy $V_{ne}(\mathbf{r}, R_{12}) = -e^2\left(|\mathbf{r} - R_{12}\,\hat{\mathbf{z}}/2|^{-1} + |\mathbf{r} + R_{12}\,\hat{\mathbf{z}}/2|^{-1}\right)$ that the electron experiences due to the nuclei of H_2^+. This function of $\mathbf{r}$ is axially symmetric around the $\hat{z}$ axis. The curves report this function along three parallel straight lines: the line through the nuclei (the $\hat{z}$ axis), and two parallel lines at distances $a_0/2$ and a_0 from the nuclei. In this example we consider an internuclear distance $R_{12} = a_0$

To estimate if any energy can be gained, we use a simple variational approach (see Appendix C.5) based on "trial" states built as linear combinations of two states only, namely the two 1s hydrogen orbitals, Eq. (2.21), $|1s\,L\rangle$ and $|1s\,R\rangle$, centered around the left and right nucleus respectively. Guided by the reflection symmetry of $V_{ne}(\mathbf{r}, R_{12})$ across the xy plane and assuming that $\langle 1s\,L|1s\,R\rangle > 0$, we build the following normalized trial electronic kets:

$$|S\rangle = \frac{1}{[2(1 + \langle 1s\,L|1s\,R\rangle)]^{1/2}} (|1s\,L\rangle + |1s\,R\rangle) \quad [\textit{Symmetric}]$$

$$|A\rangle = \frac{1}{[2(1 - \langle 1s\,L|1s\,R\rangle)]^{1/2}} (|1s\,L\rangle - |1s\,R\rangle) \quad [\textit{Antisymmetric}]. \quad (3.11)$$

Figure 3.3 sketches the wavefunctions $\psi_e^{1s\,{}^L_R}(\mathbf{r}) = \langle \mathbf{r}|{}^L_R\rangle$ and their combinations $\psi_e^{{}^S_A}(\mathbf{r}) = \langle \mathbf{r}|{}^S_A\rangle$ representing the states $|S\rangle$ and $|A\rangle$ of Eq. (3.11).

For any given inter-nuclear separation R_{12}, the variational principle guarantees that the ground-state energy $E_e^{(GS)}(R_{12}) \leq \langle S|T_e + V_{ne}|S\rangle \equiv \mathcal{E}_S$, and similarly that $E_e^{(GS)}(R_{12}) \leq \mathcal{E}_A$. For infinitely large inter-nuclear separation R_{12}, $|S\rangle$ and $|A\rangle$ are exact eigenstates, both with electronic energy $\mathcal{E}_{{}^S_A}(R_{12} = \infty) = E_{1s} = -1/2\,E_{\mathrm{Ha}}$. For finite R_{12} and omitting the 1s labels, the mean energy of these electronic states becomes

$$\mathcal{E}_{{}^S_A} = \langle {}^S_A|T_e + V_{ne}|{}^S_A\rangle = \frac{(\langle L|T_e + V_{ne}|L\rangle \pm \langle R|T_e + V_{ne}|L\rangle) + (L \leftrightarrow R)}{2(1 \pm \langle L|R\rangle)}.$$
$$(3.12)$$

The $(L \leftrightarrow R)$ terms obtained by exchanging left and right equal the previous ones, thus we can omit them, canceling the factor 2 at the denominator. We substitute

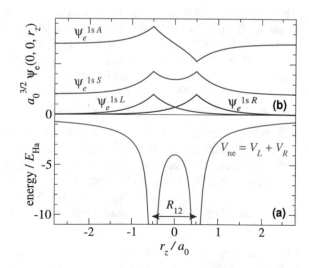

Fig. 3.3 Cuts of several functions along the molecular axis. **a** The total potential energy $V_{ne}(\mathbf{r}, R_{12})$ experienced by the electron due to the two nuclei placed at $\pm 1/2\,R_{12}$ along the $\hat{z}$ axis. **b** The ground-state (1s) eigenfunctions $\psi_e^{1s\,L}(\mathbf{r})$ and $\psi_e^{1s\,R}(\mathbf{r})$ for the electron sitting in the left- and right-nucleus potential wells respectively; shifted above: the symmetric $\psi_e^S(\mathbf{r})$ and antisymmetric $\psi_e^A(\mathbf{r})$ combinations of $\psi_e^{1s\,L}(\mathbf{r})$ and $\psi_e^{1s\,R}(\mathbf{r})$, Eq. (3.11)

$V_L + V_R$ for V_{ne}, and reorganize the matrix elements, to take advantage of the fact that $|L\rangle$ is the ground state of $(T_e + V_L)$:

$$\mathcal{E}_{\substack{S \\ A}} = \frac{\langle L|T_e + V_L|L\rangle + \langle L|V_R|L\rangle \pm \langle R|T_e + V_L|L\rangle \pm \langle R|V_R|L\rangle}{1 \pm \langle L|R\rangle}$$

$$= -\frac{E_{\text{Ha}}}{2} + \frac{\langle L|V_R|L\rangle \pm \langle R|V_R|L\rangle}{1 \pm \langle L|R\rangle}. \tag{3.13}$$

The second term contains two matrix elements of V_R. They are both real and negative functions of the inter-nuclear distance R_{12}. The $\langle L|V_R|L\rangle$ term expresses the attraction that the right nucleus exerts on the electron sitting around the left nucleus. At large R_{12} this attractive term $\langle L|V_R|L\rangle \simeq -e^2/R_{12}$, so that it balances almost exactly the inter-nuclear repulsion $V_{nn} = e^2/R_{12}$. This cancellation represents the classical result that the electrostatic interaction energy between a spherically symmetric neutral object (the H atom) and a remote point charge (the H$^+$ ion) vanishes rapidly with the distance. At large inter-nuclear distance, the cross term $\langle R|V_R|L\rangle$ is even smaller. We can approximate it as $\langle R|V_R|L\rangle \simeq -\langle L|R\rangle e^2/(R_{12}/2)$, because the distribution $\psi_L(\mathbf{r})\psi_R(\mathbf{r})$ is rather small everywhere, and peaks at the axial region between the two atoms, at the center of which $V_R \simeq -e^2/(R_{12}/2)$. Thus,

$$\mathcal{E}_{\substack{S \\ A}} + \frac{E_{\text{Ha}}}{2} = \frac{\langle L|V_R|L\rangle \pm \langle R|V_R|L\rangle}{1 \pm \langle L|R\rangle} \simeq \frac{-\frac{e^2}{R_{12}} \pm \left(-\langle L|R\rangle\frac{e^2}{R_{12}/2}\right)}{1 \pm \langle L|R\rangle} \simeq$$

$$\simeq -\frac{e^2}{R_{12}} \left(1 \pm 2\langle L|R\rangle\right)\left(1 \mp \langle L|R\rangle + \ldots\right)$$

$$\simeq -\frac{e^2}{R_{12}} \left(1 \pm \langle L|R\rangle + \ldots\right).$$

This expression, valid for $R_{12} \gg a_0$, shows that (i) the symmetric state $|S\rangle$ sits lower in energy than $|A\rangle$; (ii) in the symmetric state the *attraction* $\mathcal{E}_S + \frac{E_{\text{Ha}}}{2}$ *exceeds the repulsion* $V_{nn} = e^2/R_{12}$, thus generating *bonding*[2]; (iii) for $|A\rangle$, the negative sign in front of $\langle L|R\rangle$ lets the inter-nuclear repulsion prevails against attraction, with the result that the $|A\rangle$ state acts against bonding.

Improving beyond the large-R_{12} expansion, Eq. (3.13) can be evaluated for arbitrary R_{12}: Fig. 3.4 displays the outcome of this evaluation. Observe that

- the adiabatic potential $V_{\text{ad}}^{(S)}(R_{12})$ associated to the symmetric state exhibits a minimum at a finite separation $R_{12} = R_{\text{M}} \simeq 2.5\, a_0$;
- the adiabatic potential well is deep enough to bind the two protons together, and for this reason $|S\rangle$ is called a *bonding state*;
- the dot-dashed curve indicates that, as R_{12} is reduced, the decrease of the electronic potential energy $\langle S|V_{ne}|S\rangle$ does not compensate the raise in inter-nuclear repulsion:

---

[2]This elementary variational model predicts that, at large R_{12}, the lowering in adiabatic energy is exponentially small in R_{12}/a_0, due to $\langle L|R\rangle$ decaying with distance as the 1s atomic wavefunction.

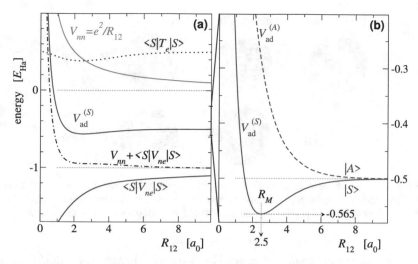

Fig. 3.4 **a** The total adiabatic potential of the $|S\rangle$ variational state of H_2^+ decomposed in its kinetic, potential and ion-ion contributions: $V_{ad}^{(S)}(R_{12}) = \langle S|T_e|S\rangle + \langle S|V_{ne}|S\rangle + V_{nn}(R_{12})$. **b** A blowup of the total adiabatic potential for the $|S\rangle$ (*solid*) and $|A\rangle$ (*dashed*) state. The indicated minimum $R_M = 2.5\,a_0$ of $V_{ad}^{(S)}(R_{12})$ corresponds to a molecular bond energy of $0.065\,E_{Ha} = 1.76\,eV$

thus our initial expectation that bonding is related to a gain in electrostatic potential energy is not confirmed by the present variational model;

- as the electron moves in the wider potential well created by the two protons rather than in the narrower well of an isolated proton, its kinetic energy (dotted line) decreases significantly, enough to dig a minimum for $V_{ad}^{(S)}(R_{12})$;
- the adiabatic potential $V_{ad}^{(A)}(R_{12})$ is monotonically decreasing, i.e. repulsive, which accounts for the qualification "*antibonding state*" for $|A\rangle$;
- as R_{12} is reduced below R_M, both adiabatic potentials shoot up, due to the divergence in V_{nn} (now compensated poorly) and to a new raise of the kinetic energy, due to the molecular potential well shrinking.

In summary, bonding of H_2^+ is to be attributed to a concurrent lowering of the kinetic energy (the electron moves in a broader well) and the potential energy (the electron screens the inter-nuclear Coulomb repulsion by spending a significant fraction of its probability density in the inter-nuclear region). In our simple variational model the molecular bond energy amounts to $V_{ad}^{(S)}(\infty) - V_{ad}^{(S)}(R_M) \simeq 1.76\,eV$. This variational estimate provides a fair approximation as long as $|\langle L|R\rangle| \ll 1$, i.e. $R_{12} \gg a_0$, but becomes especially inaccurate at smaller R_{12}, where non-1s components should be considered for an accurate description of the bonding and antibonding molecular states. The exact solution of the Schrödinger equation in the double Coulomb well of Fig. 3.2 yields an accurate $V_{ad}^{(S)}(R_{12})$, which is consistent with a bond energy of 2.79 eV at an equilibrium distance $R_M = 2.00\,a_0$, in agreement with experiment.

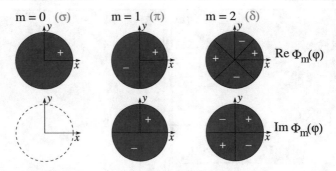

Fig. 3.5 The sign of the real and imaginary parts of the angular wavefunction, Eq. (2.11), of $m = 0, 1, 2$ electronic molecular states around the $\hat{\mathbf{z}}$ symmetry axis. For reverse-rotating $m = -1, -2$ states (not drawn), the real part is identical, while the sign of $\text{Im}\,\Phi_m(\varphi)$ is inverted

The symmetry of the field generated by two nuclei is axial (cylindrical), rather than spherical as in an atom. Accordingly, electrons in molecules have no analogue of the orbital-angular-momentum quantum number l. The electronic states are rather eigenstates of the angular-momentum *projection* L_z along the molecular symmetry axis $\hat{\mathbf{z}}$, namely the line through the nuclei. Figure 3.5 displays the φ-dependence of the molecular electronic states, labeled $\sigma, \pi, \delta, \ldots$, to identify the absolute value $|m| = 0, 1, 2, \ldots$ of their angular-momentum projection $m = 0, \pm 1, \pm 2, \ldots$ Both $|S\rangle$ and $|A\rangle$ of Eq. (3.11) are σ states, because they consist of 1s ($l = 0$, thus $m = 0$) atomic states. Standard spectroscopic notation adds a star apex to antibonding states, e.g. $|A\rangle$ would usually be labeled $1\sigma^*$. H_2^+ has further excited bonding and antibonding electronic states derived from $n > 1$ atomic levels. Certain of them possess axial angular momentum $m \neq 0$, thus labels $\pi, \pi^*, \delta, \delta^*, \ldots$ Note that the two signs of m make all non-σ states twofold degenerate orbitally.

3.2.2 Covalent and Ionic Bonding

The simple model for the H_2^+ ion is instructive because it illustrates the mechanism of chemical bonding, provides a basic picture of bonding and antibonding orbitals, and yields a numerical estimate of typical bond energies and distances in molecules. Nonetheless, the electronic ground state of H_2^+ does not fully qualify as a chemical bond. A proper *covalent bond* requires *two electrons occupying the same bonding orbital*, of course with antiparallel spins. This occurs, e.g. in neutral H_2. The precise one-electron molecular orbital $|S\rangle$ should be determined by some method (e.g. HF) which accounts—at least approximately—for electron-electron repulsion; eventually, $|S\rangle$ turns out similar to our variational guess, Eq. (3.11). Accordingly, the ground wavefunction of a neutral H_2 molecule is described approximately by

$$\psi_e^{gs}(\mathbf{r}_1\sigma_1, \mathbf{r}_2\sigma_2) = \langle \mathbf{r}_1\sigma_1, \mathbf{r}_2\sigma_2 | S \uparrow, S \downarrow \rangle^A \qquad (3.14)$$

$$= \langle \mathbf{r}_1 | S \rangle \langle \mathbf{r}_2 | S \rangle \frac{1}{\sqrt{2}} \begin{vmatrix} \chi_\uparrow(\sigma_1) & \chi_\uparrow(\sigma_2) \\ \chi_\downarrow(\sigma_1) & \chi_\downarrow(\sigma_2) \end{vmatrix}.$$

Both electrons occupy the same *spatially symmetric* bonding orbital, forming a 2-electron state which is *antisymmetric for the exchange* of electron 1 and 2 (through its spin-singlet part). The wavefunction (3.14) represents a typical *homonuclear covalent bond*. For H_2, the bond energy $V_{ad}^{(GS)}(\infty) - V_{ad}^{(GS)}(R_M) \simeq 4.8$ eV, at an equilibrium distance $R_M \simeq 1.4\, a_0 = 74$ pm. This energy is less than twice the bond energy of H_2^+, due to the significant electron-electron repulsion paid by state (3.14).

The potential well generated by the two protons has room for several excited states, mostly single-electron excitations with one electron promoted to a higher bonding/antibonding state. Excited states with both electrons in bonding orbitals, also lead to a bonding $V_{ad}(R_M)$, although usually characterized by lower binding energy and longer interatomic equilibrium distance than the electronic ground state [24]. However, other excited electronic states do not support bonding: for example, for the lowest triplet state of H_2, with electron one in $|S\rangle$ and electron two in $|A\rangle$, $V_{ad}(R_{12})$ decreases monotonically.

When pairs of larger-Z atoms interact and form dimers, to accommodate all electrons the number of involved bonding and antibonding orbitals is necessarily larger. Figure 3.6 sketches the ground-state valence electronic structure of several homonuclear diatoms (the core 1s electrons are left out). This figure prompts several remarks:

- With few exceptions, the overall ordering of the one-electron levels does not change much in passing from one molecule to another. In particular every antibonding level stands higher up in energy than the corresponding bonding level.
- Electrons fill the levels according to the same scheme as in atoms (low to higher energy, fulfilling Pauli's principle).
- The *bond order* is the number of electron *pairs* present in a bonding level and missing in the corresponding antibonding level: e.g. O_2 is double-bonded.
- The strength of covalent bonds increases with their bond order: among the dimers of Fig. 3.6, the triple bond of N_2 is the strongest, with a bond energy nearing 10 eV.
- When electrons start to fill 2p-derived molecular orbitals, the π states are lower in energy than the σ combination. In a way, this is surprising, as the overlap of the $m = 0$ p orbitals (pointing along the molecular axis) is larger than that of the $m = \pm 1$ orbitals (mostly located away from the axis). As a result, the $\sigma - \sigma^*$ splitting is generally larger than that of the corresponding $\pi - \pi^*$ orbitals. However, the Coulomb repulsion of the deeper filled σ and σ^* orbitals and the hybridization with the 2s-derived orbitals pushes the 2p-derived σ orbitals up by a substantial amount. The "natural" ordering is restored starting from O_2.
- In B_2 and O_2, two electrons sit in a degenerate π or π^* molecular orbital, which has room for 4 electrons: to minimize the residual Coulomb repulsion, the ground

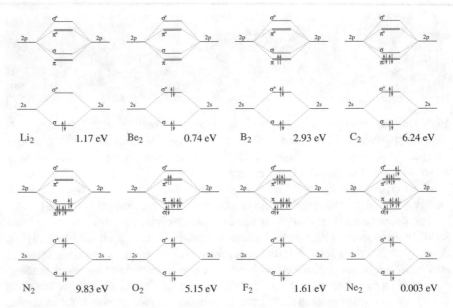

Fig. 3.6 A qualitative level scheme of the valence electronic structure of a few simple homonuclear diatomic molecules, with their observed bond energy. *Arrows* represent electrons occupying the single-particle states. Stars label antibonding orbitals

state is a spin-triplet Hund's-first-rule state ⊞ (Sect. 2.2.9.3). These dimers are magnetic, and their gases are therefore paramagnetic (see Sect. 4.3.1.3 below).

• Even dimers such as Be_2 and Ne_2, with null bond order (equally populated bonding and antibonding orbitals), exhibit a weak bonding, of the type described in Sect. 3.2.3 below.

Homonuclear diatomic molecules are very special diatoms, due to their peculiar $L \leftrightarrow R$ symmetry. Non-identical atoms bind together too. Many pairs form covalent bonds analogous to those illustrated above for the homonuclear molecules. For example, CO and N_2 have the same number of electrons and nuclei, and thus similar electronic structures. The main novelty of CO is the lack of $L \leftrightarrow R$ symmetry: electrons are more strongly attracted to the O nucleus ($Z = 8$) than to C ($Z = 6$). As a consequence, as shown in Fig. 2.21, the 2s and 2p shells of O sit deeper than the 2s and 2p shells of C: the bonding and antibonding orbitals deviate from symmetric/antisymmetric combinations of the type (3.11). Extending the variational treatment of H_2^+ with the methods of Appendix C.5 (neglecting overlaps $\langle L|R \rangle$ for simplicity), the approximate molecular orbitals are the eigenstates of the 2×2 matrix

$$\begin{pmatrix} E_R & -\Delta \\ -\Delta^* & E_L \end{pmatrix} \equiv \begin{pmatrix} \langle R|H_1|R \rangle & \langle R|H_1|L \rangle \\ \langle L|H_1|R \rangle & \langle L|H_1|L \rangle \end{pmatrix}. \tag{3.15}$$

Here $H_1 = T_e + V_{\text{eff}}$ is the effective one-electron Hamiltonian determined by the self-consistent potential acting on each electron, and $|L\rangle$, $|R\rangle$ are the relevant atomic states (e.g. the 2s or 2p) of the left and right atom (here C and O). The diagonal elements track the energy positions of the atomic shells. The off-diagonal intersite energy $\Delta > 0$ is very small at large separation R_{12}, and grows larger and larger as the atoms move closer.

Expressions for the eigenvalues $\bar{E}_1$ and $\bar{E}_2$ of the 2×2 matrix (3.15) are reported in Appendix C.5.2. In detail, Fig. C.2 shows how they change as a function of the "asymmetry ratio" $u = (E_L - E_R)/|2\Delta|$. The eigenenergies in Eq. (C.39) remain centered around $(E_L + E_R)/2$. Their splitting $\bar{E}_2 - \bar{E}_1 = \left[(E_L - E_R)^2 + (2\Delta)^2 \right]^{1/2}$ exceeds both the diagonal splitting $E_L - E_R$ and the minimum splitting $|2\Delta|$, which is recovered in the symmetric homonuclear limit $u = 0$ (i.e. for $E_L = E_R$). In the symmetric $u = 0$ limit the bonding $|\bar{E}_1\rangle$ and antibonding $|\bar{E}_2\rangle$ coincide respectively with the equally-weighted $|S\rangle$ and $|A\rangle$ combinations of Eq. (3.11). For increasing u, the eigenkets given in Eq. (C.40) resemble less and less the simple symmetric and antisymmetric combinations (3.11): $|\bar{E}_1\rangle$ acquires a prevalent $|R\rangle$ character, while $|\bar{E}_2\rangle$ acquires a mainly $|L\rangle$ character. Electrons in the $|\bar{E}_1\rangle$ state reside predominantly at the R side, thus providing a *polar character* to the dimer, unless an equal number of electrons occupies the $|\bar{E}_2\rangle$ state, which is displaced more to the L side.

As sketched in Fig. 3.7a, an intermediate value $u \simeq 1$ applies for the 2s and 2p orbitals of CO near its equilibrium separation: bonding molecular orbitals lie prevalently at the O side, antibonding ones at the C side. The bond of CO and of many similar heteronuclear diatoms is classified as *polar covalent*. It carries a nonzero average electric-dipole moment due to the charge transfer produced by the asymmetric charge distribution of the electrons in the bonding state $|\bar{E}_1\rangle$.

In the limit of very large asymmetry $u \gg 1$, the eigenenergies (C.39) $\bar{E}_1 \simeq E_R$, $\bar{E}_2 \simeq E_L$, and the eigenkets $|\bar{E}_1\rangle \simeq |R\rangle$, $|\bar{E}_2\rangle \simeq |L\rangle$. At typical interatomic separation, the highly polar bond of HF (here hydrogen fluoride, not Hartree-Fock!) is characterized by large u, see Fig. 3.7b. Recall—Fig. 2.21—that $\mathcal{E}_{2p}(F) \ll \mathcal{E}_{1s}(H)$. As a result, the relevant bonding orbital almost coincides with the 2p of the F atom. Thus an approximate description of the bond of HF invokes a complete charge transfer from H to F: fluorine therefore completes its open shell to $2p^6$. As long as the H^+ and F^-

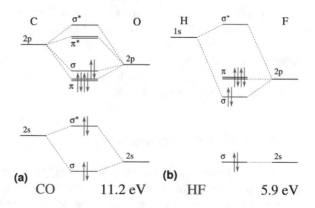

Fig. 3.7 A schematic electronic structure of two heteronuclear diatomic molecules: **a** CO and **b** HF. Stars label antibonding orbitals. The observed bond energy is indicated. Single-electron energies and splittings are purely qualitative

(a) CO   11.2 eV

(b) HF   5.9 eV

ions are separated widely, they would then attract each other like point charges, with a $-e^2/R_{12}$ attraction energy. As the proton moves inside the outer shell of F^-, electrons screen the positive charge of the F nucleus less and less, and the initial attraction turns gradually into repulsion. The origin of bonding in this simple picture is the energy gained in moving the ions from infinity to the 92 pm equilibrium distance, to which the energy paid to form the ions from the neutral atoms (the first ionization energy of the atom which turns into a cation—here H—minus the electron affinity of the atom acquiring the electron—here F) must be subtracted. Indeed, $[e^2/(92\,\text{pm}) = 15.7\,\text{eV}]$ minus the ionization energy of H (13.6 eV) plus the electron affinity of F (3.4 eV) yields 5.5 eV, in fair agreement with the observed bond energy = 5.9 eV. A molecular bond such as that of HF, where u is so large that almost complete charge transfer occurs is named an *ionic bond*.

The picture of molecular orbitals constructed as combinations of specific atomic orbitals, as in Figs. 3.6 and 3.7, is oversimplified. All orbitals with the same m in fact participate to the mix, although each molecular orbital has usually one prominent atomic-orbital component. For example, each of the orbitals labeled σ and σ^* in Fig. 3.6 is a linear combination involving some component of all $m = 0$ states of both atoms, mostly 2s and $2p_z$, but also a little 1s, 3s, $3p_z$, 4s....

This *hybrid* character of molecular orbitals, a side detail for diatomic molecules, becomes a crucial ingredient in determining the 3D structure of *polyatomic molecules* and covalent solids. Hybrid orbitals often generate adiabatic potentials which depend strongly not just on interatomic distances, but also on angles. An especially remarkable example is the mix of 2s and 2p, at the origin of the shapes of organic molecules, Fig. 3.8. As specific and important examples, sp^3 combinations intermixing 2s $2p_x$ $2p_y$ and $2p_z$ orbitals determine the ideally 109.5° angle between the bonds of tetrahedrally coordinated carbon; sp^2 combinations of 2s $2p_x$ $2p_y$ tend to bind atoms such as carbon to three other atoms in the same plane, forming ideally 120° angles. The multiple combinations of the molecular orbitals determining 3D polyatomic molecular structures, in interplay with weaker intermolecular interactions, originate the endless variety of shapes and functionalities of organic compounds, which are the microscopic building blocks and byproducts of living matter. Analogous mixtures of 3s 3p, or 4s 4p, provide Si, P, S, Ge, As, Se with a marked tendency to form directional bonds, producing the extended (rather than molecular as for N_2) covalently bound structures characterizing the solid state of these elements (see Fig. 5.4 below).

Each minimum of the adiabatic potential $V_{ad}(R)$ in the multidimensional space of configurations identifies an equilibrium molecular structure. Polyatomic molecules are often characterized by several local minima of V_{ad}, each representing a different *isomer*, such as those of Fig. 3.8i–k.

3.2.3 Weak Non-chemical Bonds

Our simplified model of covalent bonds in terms of linear combinations of single fixed atomic orbitals generates adiabatic potentials $V_{ad}(R_{12})$ that flatten out exponentially at large distance R_{12}, due to the exponential decay of the orbitals themselves. Exper-

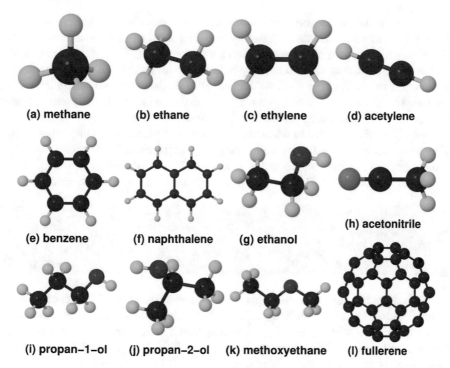

Fig. 3.8 The 3D equilibrium structures [25] of organic molecules is dictated by the hybrid orbitals of carbon. sp^3 hybridization, with C atoms at the center of a tetrahedron of 4 bonds, is relevant e.g. for **a** methane CH_4, **b** ethane C_2H_6, **g** ethanol CH_3CH_2OH, the carbon in the CH_3 group of **h** acetonitrile, and **i–k** all carbons in these isomers of C_3H_8O. sp^2 hybridization, with C atoms at the center of a triangle of 3 bonds, is relevant e.g. for **c** ethylene C_2H_4, **e** benzene C_6H_6, and **f** naphthalene $C_{10}H_8$. sp hybridization with C atoms in a linear 2-bonds configuration is relevant for **d** acetylene C_2H_2, and the carbon in the CN group of **h** acetonitrile. The hybridization of **l** fullerene C_{60} is intermediate between sp^2 and sp^3

imentally however, at large distance, the interaction energy between any two atoms is attractive and decays following a universal power law: $\propto R_{12}^{-6}$. This attraction is due to a classical-electromagnetism concept which the previous analysis completely overlooked: atomic polarizability.

On average, the electric-dipole moment of any atom vanishes. As long as spherical symmetry is unbroken, the single-electron atomic orbitals are also eigenstates of the orbital angular momentum $|\mathbf{L}|^2$. The parity of an eigenfunction $R_{nl}(r) \times Y_{l m_l}(\hat{\mathbf{r}})$ is the same as that of its spherical harmonic: $(-1)^l$; thus the angular probability distribution $|Y_{l m_l}(\hat{\mathbf{r}})|^2$ of each electron is parity-even (see also Fig. 2.5 and Sect. 2.1.9). As a result, the average electric dipole $\langle n, l, m_l | \mathbf{d} | n, l, m_l \rangle$ of each electron vanishes. However, electrons move around the nucleus and occupy instantaneously specific positions, producing a fluctuating nonzero dipole moment. According to basic electromagnetism, this dipole moment generates an instantaneous electric field $\mathbf{E}$, whose intensity decays away from the atom as distance^{-3}.

An electric field acting on a second remote atom "polarizes" it. The electrons of atom b react to minimize the total energy, now including an additional term $-\mathbf{d} \cdot \mathbf{E}$ describing the coupling to the external field, where $\mathbf{d} = -q_e \sum \mathbf{r}_i$. Atom b responds to the applied external electric field by building up an *induced dipole* in the same direction as the field: $\langle \mathbf{d} \rangle = \alpha_b \mathbf{E}$. As the field is weak, the *atomic polarizability* α_b is independent of the field (*linear response*). Due to this induced polarization, the total energy of the two atoms at separation R_{12} decreases by an amount $-1/2\langle \mathbf{d} \rangle \cdot \mathbf{E} = -1/2\alpha_b |\mathbf{E}|^2 \propto R_{12}^{-6}$ relative to when atoms a and b are placed at infinite separation. Dimension-wise, the coefficient multiplying R_{12}^{-6} is a [distance]^6 × [energy]. As only atomic physics is involved, the order of magnitude of this coefficient is $E_{Ha} a_0^6 \simeq 10^{-79}$ J m$^6 \simeq 0.6$ eV Å^6. This *van der Waals attraction* is then approximately

$$V_{ad}^{vdW}(R_{12}) \approx -E_{Ha}\left(\frac{a_0}{R_{12}}\right)^6. \tag{3.16}$$

More quantitatively, the coefficient of R_{12}^{-6} is proportional to the product of the two atomic electrical polarizabilities $\alpha_a \alpha_b$ (because the dipole fluctuation in atom a is also proportional to its polarizability α_a), or, more precisely, to the energy integral of their frequency- or energy-dependent version $\alpha_{a/b}(\hbar\omega)$ [26]. The physical dimension of polarizability is [Charge]^2[Length]^2[Energy]$^{-1}$ = [Charge]^2[Time]^2[Mass]$^{-1}$, the same as $4\pi\epsilon_0$ × [Length]^3. It is therefore natural to express atomic polarizabilities in units of $\alpha_{A.U.} = 4\pi\epsilon_0 a_0^3 \simeq 1.65 \times 10^{-41}$ C^{2}s^2/kg.

The atomic polarizability can be either measured (for example, by measuring the dielectric properties of an atomic gas inserted in a charged parallel-plate capacitor) or computed, by evaluating the distortion of the atomic state under the action of a weak external electric field. By standard perturbation theory, Appendix C.9, an external field distorts the ground state so that (i) the electronic wavefunction acquires components (proportional to $|\mathbf{E}|$) of excited states with l changed by ± 1, and (ii) the total energy decreases $\propto |\mathbf{E}|^2$. For example, in an electric field $\mathbf{E} = E_z \hat{\mathbf{z}}$, the ground state of a hydrogen atom changes to the following linear combination

$$|1, 0, 0; E_z\rangle = b_1 |1, 0, 0\rangle + \sum_{n>1} b_n |n, 1, 0\rangle. \tag{3.17}$$

Here the hydrogen-atom eigenkets $|n, l, m\rangle$ are represented by the eigenfunctions detailed in Eq. (2.16). First-order perturbation theory Eq. (C.87) provides the coefficients $b_1 = 1$, $b_{n>1} = -E_z \langle n10|d_z|100\rangle/(\mathcal{E}_1 - \mathcal{E}_n)$, where $\mathcal{E}_n$ are the energy levels of H, Eq. (2.10). As a result, the acquired electric-dipole moment $\langle 1, 0, 0; E_z|d_z|1, 0, 0;$ $E_z\rangle = 2b_1 \sum_n b_n \langle n, 1, 0|d_z|1, 0, 0\rangle$ is proportional to E_z, because the $b_{n>1}$ coefficients are. The atomic polarizability α is then given by

$$\alpha_H = \lim_{E_z \to 0} \frac{\langle 1, 0, 0; E_z|d_z|1, 0, 0; E_z\rangle}{E_z} = -2\sum_{n>1} \frac{|\langle n, 1, 0|d_z|1, 0, 0\rangle|^2}{\mathcal{E}_1 - \mathcal{E}_n}. \tag{3.18}$$

The static atomic polarizability of H amounts to $\alpha_H = 4.50\,\alpha_{A.U.}$. Taking all electrons into account, one can evaluate the same quantity for any atom. For example, He has a record-low $\alpha_{He} = 1.383\,\alpha_{A.U.}$ due to its huge 1s-2p gap. Li has a much larger $\alpha_{Li} = 164.0\,\alpha_{A.U.}$ due to the comparably small 2s-2p gap. Cs boasts the record-high $\alpha_{Cs} = 401.0\,\alpha_{A.U.}$ mostly due to the very small 6s-6p gap. Indeed, Eq. (3.18) suggests that α depends inversely on the energy separation between the highest filled state and the lowest empty state with l differing by ± 1.

The dipole–induced-dipole mechanism is perfectly general, and it accounts for the leading R_{12}^{-6} long-range attraction of all dimers of neutral atoms. As distance R_{12} decreases, the dimer can evolve along two distinct paths: for pairs of close-shells atoms (Be_2, Ne_2) the only attraction comes from this weak van der Waals mechanism; in contrast, pairs of open-shell atoms, e.g. H_2 and the diatoms of Fig. 3.7, modify gradually their orbitals to form robust covalent bonds.

As R_{12} is further reduced, repulsion sets in. The short-distance repulsive Coulombic divergence of $V_{ad} \propto R_{12}^{-1}$ is peculiar to H_2^+, H_2, and few other dimers involving H. For dimers of $Z \geq 3$ atoms, long before the nucleus-nucleus repulsion $V_{nn}(R_{12})$ becomes relevant, the electronic energy $E_e^{(a)}(R_{12})$ blows up because of Pauli's principle: when the core electrons are brought together in the same region of space, their wavefunctions become less and less orthogonal. Part of these electrons are then pushed up into some orthogonal empty valence level, which makes tiny reductions of R_{12} cost tens or even hundreds E_{Ha}. The resulting "hard-core" repulsion as two atoms come into intimate contact is responsible for the "impenetrability" of matter, i.e. the sharp increase of pressure of a sample whose volume is reduced so much that each atom is squeezed into less than the volume occupied by its core shells ($\sim$ a few Å^3). We see that a combined effect of electrons indistinguishableness and quantum kinetic energy sustains matter against collapse due to electromagnetic attraction. In modeling, the $V_{ad}(R_{12})$ blowup at short distance is often parameterized phenomenologically with a R_{12}^{-12} power law.

A simple functional form capturing the long-distance van-der-Waals attraction and the rapid short-range repulsion is the popular Lennard-Jones potential

$$V_{LJ}(R_{12}) = 4\varepsilon \left[\left(\frac{\sigma}{R_{12}} \right)^{12} - \left(\frac{\sigma}{R_{12}} \right)^6 \right]. \tag{3.19}$$

This expression is nothing but a phenomenological approximation of the actual $V_{ad}(R_{12})$ of a dimer. Values for its two parameters ε (the depth of the potential well at R_M) and σ (the distance where $V_{LJ}(R_{12})$ changes sign) are listed in Table 3.1 for the noble-gas dimers. The Lennard-Jones potential $V_{LJ}(R_{12})$ is a fair approximation for $V_{ad}(R_{12})$ for noble-gas atoms, and is also used for describing the dynamics of a collection of more than two of these atoms as a *sum of pair potentials*). This *two-body* approximation is fair for collections of close-shell atoms: the phase diagram and correlation properties of the Lennard-Jones model exhibit close qualitative and semi-quantitative agreement to the observed properties of noble-gas fluids and solids. In contrast, the two-body Lennard-Jones model is entirely inappropriate to describe atoms forming strongly-directional covalent bonds.

Table 3.1 Parameters, obtained fitting atom-atom scattering data [2], for the Lennard-Jones pair potential, Eq. (3.19), of the dimers of the noble-gas elements; note the increase of atomic size (i.e. σ) with Z and the rapid increase in atomic polarizability in progressing from helium to xenon [reflected by the coefficient $4\varepsilon\sigma^6 \propto \alpha^2$ of $-R_{12}^{-6}$ in Eq. (3.16)]

Element	σ [pm]	ε [meV]	$4\varepsilon\sigma^6$ [$E_{Ha}\,a_0^6$]
He	256	0.879	1.7
Ne	275	3.08	8.9
Ar	340	10.5	109
Kr	368	14.4	239
Xe	407	19.4	590

3.2.4   A Classification of Bonding

We have clarified the mechanism for the general tendency, pictured in Fig. 3.1, of atoms to mutually attract at large distance and repel when coming into contact. Contrasting this qualitative likeness, significant differences in the equilibrium distances and huge differences in well depths differentiate various bonding mechanisms.

When either atom or both is a noble gas, little or no covalency occurs, and only the dipole–induced-dipole *van der Waals* mechanism provides attraction. Consequently the atom-atom equilibrium distance R_M is relatively large ($R_M \sim 250$–400 pm) and the bond energy [$V_{ad}(+\infty) - V_{ad}(R_M)$] is small (few meV, see Table 3.1). The noble-gas elements usually retain the monoatomic gas phase down to relatively low temperature, and show a scarcer tendency to form dimers than to eventually condense to extended liquid and solid phases.

In contrast, a few open-shell atoms (O, N, F) exhibit a prominent tendency to form diatomic molecules, with short strong *covalent bonds*, characterized by several-eV bond energies. These molecular units are retained in the low-temperature liquid and solid phases. For most other elements (e.g. Li, Be, B, C…), the extra energy gain in forming multiple chemical bonds per atom makes it energetically convenient to form extended *metallic* or covalent solids, rather than diatoms.

When different atoms are bound together covalently, their valence electrons move asymmetrically closer to an ion than to the other, because of differences in atomic-shell energies, as illustrated in the examples of Fig. 3.7. This charge displacement provides a nonzero average electric dipole to *polar* heteroatomic bonds. In the extreme limit of complete or almost complete charge transfer (e.g. HF, LiF), *ionic bonds* are formed. Equilibrium distances ($R_M \sim 80$–250 pm) and energies involved in metallic, covalent, and ionic bonds are in the same range: 1–10 eV per bond.

In addition to van der Waals, covalent, ionic, and metallic bonds, we just mention a fifth bonding mechanism which plays a relevant role under specific conditions: *hydrogen bonding*.

3.3 Intramolecular Dynamics and Spectra

In the previous Section, we have discussed the general properties of the solutions of the electronic equation (3.2) for a diatom, thus acquiring information on the typical shape of the adiabatic potential $V_{ad}(R_{12})$. Time has come to consider the motion of the two nuclei in the adiabatic "force field" described by V_{ad}, and its spectroscopic implications. This motion is described by Eq. (3.9) and its solutions.

As remarked in Sect. 3.1, the adiabatic potential is independent of the center-mass position of the molecule (translational invariance). It is therefore natural to switch to center-mass and relative coordinates. Precisely the same transformation (2.1) applied to the two-body problem of the one-electron atom separates the center-mass motion of the two-body problem of the diatomic molecule. Like for atoms, the molecular center of mass translates freely: the random thermal translational motion in a gas-phase sample originates Doppler and collisional broadening of the spectra.

The diatom energy is also independent of the orientation in space of the straight line through the two nuclei (rotational invariance). By adopting spherical coordinates for the vector joining the two nuclei, we separate the Schrödinger equation for the internal motion into angular (2.5) (2.6) and radial (2.7) equations, exactly like for the one-electron atom. The angular equations are universal, so that the angular eigenfunctions, describing the orientation in space of the molecule (thus molecular rotations) are standard spherical harmonics $Y_{l m_l}$. Rotational states $|l, m_l\rangle$ are labeled by the molecular angular momentum l,[3] plus its z component m_l.

When applying this method to diatoms, the nucleus-nucleus separation R_{12} replaces the electron-nucleus distance r of the one-electron atom and the distance-dependent adiabatic potential $V_{ad}(R_{12})$ replaces $U(r)$ in Eq. (2.7). The formal structure of the radial equation is the same as for the one-electron atom. The substantial physical difference stands in the equilibrium distance (i.e. the separation where the potential is the most attractive) which for the diatom is finite $R_{12} = R_M > 0$, rather than null as for the one-electron atom. The main consequence is that the radial motion remains localized mostly near R_M, in a region where the centrifugal term $\hbar^2 l(l + 1)/(2\mu R_{12}^2)$ in Eq. (2.7) is usually fairly small. If we neglect the variations of R_{12}^{-2} in the region around R_M where the radial wavefunction differs significantly from zero, then the radial motion is approximately independent of the rotation, i.e., the radial solutions are independent of l. The quantum number $v = 0, 1, 2, \ldots$ for the radial motion of the diatom counts the radial nodes, matching $n - l - 1$ in the one-electron atom. We Taylor-expand the adiabatic potential around its minimum

$$V_{ad}(R_{12}) = V_{ad}(R_M) + \frac{1}{2} \left. \frac{d^2 V_{ad}(R')}{dR'^2} \right|_{R'=R_M} (R_{12} - R_M)^2 + \cdots . \qquad (3.20)$$

---

[3] In the molecular-spectroscopy literature, the angular-momentum quantum number l describing the molecular rotation is occasionally labeled r, or j.

Retaining terms up to second order, the radial motion is approximately a *harmonic* vibration. In this scheme, the total energy consists of the sum of three contributions, with decreasing magnitude:

- An *electronic* term $V_{ad}(R_M)$. Its lowering at the equilibrium distance $V_{ad}(\infty) - V_{ad}(R_M)$ represents the depth of the energy well responsible for the bond formation, typically in the few-eV range, see Figs. 3.6 and 3.7.
- A *vibrational* term measuring the energy of the radial vibration around the equilibrium position R_M. In the harmonic approximation, Eq. (C.65), it amounts to

$$\mathcal{E}_{vib}(v) = \hbar\omega \left(v + \frac{1}{2} \right).$$
(3.21)

Here

$$\omega = \sqrt{\frac{K}{\mu}}, \quad \text{with } K = \left. \frac{d^2 V_{ad}(R_{12})}{d R_{12}^2} \right|_{R_{12}=R_M} \quad \text{and } \mu = \frac{M_1 M_2}{M_1 + M_2}.$$
(3.22)

μ is the reduced mass of the two-body oscillator consisting of the two nuclei with mass M_1 and M_2, see also Eq. (2.4). Typical vibrational energies $\hbar\omega \approx 20\text{--}400$ meV are observed.

- A rotational contribution from the rotational term $\hbar^2 l(l+1)/(2\mu R_{12}^2)$ in the radial equation (2.7). Approximating $R_{12}^{-2} \simeq R_M^{-2}$ yields:

$$\mathcal{E}_{rot}(l) = \frac{\hbar^2 l(l+1)}{2\mu R_M^2} = \frac{|\mathbf{L}|^2}{2I}.$$
(3.23)

I stands for the classical momentum of inertia μR_M^2 of the diatom relative to its center of mass, described as if it was a rigid dimer with interatomic separation $R_{12} \equiv R_M$. Typical molecular rotational energies $\hbar^2/(2I)$ stand in the meV region, with H_2 exhibiting the largest one: 7 meV.

3.3.1 Rotational and Rovibrational Spectra

In an "adiabatic" transition with the electrons remaining in the electronic ground state, transitions observed in spectroscopy fulfill the standard electric-dipole selection rule: $\Delta l = \pm 1$. The involved dipole operator is the product of the inter-nuclear

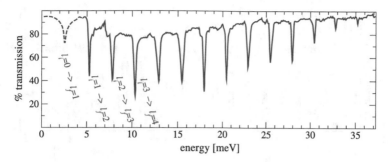

Fig. 3.9 Observed purely rotational absorption spectrum of gas-phase HCl, with a few $|l_i\rangle \rightarrow |l_f\rangle$ assignments. In this spectrum, the $|l_i = 0\rangle \rightarrow |l_f = 1\rangle$ peak at ~ 2.5 meV (*dashed*) is guessed rather than observed, due to range limitations of the spectrometer (Data from Ref. [3])

separation times the charge difference permanently attached to the two atoms. This charge difference certainly vanishes for equal nuclei: as a result, no dipole transition occurs for homonuclear molecules.[4] In contrast, the large permanent electric-dipole moments of strongly polar molecules, such as HF and HCl, support intense dipole transitions. The dipole moment of CO, a moderately polar molecule, is only $\sim 10\%$ of that of HCl, thus producing much weaker IR absorption.

Rotational and vibrational molecular spectra are mostly observed in *absorption* rather than emission. The reason is the small spontaneous emission rate of such low-energy IR transitions, due to the $\mathcal{E}_{if}^3$ dependence of the electric-dipole decay rate, Eq. (2.45). Competing radiationless decay phenomena (typically related to molecular collisions) occur over far shorter time scales, thus making the observation of emission spectra in this spectral region practically unfeasible.

Purely rotational spectra, usually in the far IR region, are associated to $\Delta v = 0$, $\Delta l = 1$ transitions. The energy difference between the $|l_i\rangle$ and $|l_f\rangle = |l_i + 1\rangle$ states is

$$\Delta \mathcal{E}_{\rm rot}(l_i) = \frac{\hbar^2}{2I}[(l_i + 1)(l_i + 1 + 1) - l_i(l_i + 1)] = \frac{\hbar^2}{I}(l_i + 1) . \qquad (3.24)$$

Accordingly, the rotational spectrum of a sample composed of molecules in several initial rotational states consists of an array of equally-spaced lines. The energy spacing is twice as large as the typical rotational energy quantum $\hbar^2/(2I)$. Figure 3.9 reports a typical purely-rotational absorption spectrum. The observed separation of the lines permits us to estimate the interatomic equilibrium separation $R_{\rm M}$ through Eqs. (3.24) and (3.23).

Rovibrational spectra are usually observed by means of near-IR absorption spectroscopy. The $\Delta v = 1$ transition concentrates most of the dipole intensity, but weaker $\Delta v > 1$ *overtone* transitions are observed routinely too. Figure 3.10 reports a char-

---

[4]This is the reason why clean air (consisting mostly of homonuclear N_2 and O_2) is highly transparent to IR light. Transparency in the visible and near-UV range is associated to the wide gaps (several eV) separating the electronic ground state to the first allowed electronic excitation.

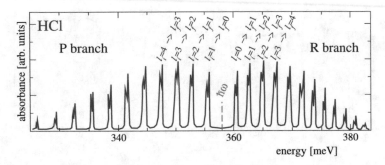

Fig. 3.10 Observed rovibrational spectrum of gas-phase HCl: absorption in this region is associated to the "fundamental" vibrational transition $|v_i = 0\rangle \rightarrow |v_f = 1\rangle$. The P- and R-branch peaks involve $|l_i\rangle \rightarrow |l_i - 1\rangle$ and $|l_i\rangle \rightarrow |l_i + 1\rangle$, respectively. Each peak exhibits an isotopic duplication: $H^{35}Cl$ is responsible for the stronger peaks, and the less abundant $H^{37}Cl$ for the weaker ones

Fig. 3.11 A scheme of the rotational levels (*horizontal lines*) and the relative electric-dipole-allowed absorption transitions (*vertical arrows*) around the $v = 0 \rightarrow 1$ fundamental vibrational excitation (at energy $\hbar\omega$) of a diatomic molecule

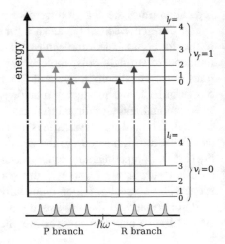

acteristic rovibrational spectrum. Like for purely-rotational spectra, the gas sample consists of molecules in several initial l states: as a consequence, the purely vibrational peak is "decorated" by rotational transitions, as illustrated in Fig. 3.11. In the transitions at energies $< \hbar\omega$, the rotational energy decreases, $\Delta l = -1$: this sequence is called P branch. In the transitions at energies $> \hbar\omega$, l increases, $\Delta l = +1$ (R branch). Equally-spaced rotational peaks are expected according to Eq. (3.24) (for the P branch a similar result holds). The rovibrational spectra are characteristic for the absence of a purely vibrational peak at energy $\hbar\omega$, which could only occur if $\Delta l = 0$ transitions were dipole allowed (which are not).

Alternatively, rovibrational spectra are investigated by means of Raman spectroscopy, a technique based on non-resonant electronic excitations of the molecule, which rapidly decay back to the electronic ground state. When this light-scattering process leaves a vibrational and/or rotational excitation, the re-emitted photon has a lower energy than the original one: this downshift measures the excitation energy.

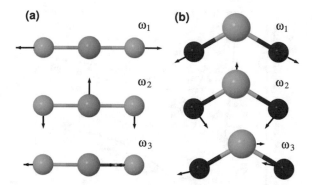

Fig. 3.12 The vibrational normal-mode displacements of two triatomic molecules. **a** CO_2, a molecule whose equilibrium geometry is linear. Its normal-mode energies are $\hbar\omega_1 = 166$ meV, $\hbar\omega_2 = 83$ meV, $\hbar\omega_3 = 291$ meV. **b** H_2O, a molecule with a bent equilibrium configuration. Its normal-mode energies are $\hbar\omega_1 = 453$ meV, $\hbar\omega_2 = 198$ meV, $\hbar\omega_3 = 466$ meV

This technique involves two photons, with selection rule $\Delta l = 0, \pm 2$, different from the electric-dipole transitions of single-photon absorption.

The rovibrational dynamics of polyatomic molecules involves further intricacies. Once translations and rotations have been accounted for, $3N_n - 6$ (or $3N_n - 5$ for linear molecules) internal degrees of freedom account for vibrations, approximately described in terms of small-amplitude harmonic oscillations (the *normal modes* of classical mechanics) around the multi-dimensional minimum of V_{ad}. Each normal mode i behaves as an independent quantum harmonic oscillator, with its own vibrational frequency ω_i (and thus ladder energy step $\hbar\omega_i$). Figure 3.12 sketches such vibrational modes for CO_2 and H_2O.

3.3.2 Electronic Excitations

Visible and UV resonant photons can excite the electronic states of molecules. Like in atoms, these transitions can roughly be understood as promotions of an electron from a filled molecular orbital to an empty one, e.g. from a bonding to an antibonding orbital. An electronic transition $\psi_e^{(a)} \to \psi_e^{(b)}$ in a molecule leads from one adiabatic potential surface to another, as illustrated in Fig. 3.13.

During the extremely short timescale of the electronic transition, the ions have no time to move, thus the transition is "vertical" in R_{12}. In general, the shapes of different adiabatic potentials $V_{ad}^{(a)}(R_{12})$ and $V_{ad}^{(b)}(R_{12})$ associated to different electronic states are often well distinct; in particular $R_M^{(b)} \neq R_M^{(a)}$. This means that electronic transitions are usually accompanied by vibrational "shakeup" transitions, excited by the displacement of the equilibrium geometry. For this reason, an electronic transition usually gets decorated by approximately equally-spaced vibrational satellites, separated by the harmonic frequency $\hbar\omega^{(b)}$ of the final adiabatic potential, as sketched in Fig. 3.14.

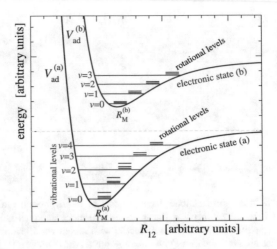

Fig. 3.13 Two adiabatic potentials $V_{ad}^{(a)}(R_{12})$ and $V_{ad}^{(b)}(R_{12})$ associated to different electronic eigenfunctions $\psi_e^{(a)}$ and $\psi_e^{(b)}$. Each potential-energy well holds a ladder of vibrational states (*red horizontal lines*), whose spacing is related to the curvature of the corresponding well. On top of each vibrational state, rotational states build up a tiny nonuniformly-spaced ladder (*black horizontal lines*)

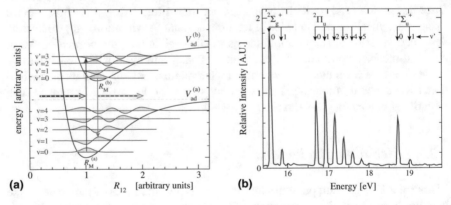

Fig. 3.14 **a** A sketch of the adiabatic potentials of two different molecular electronic states, with the respective vibrational levels, the ground and a few excited vibrational wavefunctions. *Arrows* highlight the most intense vertical (Franck-Condon) transitions in absorption and in emission (fluorescence), illustrating the photon-frequency reduction usually occurring in molecular fluorescence, see also Appendix B.2. **b** The observed photoemission spectrum of N_2, with the transition between the electronic ground state of N_2 and three different electronic states of N_2^+ labeled $^2\Sigma_g^+$, $^2\Pi_g$, $^2\Sigma_u^+$. This spectrum exhibits sequences of regularly-spaced vibrational satellites built on top of each electronic state of N_2^+ (Data from Ref. [27])

The intensities of the different satellites are distributed proportionally to the overlaps $|\langle v=0|v'\rangle|^2$, expressing the projection of the vibrational ground state $|v=0\rangle$ of the initial adiabatic potential $V_{ad}^{(a)}$ on each of the final vibrational eigenstates $|v'\rangle$

of $V_{\rm ad}^{(b)}$. The most probable transitions involve those $|v'\rangle$ states with large amplitude in the region of the initial minimum $R'_{\rm M}$, as illustrated in Fig. 3.14a (*Franck-Condon principle*). Accordingly, the spectral intensity shared by a large number of vibrational satellites indicates a large displacement of the equilibrium position in the electronic transition. For example, the spectrum of Fig. 3.14b indicates that, in the electronic transition from the ground state of N_2 to N_2^+, $R_{\rm M}$ shifts more in going to the $^2\Pi_u$ state than to either of $^2\Sigma_g^+$ or $^2\Sigma_u^+$. Rotational structures accompanying the electronic-vibrational structures are complicated by the change in momenta of inertia and by occasional changes in the electronic angular momentum projection.

3.3.3 Zero-Point Effects

We conclude this Section with an intriguing detail: the bond energy of a diatom is slightly less than the depth $V_{\rm ad}(+\infty) - V_{\rm ad}(R_{\rm M})$ of the adiabatic potential well. The bond energy would coincide with the well depth $V_{\rm ad}(+\infty) - V_{\rm ad}(R_{\rm M})$ if the nuclear masses were infinite, or equivalently if the ions moved according to classical mechanics. In fact, due to the quantum zero-point motion associated to Heisenberg's uncertainty principle (see Appendix C.2.1), the actual ground-state energy includes a vibrational contribution, that equals $\mathcal{E}_{\rm vib}(0) = \hbar\omega/2$ in the harmonic approximation, see Eq. (3.21). As illustrated in Fig. 3.15, when zero-point energy is accounted for, the actual binding energy decreases to $E_b = V_{\rm ad}(+\infty) - [V_{\rm ad}(R_{\rm M}) + \mathcal{E}_{\rm vib}(0)]$.

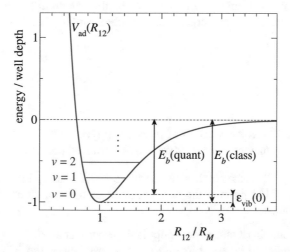

Fig. 3.15 The quantum zero-point vibrational energy $\mathcal{E}_{\rm vib}(0)$ reduces the bond energy E_b of a diatomic molecule relative to $V_{\rm ad}(+\infty) - V_{\rm ad}(R_{\rm M})$. Quantum confinement produces a zero-point motion which adds extra kinetic and potential energy to the well bottom. The zero-point energy $\mathcal{E}_{\rm vib}(0)$ decreases for heavier atomic masses M_α, and would eventually vanish in the hypothetical $\mu \to \infty$ limit where Heisenberg's uncertainty principle plays no relevant role

Experimentally, this usually small effect can be probed by changing the nuclear isotopic masses, thus modifying the vibrational frequency $\omega \propto \mu^{-1/2}$, without affecting $V_{\rm ad}(R_{12})$.

^{4}He$_2$ exhibits a spectacular zero-point effect. This dimer is so extremely weakly bound (the well depth is approximately 900 μeV, see Table 3.1) that just one bound $v = 0$ "vibrational" level exists [28], with a zero-point energy $\mathcal{E}_{\rm vib}(0)$ that balances the adiabatic attraction almost entirely. The resulting total net binding energy is $E_b \simeq 0.1$ μeV only! Of the 25% lighter ^{3}He$_2$, no bound state is observed.

Final Remarks

The present Chapter summarizes basic concepts and experimental evidence in the field of molecular physics. Adiabatic separation and chemical bonding stand at the heart of all physics and chemistry of matter: they open the way to understanding the structure and dynamics of systems composed of more than a single atom, namely molecules, as well as polymers, clusters, solids... These pages only scratch the surface of the vast field of molecular spectroscopy, which provides detailed quantitative information about the geometry and dynamics of diatomic and polyatomic molecules [6, 8, 9]. Before being subjected to spectroscopic characterization, molecules are synthesized transformed and probed by means of all sorts of chemical reactions: these reactions can mostly be analyzed conceptually in terms of the dynamics of atoms guided by appropriate multi-dimensional potential energy surfaces $V_{\rm ad}(R)$, across regions straddling far from their minima. The qualitative understanding and quantitative study of these reactions constitutes the hard core of chemistry and chemical physics.

Problems

Problems labeled $^{\rm S}$ involve concepts of statistical physics introduced in the following Chapter 4, mainly in Sect. 4.3.1.2. Stars* mark advanced problems.

3.1$^{\rm S}$ Consider a H^{80}Br gas sample at temperature 400 K. The equilibrium interatomic distance is $R_M = 141$ pm and the wave number of the vibrational transition $\bar{\nu} = 2650$ cm^{-1}. Evaluate:

- the quantum number l of the most populated rotational level;
- the wave number of the most intense transitions of the P and R branches of the absorption spectrum.

3.2$^{\rm S}$ The vibrational wave number of the isotopically pure H^{79}Br molecule is $\bar{\nu} = 2650$ cm^{-1}. Evaluate the force needed to stretch the bond, maintaining the two nuclei farther than their equilibrium distance R_M by 1 pm. Evaluate also the vibrational contribution to the specific heat capacity at temperature $T = 350$ K.

3.3[S] The figure below records an infrared absorption spectrum of gas-phase $H^{80}Br$.

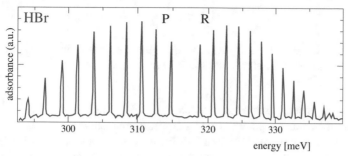

Based on this experimental spectrum, determine:

- The elastic constant of the adiabatic potential in the harmonic approximation;
- the equilibrium distance between the H and Br nuclei;
- an estimate (with a 30% precision) of the sample temperature.

3.4[S] Consider fluorine gas at temperature 5,000 K. Using the harmonic approximation, estimate the number of bound vibrational levels for a F_2 molecule, knowing that the vibrational transition is observed at $\bar{\nu} = 892$ cm^{-1} and that the molecular binding energy is 1.5 eV. Write the vibrational partition function in the harmonic approximation, and estimate the fraction of bound molecules and the fraction of dissociated molecule. For the latter, assume simply that the dissociated molecules are represented by the harmonic vibrational levels above the dissociation threshold.

3.5 Evaluate the force (in Newton) necessary to keep the two nuclei of a N_2 molecule 1 pm farther apart than their equilibrium separation $R_M = 109$ pm. The vibrational fundamental excitation is observed at a wave number 2360 cm^{-1}. Evaluate also the excitation energy of the rotational state with angular momentum $l = 2$ relative to the ground state $l = 0$.

3.6 In the rovibrational spectrum of gas-phase HCl, the transition $(v = 0, l = 2) \rightarrow (v = 1, l = 3)$ is observed at frequency 88,380 GHz. Evaluate the frequency of the homologous line in the DCl spectrum, knowing that the molecular equilibrium separation is 127 pm. Additionally, estimate the difference in binding energy between the two molecular species.

3.7[S] Evaluate the total molar specific heat capacity of a gas of CO molecules at temperature 500 K, knowing that the lowest pure rotational line is observed at frequency 115.27 GHz, and that the vibrational frequency is 64,100 GHz.

3.8 Model the adiabatic potential energy as a function of the distance R_{12} between a carbon and a oxygen nucleus with the Lennard-Jones potential:

$$V(R_{12}) = \epsilon \left[\left(\frac{\sigma}{R_{12}} \right)^{12} - \left(\frac{\sigma}{R_{12}} \right)^{6} \right].$$

Evaluate the parameters ϵ and σ that reproduce the spectroscopic values of the vibrational quantum $\bar{\nu} = 2162$ cm^{-1}, and the 3.8 cm^{-1} separation between the rotational lines of the ^{12}C^{16}O molecule.

3.9 ^{S} The equilibrium interatomic distance of the HCl molecule is 127 pm, and the curvature of the adiabatic potential at this equilibrium distance amounts to 590 kg/s^2. For a gas-phase sample of ^{1}H^{37}Cl at equilibrium at temperature 390 K, determine the frequency in Hz of the most intense rovibrational absorption line in the P branch.

3.10 * Refining the standard rigid-rotor approximation, the actual equilibrium distance of a diatomic molecule does not coincide with the adiabatic minimum of $V_{ad}(R_{12})$, but with the minimum of $V_{ad}(R_{12}) + V_{centrif}(R_{12})$, where the centrifugal contribution to the radial equation depends on the interatomic separation R_{12} and on the rotational quantum number l as well. As a consequence, the momentum of inertia depends on l, so that the purely rotational spectrum gets distorted compared to the rigid-rotor spectrum. Consider the ^{1}H^{127}I molecule and assume

$$V_{ad}(R_{12}) = \epsilon \left[\left(\frac{\sigma}{R_{12}} \right)^{4} - 2 \left(\frac{\sigma}{R_{12}} \right)^{2} \right],$$

with $\epsilon = 3.0$ eV and $\sigma = 161$ pm. Evaluate the l value for which the corrected rotational energy $\mathcal{E}_{rot}(l)$ differs by 1% compared to the prediction of the rigid-rotor formula.

3.11 * Evaluate the mean distance between the two protons in the H$_2$ molecule and the mean square fluctuation of that distance due to the zero-point radial motion in the ground state, assuming the harmonic approximation. Also compute the ratio of these lengths. The vibrational wave number (4395 cm^{-1}) and the rotational spectroscopic constant ($\hbar^2/2I = 61.0$ cm^{-1}) have been determined experimentally.
[Recall the expression $x_0 = [\hbar/(\omega\mu)]^{1/2}$ for the characteristic length scale of a harmonic oscillator with angular frequency ω and mass μ.]

3.12 * Consider the transition between two electronic states 1 and 2 of the diatomic molecule NH. Assume that the corresponding adiabatic potential energies $V_{ad}^{(i)}(R_{12})$ [for $i = 1, 2$] as a function of the interatomic distance R_{12} can be represented by the (Morse) form:

$$V_{ad}^{(i)}(R_{12}) = E_{bi} \left[e^{-2a_i(R_{12}-R_{0i})} - 2e^{-a_i(R_{12}-R_{0i})} \right].$$

Assume the following values of the parameters: $E_{b1} = 5.9$ eV, $R_{01} = 95$ pm, $a_1 = 19.0\,(\text{nm})^{-1}$, $E_{b2} = 3.2$ eV, $R_{02} = 107$ pm, $a_2 = 19.0\,(\text{nm})^{-1}$. Adopting the harmonic approximation for the vibrational energies, estimate the wavelength of the radiation necessary to excite the transition from level 1 (ground vibrational state $v_1 = 0$) to level 2 (vibrational level $v_2 = 3$).

3.13 Three measurements of the molar heat capacity of gas-phase HD provide the following values:

T (K)	C_V [J mol^{-1} K^{-1}]
500	20.81
1000	21.83
1500	23.80

Given the energy $\hbar^2/I = 5.69$ meV of the first rotational excited state, estimate the vibrational energy quantum $\hbar\omega$ of the HD molecule with a 10% precision. [Hint: the approximate solution of a transcendental equation $f(x) = $ constant can be determined by bisection when $f(x)$ is a monotonic function.]

3.14 Consider the molecular electronic levels derived from the 2p atomic orbitals of the O_2 molecule. Such orbitals are classified as σ and π according to their axial angular momentum $\hbar m$. Assume that the energies of these $\sigma - \sigma^*$ and $\pi - \pi^*$ molecular orbitals result from the diagonalization of the following matrices:

$$\begin{pmatrix} E_{2p} & -\Delta_m \\ -\Delta_m & E_{2p} \end{pmatrix}.$$

Here $E_{2p} = -8$ eV and $\Delta_m = \Delta_m(R) = \epsilon_m \exp(-R/\lambda_m)$, where $\epsilon_m = (1 + |m|) \times 12$ eV and $\lambda_m = (2 - |m|) \times 72$ pm. Taking the occupations of such molecular levels and Hund's rules into account, show that at the equilibrium distance $R_M = 121$ pm the molecule is paramagnetic ($S = 1$). Determine also the interatomic spacing below which the ground-state occupations of the electronic levels imply a diamagnetic state with all paired spins.

3.15 In the rovibrational absorption spectrum of hydrogen fluoride HF, consecutive lines are observed at the following wavelengths in nm: 870.301, 873.254, 876.226, 882.232, 885.266, and 888.321. Knowing that these lines belong to the $v = 0 \rightarrow 3$ excitation, assign the rotational quantum numbers of such lines and evaluate the vibrational-energy contribution to these transitions.

Chapter 4
Statistical Physics

The purpose of statistical physics is to relate average properties of *"macroscopic" objects* (thermodynamic quantities) to the fundamental interactions governing their *microscopic dynamics*. The previous chapters have discussed the way many mechanical properties of individual atoms and molecules are accounted for based on the fundamental electromagnetic interactions driving the motion of the composing electrons and nuclei. The number of electrons and nuclei in atoms and small molecules does not exceed few hundred or, at most, few thousand. Compared to such microscopic systems, macroscopic objects are characterized by huge numbers of degrees of freedom. For example, a sodium-chloride crystal weighting 1 g is composed by $\sim 3 \times 10^{23}$ electrons plus approximately 10^{22} Na and 10^{22} Cl nuclei. One may as well conceive some (possibly approximate) wavefunction for the dynamics of such a huge number of degrees of freedom, but must also readily give up any hope to ever record or process the huge amount of information stored by the wavefunction describing the full microscopic detail of such an object. On the other hand, this limitation is not so bad because the motion of individual electrons and nuclei and other intimate details of the dynamics of this crystal are probably boring and of little practical use. Physicists and materials scientists are more interested in the measurable *average macroscopic properties* of objects and substances, e.g. stiffness, tensile strength, heat capacity, heat and electric conductivity, dielectric and magnetic susceptibilities, phase transitions... and here statistics does help.

4.1 Introductory Concepts

To draw a link between the microscopic dynamics and the macroscopic thermodynamic properties, equilibrium *statistical physics* borrows its mathematical tools from *the theory of probability* and from general *statistics*.

N. Manini, *Introduction to the Physics of Matter*, Undergraduate Lecture Notes
in Physics, https://doi.org/10.1007/978-3-030-57243-3_4

4.1.1 Probability and Statistics

Statistics is rooted in the notion of probability. The naive notion of probability coincides with the relative number of observations. For example, after rolling a dice a large number N of times, we observe "two" N_2 times: the ratio N_2/N is an estimate of the probability P_2 of obtaining "two" in a single roll. However, we have also an *a-priori* concept of probability. We will assert that $P_2 = 1/6$, even against contradictory observation, unless we have evidence that the dice is loaded. Such *a-priori* notion of probability is an example of a rigorous definition of probability as a measure defined on a "space of events" (here all possible dice outcomes), such that the measure of all space equals unity. The basic assumptions are:

1. for non-intersecting sets of events A and B, the probability of $A \cup B$ (i.e. that any event in either A or B is realized) equals $P_A + P_B$;
2. given two spaces of events (which could intersect, or even coincide), two sets of events A and B belonging to the first and second space are called *independent* if the probability of (A and B) (i.e., an event occurs that satisfies the conditions for belonging to A and at the same time to B) is the product $P(A) \cdot P(B)$.

Both these properties are trivial when probability and relative number of observations are identified.

The two basic properties of probability sketched above allow us to relate the probabilities of complicated events to those of elementary events. For example, the probability of obtaining two ones when two dices are rolled independently is $1/6 \times 1/6 = 1/36$ (property 2); the probability for a four and a five is $1/36 + 1/36 = 1/18$, since both (a four on dice 1 and a five on dice 2) and (a five on dice 1 and a four on dice 2) are mutually exclusive events (property 1).

Statistics makes a wide use of *probability distributions*: these are lists of probabilities of mutually exclusive sets of events which cover the whole space of events. For example, when the two rolled dices are considered, the outcome may be grouped in equal numbers (6 mutually exclusive possibilities, each of probability $1/36$) or different numbers ($6 \times 5/2 = 15$ mutually exclusive possibilities, each of probability $1/18$): using property 1, this leads to a distribution $P_{\text{equal}} = 1/6$, and $P_{\text{different}} = 5/6$. Clearly, the events considered exhaust all space, and the sum of all the probabilities in the distribution equals unity. It is a useful exercise to work out the distribution $P_{\text{both even}}$, $P_{\text{even odd}}$, $P_{\text{both odd}}$ on the same space of events. As in this example, a probability distribution makes probability a function of certain properties of the events considered. The dice outcomes provide an example of a discrete space of events. Continuous spaces of events are common, too. Examples of popular statistical distributions are the binomial and the Poisson distribution (both for discrete events), and the Gaussian distribution (for a continuous space of events).

QM, even at the level of one or few particles, involves a statistical interpretation, in the probabilistic postulate of observation. When observable A is measured on a quantum system initially in some state $|i\rangle$, the system will be found in the eigenstate $|a\rangle$ of A associated to eigenvalue a, with probability $P_a(|i\rangle) = |\langle a|i\rangle|^2$. The ket $|i\rangle$ contains all probability distributions corresponding to all possible operators associated to conceivable measurements that could be carried out on the system.

Statistical physics is concerned only marginally with this statistical interpretation of QM because, as long as the mechanical system is left undisturbed, its ket evolves from any given initial state according to the (deterministic) Schrödinger equation (1.7). The statistical description of a macroscopic system attempts to rather address its average properties without the need of a precise specification of its initial conditions, rather assuming a "typical" situation, where any of all "plausible" initial conditions can occur with equal probability. Banning the unlikely situation of an highly untypical initial condition, a macroscopic system is expected to explore successively a sequence of many standard microscopic conditions, so that it should be possible to identify its macroscopic properties with time averages of the relevant observables along the time evolution driven by the internal dynamics.

Two basic assumptions reconcile the irrelevance of the initial conditions with the time-average approach: equilibrium and ergodicity. *Equilibrium* requires that long ago the system has undergone some initial transient, and that now, at the time of interest, no systematic drift is occurring any more, all collective quantities (e.g. pressure) fluctuating in time around some well-defined average value. *Ergodicity* assumes that all kinds of states are randomly explored in a period of time short compared to the typical duration of measurements: subsequent times provide *independent random* realizations of an underlying probability distribution. A dishonest dice roller violates ergodicity by controlling accurately the initial conditions, rather than rolling blindly to generate successive truly random independent numbers. A similar violation may occasionally occur in statistical physics, as illustrated in the example of H_2 nuclear spin. H_2 molecules occur with total nuclear spin 1 (orthohydrogen) or 0 (parahydrogen). These states have nearly the same energy, thus they should all occur with equal probability: one expects to observe a ratio of ortho- to parahydrogen matching their degeneracy ratio $3 : 1$. However ortho-para inter-conversion is rather difficult: ordinary molecular collisions conserve nuclear spins, thus leaving the total abundance of each species unaltered. Inter-conversions can occur at the walls, in the neighborhood of magnetic impurities. However, one can store the H_2 sample in a vessel whence all magnetic impurities have been carefully removed. In this condition, an anomalous abundance ratio of ortho- to parahydrogen can be stabilized for an extended time. Instead, if magnetic impurities are present, then careless "shuffling" of the nuclear spins occurs, the expected $3 : 1$ equilibrium distribution is readily recovered, and the system is ergodic. In brief, the ergodic hypothesis assumes that the system is sufficiently random that no conserved quantities prevent the access (in periods of time short compared to the duration of the experiment) to some major subset of the space of states.

When a system is at equilibrium and ergodic, time averages can safely be replaced by averages over a suitable probability distribution of the microscopic states. We now sketch the standard formalism for a random distribution of quantum states.

4.1.2   Quantum Statistics and the Density Operator

We formalize our ignorance about the precise quantum state of a system, while describing correctly its statistical properties [17]. A system may be found in any of a number of quantum states $|a_1\rangle$, $|a_2\rangle$,... with probability w_1, w_2,... respectively. The states $|a_1\rangle$, $|a_2\rangle$,... need not be eigenstates of any observable, they need not be orthogonal, and in number they could even exceed the dimension of the Hilbert space of states. The normalization of probability requires that $\sum_i w_i = 1$.

On average, the measurement of an observable B on such a system should provide

$$[B] = \sum_i w_i \langle a_i|B|a_i\rangle = \sum_i w_i \sum_b b \, |\langle a_i|b\rangle|^2, \tag{4.1}$$

where the $\{|b\rangle\}$ is the basis of eigenkets of B, with eigenvalues b, allowing us to write $B = \sum_b b \, |b\rangle\langle b|$. We introduce the square brackets [ ] to indicate the statistical ensemble average, i.e. the statistical mean of the quantum average values.

An average (4.1) can also be computed using the *statistical density operator*

$$\hat{\rho} = \sum_i w_i \, |a_i\rangle\langle a_i|, \tag{4.2}$$

as

$$[B] = \sum_i w_i \langle a_i|B|a_i\rangle = \sum_i w_i \sum_b b \, \langle b|a_i\rangle\langle a_i|b\rangle \tag{4.3}$$

$$= \sum_b b \, \langle b| \left(\sum_i w_i \, |a_i\rangle\langle a_i| \right) |b\rangle = \sum_b b \, \langle b|\hat{\rho}|b\rangle = \sum_b \langle b|\hat{\rho}B|b\rangle = \mathrm{Tr}(\hat{\rho}B).$$

The density operator collects all dynamical and statistical properties of the system. $\hat{\rho}$ is self-adjoint: it can therefore be diagonalized. On its diagonal basis, the density operator is expressed as:

$$\hat{\rho} = \sum_m P_m \, |\rho_m\rangle\langle\rho_m|, \tag{4.4}$$

where the kets $|\rho_m\rangle$ form a complete orthonormal basis of the Hilbert space [unlike the $|a_i\rangle$ of the definition (4.2)], associated to nonnegative eigenvalues P_m. Note the normalization

$$\mathrm{Tr}(\hat{\rho}) = \sum_b \langle b|\hat{\rho}|b\rangle = \sum_{b\,i} w_i \langle b|a_i\rangle\langle a_i|b\rangle = \sum_i w_i \sum_b \langle a_i|b\rangle\langle b|a_i\rangle$$

$$= \sum_i w_i \langle a_i| \left(\sum_b |b\rangle\langle b| \right) |a_i\rangle = \sum_i w_i \langle a_i|a_i\rangle = 1, \tag{4.5}$$

implying that also $\sum_m P_m = 1$: the eigenvalues P_m can be interpreted as probabilities.

Fig. 4.1 The whole
"universe" U partitioned into
two weakly-interacting
regions: a "system" S plus
the "rest of the universe" W

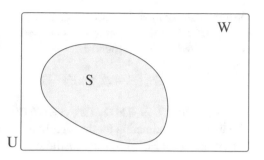

The special case of a statistical operator describing a single *pure state* $\hat{\rho} = |a_1\rangle\langle a_1|$ covers standard deterministic QM: $[B] = \mathrm{Tr}(\hat{\rho}B) = \langle a_1|B|a_1\rangle = \langle B\rangle$. In this case (and only in this case) $\hat{\rho}$ has the property of a projector $\hat{\rho}^2 = \hat{\rho}$.

4.2 Equilibrium Ensembles

Consider a macroscopic isolated system (the "universe") U, whose energy E is fixed (*microcanonical ensemble*) within an uncertainty ΔE. The fundamental *postulate of equilibrium statistical physics* follows from the ergodic hypothesis: *all quantum states of U are equally likely*. In other words, the probability of any state $|i\rangle$ equals a constant $P_i^U = 1/\Omega_U(E; \Delta E)$ if the energy E_U of $|i\rangle$ is in the range $E \le E_U \le E + \Delta E$, or vanishes otherwise. $\Omega_U(E; \Delta E)$ equals the number of individual states of U in the selected energy range $E \le E_U \le E + \Delta E$. In this way, the probability P_i^U is correctly normalized to unity. From this very "democratic" postulate, we derive the probability distribution of a system S in thermal equilibrium with the rest of the universe W, Fig. 4.1. The total Hamiltonian $H_U = H_S + H_W + H_{SW}$ is assumed to involve a negligibly weak interaction term $H_{SW} \approx 0$, so that $E_U \approx E_S + E_W$.

We address this specific problem: What is the probability P_m of finding the system S in a given quantum state $|m\rangle$ with energy E_m? As the coupling between S and W is weak, we can assume that the states in S and in W are distributed *independently* at random. The only correlations between S and W are induced by the energy conservation. Accordingly,

$$P^U = P_m^S P^W. \tag{4.6}$$

Microcanonical distributions govern both U and W individually: like U-states occur with probability $P^U = 1/\Omega_U(E; \Delta E)$, in W each state has probability $P^W = 1/\Omega_W(E - E_m; \Delta E)$. Here $E - E_m$ is the residual energy of the rest of the universe W, once the energy E_m of the system S has been removed from E. From Eq. (4.6) we extract the probability distribution of S:

$$P_m^S = \frac{P^U}{P^W} = \frac{\frac{1}{\Omega_U(E; \Delta E)}}{\frac{1}{\Omega_W(E-E_m; \Delta E)}} = \frac{\Omega_W(E - E_m; \Delta E)}{\Omega_U(E; \Delta E)}. \tag{4.7}$$

Observe that for most U-states the energy E_m in the system S is a tiny fraction of the total energy E of the universe: one can then Taylor-expand the numerator—or, even better, its logarithm[1]—around E:

$$\ln \Omega_W(E - E_m; \Delta E) = \ln \Omega_W(E; \Delta E) - \beta E_m + \cdots, \tag{4.8}$$

where the slope $\beta = \partial \ln \Omega_W(E'; \Delta E)/\partial E'|_{E'=E}$ has dimension of inverse energy. The linear approximation for the logarithm is extremely good as long as $E_m \ll E_W$. We exponentiate Eq. (4.8) and substitute it back into Eq. (4.7), obtaining

$$P_m^S = \frac{\Omega_W(E; \Delta E)}{\Omega_U(E; \Delta E)} e^{-\beta E_m}. \tag{4.9}$$

This is a remarkable result for at least two reasons:

- the whole dependence on the state $|m\rangle$ of S is confined to the exponential factor, because the prefactor $Z^{-1} = \Omega_W(E; \Delta E)/\Omega_U(E; \Delta E)$ is independent of $|m\rangle$, and is therefore a pure normalization factor;
- the *probability* of a state $|m\rangle$ of S *is a function of its energy E_m only*, and a simple one too: an exponential.

We rewrite Eq. (4.9) inserting the normalization factor Z^{-1} and omitting the label S:

$$P_m = P(E_m) = \frac{e^{-\beta E_m}}{Z}. \tag{4.10}$$

Equation (4.10) describes the *Boltzmann equilibrium probability distribution* for the states of a generic system in weak thermal contact with a huge environment. This distribution is associated to the Gibbs *canonical ensemble*.

The probability distribution (4.10) is necessarily normalized, $\sum_m P_m = 1$. Therefore, Z can alternatively be expressed entirely in terms of properties of the system S under study, without any reference to U and W. Explicitly:

$$Z = \sum_m e^{-\beta E_m} = \mathrm{Tr}\left(e^{-\beta H}\right), \tag{4.11}$$

where we write H in place of H_S. Z is called the *partition function*. Note that the sum in Eq. (4.11) involves all microstates including, in particular, all same-energy components of degenerate levels. An equivalent formulation of the same sum is:

---

[1]The number Ω_W of "microstates" in a given small energy interval ΔE of a macroscopic "rest of the universe" changes roughly with a macroscopically large power of its energy level E'. Thus the expansion of $\ln \Omega_W$ is far more accurate and better convergent than that of Ω_W.

$$Z = \sum_E n_{\deg}(E)e^{-\beta E}. \tag{4.12}$$

where $n_{\deg}(E)$ is the number of degenerate states with energy E. Note also that the trace in Eq. (4.11) can be taken on any basis, as convenient.

The density operator $\hat{\rho}_{eq}$ associated to the Boltzmann equilibrium probability distribution is *diagonal in the energy representation*:

$$\hat{\rho}_{eq} = \sum_m \frac{e^{-\beta E_m}}{Z} |m\rangle \langle m| = \frac{1}{Z} e^{-\beta H}. \tag{4.13}$$

This density operator can be written on any basis, by applying a suitable unitary transformation.

4.2.1 Connection to Thermodynamics

We assume now that the system S can be partitioned into two weakly interacting subsystems S_1 and S_2, so that $E_{m_1 m_2} \approx E_{1m_1} + E_{2m_2}$. When thermal equilibrium is established between S_1, S_2, and the rest of the world W, the distribution

$$P_{m_1 m_2} = \frac{e^{-\beta E_{m_1 m_2}}}{Z} = \frac{e^{-\beta(E_{1m_1} + E_{2m_2})}}{\sum_{n_1 n_2} e^{-\beta(E_{1n_1} + E_{2n_2})}}$$

can be factorized into

$$P_{m_1 m_2} = \frac{e^{-\beta E_{1m_1}}}{\sum_{n_1} e^{-\beta E_{1n_1}}} \cdot \frac{e^{-\beta E_{2m_2}}}{\sum_{n_2} e^{-\beta E_{2n_2}}} = \frac{e^{-\beta E_{1m_1}}}{Z_1} \cdot \frac{e^{-\beta E_{2m_2}}}{Z_2}. \tag{4.14}$$

This decomposition shows that S_1 and S_2 follow independent equilibrium Boltzmann distributions $P_{m_i}^{S_i} = e^{-\beta E_{im_i}}/Z_i$, with the *same* β parameter. This observation suggests that the intensive quantity β might be a function of temperature T.

The natural statistical definition for the *internal energy* U of thermodynamics is the average of the energy operator H:

$$U = [H] = \text{Tr}(\hat{\rho}H) = \sum_m E_m P_m = \frac{1}{Z} \sum_m E_m e^{-\beta E_m}. \tag{4.15}$$

Deriving U with respect to β yields minus the squared energy fluctuation:

$$\frac{\partial U}{\partial \beta} = \frac{-Z \sum_m E_m^2 e^{-\beta E_m} + \left(\sum_m E_m e^{-\beta E_m}\right)^2}{Z^2} = -[H^2] + [H]^2 = -[(H - [H])^2].$$
$$\tag{4.16}$$

Thus, $\partial U/\partial\beta \leq 0$, i.e. the internal energy decreases when β increases. This observation suggests that β and temperature T may be inversely related. We work out now the explicit relation between β and T.

Firstly, note that Z is a multiplicative function [$Z = Z_1 Z_2$ for a system composed of two subsystems, see Eq. (4.14)], thus its logarithm is an *additive function* ($\ln Z = \ln Z_1 + \ln Z_2$). $\ln Z$ must then represent an *extensive* thermodynamic quantity. Thus

$$F = -\frac{\ln Z}{\beta} \tag{4.17}$$

defines an extensive quantity with energy dimension. Starting from this definition, we derive a relation between the derivative of $\ln Z$ w.r.t. β and the internal energy U as defined in Eq. (4.15):

$$\frac{\partial(\beta F)}{\partial\beta}\bigg|_{V,N} = -\frac{\partial(\ln Z)}{\partial\beta} = -\frac{1}{Z}\frac{\partial Z}{\partial\beta} = -\frac{\frac{\partial}{\partial\beta}\sum_m e^{-\beta E_m}}{Z}$$

$$= -\frac{\sum_m \frac{\partial}{\partial\beta}e^{-\beta E_m}}{Z} = \frac{\sum_m E_m e^{-\beta E_m}}{Z} = [H] = U . \tag{4.18}$$

Equation (4.18) identifies the derivative of βF w.r.t. β with the internal energy U. This identification reminds us of a thermodynamic identity involving the free energy F and temperature: starting from the basic definition $F = U - TS$ (where S is entropy), one finds

$$U = F + TS = F - T\frac{\partial F}{\partial T} = -T^2\left(-T^{-2}F + T^{-1}\frac{\partial F}{\partial T}\right) = -T^2\frac{\partial(T^{-1}F)}{\partial T} . \tag{4.19}$$

This equality can be written more compactly in terms of a derivative w.r.t. the inverse temperature:

$$\frac{\partial(T^{-1}F)}{\partial(T^{-1})} = U . \tag{4.20}$$

In Eqs. (4.19) and (4.20) all derivatives are carried out at constant number of particles N and volume V. The comparison of Eq. (4.18) and Eq. (4.20) suggests that

- it is perfectly natural to identify the extensive quantity F defined statistically in Eq. (4.17) with the free energy F of thermodynamics;
- β must then be proportional to the inverse temperature T^{-1}.

The proportionality constant between T^{-1} and β is known as the Boltzmann constant k_B, and represents the numerical conversion factor between temperature [K] and energy [J]:

$$\beta = \frac{1}{k_B T} . \tag{4.21}$$

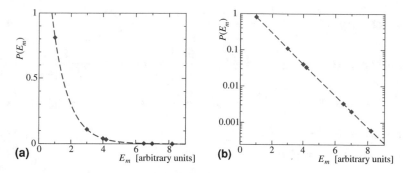

Fig. 4.2 The statistical meaning of temperature: **a** in a system at equilibrium with the "rest of the universe" the probability $P(E_m)$ of its individual microstates $|m\rangle$ decays exponentially with their energy E_m divided by $k_B T$, Eq. (4.10). **b** In a lin-log scale, the slope of the straight line representing $P(E_m)$ as a function of energy is precisely $-\beta = -(k_B T)^{-1}$

The comparison of Eq. (4.47) below with the empirical definition of absolute temperature through the ideal-gas thermometer provides the relation $k_B = R/N_A$ with the gas constant R and the Avogadro constant N_A. The resulting value $k_B = 1.38065 \times 10^{-23}$ J/K $= 86.1733$ μeV/K. In statistical physics, temperature usually appears in the energy combination $k_B T$, or as its inverse β.

The physical meaning of temperature is now clear: *the probability distribution* (4.10) *of the microstates of any system in thermal equilibrium is determined* uniquely *by the energies of these states*, in a straightforward way: at a given temperature T, the probability $P(E_m)$ *of a state with energy* E_m *is proportional to* $\exp(-\beta E_m)$. Thus, $-\beta \equiv -(k_B T)^{-1}$ is the slope of the straight line representing $P(E_m)$ as a function of E_m in a lin-log plot, as in Fig. 4.2.

Note that, due to the normalization factor Z^{-1} in Eq. (4.10), the energies of all states in the system determine the equilibrium probability P_m for the occurrence of each energy eigenstate $|m\rangle$. However, only the energy *difference* between two states $|m\rangle$ and $|n\rangle$ determines their probability *ratio* $P_m/P_n = \exp[\beta(E_n - E_m)]$.

The discussed relations, in particular Eqs. (4.17) and (4.21), establish an explicit link between statistical physics and thermodynamics, i.e. between the microscopic dynamics and macroscopic observable average properties. Based on Eq. (4.17), one can link all thermodynamic properties of a system at equilibrium purely to its partition function Z. We list here the expressions for several fundamental extensive and intensive thermodynamic quantities:

$$\text{free energy } F = -k_B T \ln Z = -\frac{\ln Z}{\beta} \qquad (4.22)$$

$$\text{internal energy } U = F + TS = \left.\frac{\partial(\beta F)}{\partial \beta}\right|_{V,N} = -\left.\frac{\partial(\ln Z)}{\partial \beta}\right|_{V,N} \tag{4.23}$$

$$\text{heat capacity} C_V = \left.\frac{\partial U}{\partial T}\right|_{V,N} = -k_B \beta^2 \left.\frac{\partial U}{\partial \beta}\right|_{V,N} = k_B \beta^2 \left.\frac{\partial^2 \ln Z}{\partial \beta^2}\right|_{V,N} \tag{4.24}$$

$$\text{entropy } S = -\left.\frac{\partial F}{\partial T}\right|_{V,N} = k_B T \frac{\partial \ln Z}{\partial T} + k_B \ln Z = \frac{U}{T} + k_B \ln Z \tag{4.25}$$

$$\text{free enthalpy } G = F + PV \tag{4.26}$$

$$\text{enthalpy } H = F + PV + TS \tag{4.27}$$

$$\text{pressure } P = -\left.\frac{\partial F}{\partial V}\right|_{\beta,N} = \left(\frac{N}{V}\right)^2 \left.\frac{\partial(F/N)}{\partial(N/V)}\right|_{\beta}. \tag{4.28}$$

4.2.2 Entropy and the Second Principle

The expression (4.25) for entropy in statistics holds at equilibrium only. A far more *general definition of entropy* is also available, which applies for any statistical distribution of the quantum states defined by an arbitrary density operator $\hat{\rho}$, whether it represents an equilibrium ensemble or not:

$$S = -k_B \operatorname{Tr}(\hat{\rho} \ln \hat{\rho}). \tag{4.29}$$

This definition conforms to the intuitive idea of entropy as a measure of disorder. Consider the basis $|\rho_m\rangle$ where $\hat{\rho}$ is diagonal. Here, $\hat{\rho} = \sum_m P_m |\rho_m\rangle\langle\rho_m|$. In this basis

$$S = -k_B \sum_m P_m \ln P_m. \tag{4.30}$$

If we compute S for the most ordered distribution, namely a pure state (i.e. $P_{\bar{m}} = 1$ for a single state $|\bar{m}\rangle$ and $P_m = 0$ for all other states), we obtain $S = 0$ since all terms in the sum vanish. S increases to some positive value when several states have nonzero probabilities. For the extreme limit of a completely random distribution with probability $P_m = \Omega^{-1}$ for a number Ω of equally-likely states, we obtain $S = -k_B \sum_m^{\Omega} \left[\Omega^{-1} \ln(\Omega^{-1})\right] = k_B \ln(\Omega)$. We conclude that *entropy* provides a *logarithmic measure of the effective number of states* that the system accesses.[2]

When the system is at equilibrium, the general statistical definition (4.29) coincides with Eq. (4.25). This is readily verified by substituting the equilibrium density operator $\hat{\rho}_{eq}$ of Eq. (4.13) into Eq. (4.29), i.e. Eq. (4.10) into Eq. (4.30):

---

[2] The definition (4.29), practically $S = -k_B[\ln \hat{\rho}]$, is related to Shannon's definition of entropy $S_{Shannon} = -[\log_2 \hat{\rho}]$ adopted in fundamental information theory by $S = \ln(2) k_B S_{Shannon}$.

$$S = -k_B \sum_m \frac{e^{-\beta E_m}}{Z} \ln \frac{e^{-\beta E_m}}{Z} = -k_B \frac{\sum_m e^{-\beta E_m} (\ln e^{-\beta E_m} - \ln Z)}{Z}$$

$$= k_B \frac{\sum_m e^{-\beta E_m} (\beta E_m + \ln Z)}{Z} = k_B \beta [H] + k_B \frac{Z \ln Z}{Z} = \frac{U}{T} + k_B \ln Z . \quad (4.31)$$

Moreover, it is possible to show [29] that *if $\hat\rho$ coincides with the* Gibbs–Boltzmann *equilibrium density operator* $\hat\rho_{eq}$, Eq. (4.13), *the system entropy*, defined in Eq. (4.29), *is maximum* under the constraint of assigned average energy $\mathrm{Tr}(\hat\rho H) \equiv U$, volume V, and number of particles N. Any generic density operator $\hat\rho_{gen}$, respecting the constraints on U, V, and N, yields an entropy S_{gen} not exceeding the S_{eq} of the equilibrium distribution $\hat\rho_{eq}$, Eq. (4.13). Any isolated system is observed to evolve spontaneously toward equilibrium: accordingly, this maximum-entropy result shows the consistency of statistical physics with the *second principle of thermodynamics*. In a spontaneous transformation toward equilibrium $\hat\rho_{gen} \to \hat\rho_{eq}$, the entropy increases from S_{gen} up to S_{eq}. This result can be extended to conditions where U, V, or N vary (not-isolated systems), but for our purposes it suffices to retain that the experimental fact that entropy increases in approaching equilibrium is rooted in statistics.

For deeper insight in the fundaments of statistics sketched in the present Section, we refer to specialized textbooks, e.g. Refs. [17, 29, 30].

4.3 Ideal Systems

Before attempting any understanding of the complicated macroscopic properties of structured objects such as a bicycle or a bowl of soup, scientists have wisely chosen to investigate simpler systems: (usually) macroscopically homogeneous pure (or controllably "doped") bunches of atoms or molecules all of the same kind, or of few kinds. These simpler systems constitute the broad class of *materials*. The rationale behind studying materials is that the properties of a complex structured object can be understood in terms of the functionality of the individual pieces it is composed of. This functionality, in turn, depends on their shapes and material properties. For example, the total heat capacity of a bicycle is practically identical to the sum of the heat capacities of the composing parts.

The macroscopic properties of a material are often studied in the limit of an infinitely large sample, for which the interactions with the surrounding environment (e.g. the containing vessel) are sufficiently weak to justify the assumptions of Sect. 4.2. As the surface atoms/molecules directly interacting with the environment usually involve a layer about $\sim$1 nm thick, for any sample whose linear dimensions (all three of them) are larger than, say, $\sim$1 μm, the error induced by this *bulk approximation* should not be too bad.[3]

Even within the idealization of a homogeneous bulk sample, the recipe

---

[3]The bulk approximation fails for submicrometric objects: in "nanoparticles" new properties often emerge, e.g., due to quantum confinement of electrons in a small volume.

1. compute the spectrum of energies E_m and eigenstates $|m\rangle$ of the system;
2. use Eq. (4.11) to evaluate the partition function Z for given T, or $\beta = 1/(k_B T)$;
3. generate the equilibrium density operator $\hat{\rho}$ using Eq. (4.13);
4. compute macroscopic average quantities as $[B] = \text{Tr}(\hat{\rho} B)$, Eq. (4.3);

is not really applicable for any realistic system, due to the difficulty of step 1.

Luckily, for a few *ideal systems* the programme of statistical physics can be carried to satisfactory conclusion. By ideal systems we mean systems composed of individual components (particles) whose mutual interactions are negligible. Ideal systems have

$$H = \sum_{i=1}^{N} H_i , \tag{4.32}$$

where H_i governs the dynamics of a small set of degrees of freedom, e.g. the position and spin of one particle. The simplicity of ideal systems allows us to obtain exact partition functions, and thus to predict their thermodynamic properties by means of statistical methods.

No ideal system exists in nature (although photons and neutrinos are very good approximations to non-interacting particles). If one did exist, strictly speaking the methods of equilibrium statistical physics would be irrelevant for that system since it would not be ergodic, and could never reach equilibrium. However, several properties of *weakly-interacting* systems follow semi-quantitatively those of ideal systems. The reader should be cautioned of the limited applicability and the risk of artifacts due to the idealized nature of non-interacting systems: a quantitative understanding of the thermodynamic properties of real materials usually requires more sophisticated methods which account for interparticle interactions.

Consider an ideal system composed of a single type of identical noninteracting particles, e.g., electrons, or atoms, or molecules... The natural basis of states is a factored basis, Eqs. (2.60), (2.61), where for particle i the set α_i of quantum numbers identifies the state $|\alpha_i\rangle$, with energy $\mathcal{E}_{\alpha_i}$. Ignoring briefly the permutation-symmetry requirements (Sects. 2.2.1 and 2.2.2), the partition function

$$Z \stackrel{?}{=} \sum_{\alpha_1 \alpha_2 \ldots \alpha_N} \exp[-\beta(\mathcal{E}_{\alpha_1} + \mathcal{E}_{\alpha_2} + \cdots + \mathcal{E}_{\alpha_N})] \tag{4.33}$$

$$= \sum_{\alpha_1 \alpha_2 \ldots \alpha_N} e^{-\beta \mathcal{E}_{\alpha_1}} e^{-\beta \mathcal{E}_{\alpha_2}} \ldots e^{-\beta \mathcal{E}_{\alpha_N}} .$$

factorizes into N single-particle terms. However, due to (anti)symmetrization, the exchange of any two quantum numbers α_i and α_j leads to the same state (up to a sign) for the N particles. For identical bosons, the summation is unconstrained but overcomplete: the same states are retrieved over and over again, namely $N!$ times whenever all α_i's differ, and fewer times when some α_i coincide.[4] For identical

---

[4]To evaluate how many times the same N-particles state reappears in the sum of Eq. (4.33), we need the number of permutations of a given set of indexes $\alpha_1, \alpha_2, \ldots, \alpha_N = \{\alpha_i\}$. First, we sort the

fermions, one must first of all constrain the multiple summation to all different α_i's. One then needs to correct for the $N!$ repetitions of the same N-particle state.

We can now correct Eq. (4.33) for those multiple counts of the same states:

$$Z = \sum_{\alpha_1 \alpha_2 \ldots \alpha_N} \frac{n_0! \, n_1! \, n_2! \, \ldots}{N!} \, \exp\left(-\beta \sum_{i=1}^{N} \mathcal{E}_{\alpha_i}\right). \tag{4.34}$$

The sum of Eq. (4.34) extends over all possible α_i's for bosons, and over different α_i's respecting Pauli's principle in the case of fermions. Unfortunately, the exact expression (4.34) for Z *does not factorize*, due to the permutation correction for bosons, and to the Pauli constraint for fermions. In Sect. 4.3.2 we will address this problem. Presently, we focus on the simpler high-temperature limit, where this factorization can be carried out, at least approximately.

4.3.1 The High-Temperature Limit

In the following, we assume that the single-particle spectrum has no upper bound, as is usually the case. At low temperature, the Boltzmann exponential factor tends to privilege those states with a total energy $\sum_{i=1}^{N} \mathcal{E}_{\alpha_i} = \sum_{\alpha} n_\alpha \mathcal{E}_\alpha$ as small as possible: relatively few low-energy single-particle states are then significantly occupied, see Fig. 4.3a. We shall come back to this low-temperature regime in Sect. 4.3.2.

In contrast, at high temperature (small β), all single-particle states with energy $\lesssim k_B T$ above the single-particle ground state, have similar chances of being occupied. For large T, there is a huge number ($\gg N$) of these single-particle states, and therefore the probability that each individual state is occupied becomes very small, see Fig. 4.3c. N-particle states with all different quantum numbers are overwhelmingly more numerous than those with two or more of them equal. As a result, at high temperature the main contribution to the boson partition function comes from states which have $n_\alpha = 0$ or 1 at most. All terms with $n_\alpha > 1$ in the sum of Eq. (4.34) contribute negligibly to a huge Z. Thus in Eq. (4.34) we replace $n_0! \, n_1! \, n_2! \, \ldots \to 1$. In the fermionic case we include even the unphysical terms with some equal α_i, thus violating the Pauli-principle constraint. We obtain an approximate expression for the high-temperature partition function valid for bosons and fermions alike:

$$Z \simeq \frac{1}{N!} \sum_{\alpha_1 \alpha_2 \ldots \alpha_N}^{\text{unrestricted}} \exp\left(-\beta \sum_{i=1}^{N} \mathcal{E}_{\alpha_i}\right) = \frac{1}{N!} \, Z_1 \cdot Z_2 \cdot \ldots \cdot Z_N = \frac{(Z_1)^N}{N!}. \tag{4.35}$$

---

single-particle states $|\alpha\rangle$, e.g. for increasing energy, say $|0\rangle, |1\rangle, |2\rangle, \ldots$ Then, in each term in the sum, we count the numbers $n_0, n_1, n_2, \ldots$ of times that each of these single-particle states appears. Clearly, the sum of these *occupation numbers* $n_0 + n_1 + n_2 + \cdots = N$. In the sum, the number of equivalent terms equals the number of permutations of the $\{\alpha_i\}$, which is $N!/(n_0! \, n_1! \, n_2! \, \ldots)$. This expression holds for fermions as well, because all $n_\alpha = 0$ or 1, and $0! = 1! = 1$.

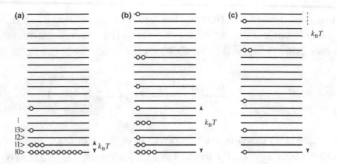

Fig. 4.3 Typical boson occupancies of single-particle energy eigenstates in a spectrum with no upper bound. **a** Low temperature. **b** Intermediate temperature. **c** High temperature, where most states are empty, occasional states have $n_\alpha = 1$ particle, and essentially none has $n_\alpha > 1$

Here the single-particle partition function $Z_1 = \sum_\alpha \exp(-\beta \mathcal{E}_\alpha)$ sums the Boltzmann weights $\exp(-\beta \mathcal{E}_\alpha)$ for all one-particle states $|\alpha\rangle$. Z_1 should not be confused with the full Z of Eq. (4.11) which involves entire-system states $|m\rangle$ instead.

In this high-temperature limit, the free energy is

$$F = -\frac{\ln Z}{\beta} \simeq -\frac{1}{\beta} \ln \frac{(Z_1)^N}{N!} = -k_{\mathrm{B}}T \left(\ln(Z_1^N) - \ln N! \right) \qquad (4.36)$$

$$\simeq -k_{\mathrm{B}}T \left(N \ln Z_1 - N \ln \frac{N}{e} \right) = -N k_{\mathrm{B}}T \ln \frac{e\, Z_1}{N},$$

where we made use of the Stirling approximation $\ln(N!) \simeq N \ln(N/e)$, which is very accurate for large N.

Ideal particles are certainly in gas phase. At very large T even real atoms are affected weakly by their mutual interactions, so that hot real atoms attain their gas phase too. The center of mass of each particle translates freely. This translational degree of freedom can be separated from the internal ones, as sketched for diatomic molecules in Sect. 3.3. In the single-particle Hamiltonian $H_{1\,\mathrm{tr}} + H_{1\,\mathrm{int}}$ the two terms commute. Accordingly, the single-particle partition function Z_1 can be factored

$$Z_1 = Z_{1\,\mathrm{tr}} \, Z_{1\,\mathrm{int}} \qquad (4.37)$$

into a translational times an internal part. The latter describes the statistical dynamics of the internal degrees of freedom of the particle (including, e.g., its spin, and/or its rotations), and depends therefore on the specific spectrum of its excitations. In contrast, the *translational part is universal*: it depends uniquely on the particle mass M and the containing volume V.

4.3.1.1 Translational Degrees of Freedom

To obtain $Z_{1\,\text{tr}}$, recall the spectrum of a freely translating particle contained in a macroscopically large cubic box of volume $V = L \times L \times L$. Impose periodic boundary conditions, i.e. the eigenfunction is the same at opposite faces of the box. The results [Eq. (4.42) onward] would not change if, slightly more realistically, we assumed that the wavefunction vanishes at the boundary instead. The allowed values of the $u = x, y, z$ momentum components

$$p_u = \hbar k_u = \hbar \frac{2\pi}{L} n_u , \qquad n_u = 0, \pm 1, \pm 2, \pm 3, \ldots \qquad (4.38)$$

are associated to plane-wave eigenfunctions $\psi_{\mathbf{k}}(\mathbf{r}) = L^{-3/2} \exp(i\,\mathbf{k} \cdot \mathbf{r})$, with translational kinetic energy

$$\mathcal{E}_{\mathbf{n}} = \frac{|\mathbf{p}|^2}{2M} = \frac{\hbar^2 |\mathbf{k}|^2}{2M} = \frac{(2\pi\hbar)^2 (n_x^2 + n_y^2 + n_z^2)}{2ML^2} . \qquad (4.39)$$

For macroscopically large L, the translational energy levels form a "continuum": the triple sum in the translational single-particle partition function

$$Z_{1\,\text{tr}} = \sum_{\mathbf{n}} \exp(-\beta \mathcal{E}_{\mathbf{n}}) = \sum_{n_x n_y n_z} \exp\left(-\beta \frac{[2\pi\hbar]^2 [n_x^2 + n_y^2 + n_z^2]}{2ML^2}\right) \qquad (4.40)$$

is conveniently approximated by the triple integral

$$Z_{1\,\text{tr}} = \int_{-\infty}^{\infty} \int_{-\infty}^{\infty} \int_{-\infty}^{\infty} \exp\left(-\frac{\beta[2\pi\hbar]^2}{2ML^2}[n_x^2 + n_y^2 + n_z^2]\right) dn_x \, dn_y \, dn_z . \qquad (4.41)$$

This expression factorizes into the product of three identical Gaussian integrals.[5] In total,

$$Z_{1\,\text{tr}} = \left(\frac{L}{\hbar}\sqrt{\frac{Mk_{\text{B}}T}{2\pi}}\right)^3 = \left(\frac{L}{\Lambda}\right)^3 = \frac{V}{\Lambda^3} , \qquad (4.42)$$

where we have introduced the *thermal length*

$$\Lambda = \hbar \sqrt{\frac{2\pi}{Mk_{\text{B}}T}} . \qquad (4.43)$$

We can now substitute the result (4.42) for the single-particle partition function into the global partition function (4.35) of the gas

---

[5]Recall that $\int_{-\infty}^{\infty} \exp\left(-x^2/[2a^2]\right) dx = \sqrt{2\pi}\, a$.

$$Z = \frac{(Z_{1\,\mathrm{tr}})^N}{N!} (Z_{1\,\mathrm{int}})^N = \frac{V^N}{N! \Lambda^{3N}} (Z_{1\,\mathrm{int}})^N . \tag{4.44}$$

Derivation yields the translational contribution to the internal energy

$$U_{\mathrm{tr}} = -\frac{\partial}{\partial \beta} \ln Z_{\mathrm{tr}} = -\frac{\partial}{\partial \beta} \ln \frac{V^N}{N! \Lambda^{3N}} = 3N \frac{\partial \ln \Lambda}{\partial \beta} = 3N \frac{\partial \ln \beta^{1/2}}{\partial \beta} = \frac{3}{2} \frac{N}{\beta} = \frac{3}{2} N k_{\mathrm{B}} T . \tag{4.45}$$

This result implies the experimentally well-established translational contribution $3/2 \, k_{\mathrm{B}}$ to the heat capacity per molecule of high-temperature gases. Likewise, we obtain this expression for the free energy:

$$F = -N k_{\mathrm{B}} T \ln \frac{e \, Z_{1\,\mathrm{tr}} Z_{1\,\mathrm{int}}}{N} = -N k_{\mathrm{B}} T \left(\ln \frac{e \, V}{N \Lambda^3} + \ln Z_{1\,\mathrm{int}} \right) . \tag{4.46}$$

$Z_{1\,\mathrm{int}}$ describes internal degrees of freedom: thus it can depend on T, but certainly not on the volume V of the containing vessel. This observation and the definition (4.28) of pressure allow us to derive a remarkably general equation of state for the ideal gas:

$$P = -\left. \frac{\partial F}{\partial V} \right|_{\beta,N} = -\frac{\partial}{\partial V} \left[-N k_{\mathrm{B}} T \left(\ln \frac{e \, V}{N \Lambda^3} + \ln Z_{1\,\mathrm{int}} \right) \right]_{T,N} = N k_{\mathrm{B}} T \frac{\partial}{\partial V} \ln \frac{e \, V}{N \Lambda^3} ,$$

i.e.

$$P = \frac{N k_{\mathrm{B}} T}{V} . \tag{4.47}$$

This relation, obtained on purely statistical grounds, is equivalent to the well-established equation of state of perfect gases $PV = nRT$, universally and quantitatively valid for atomic and molecular gases at high temperature and low density. Given that n moles of gas contain $n N_{\mathrm{A}}$ particles, by identifying $N k_{\mathrm{B}} T = n R T$, we determine the numerical value of the Boltzmann constant $k_{\mathrm{B}} = R / N_{\mathrm{A}}$.

In addition to thermodynamic functions, other quantities are accessible to statistics and to experiment. For example, the *kinetic-energy distribution* for the center-mass translations yields the probability to find a particle in a given energy interval. To obtain this distribution, we need the energy density of translational states

$$g_{\mathrm{tr}}(\mathcal{E}) = \frac{M^{3/2} V}{\sqrt{2} \, \pi^2 \, \hbar^3} \, \mathcal{E}^{1/2}, \tag{4.48}$$

Fig. 4.4 The distribution—Eq. (4.49)—of the translational kinetic energy of the molecules in an ideal gas at a temperature high enough that Bose/Fermi statistics effects can be neglected

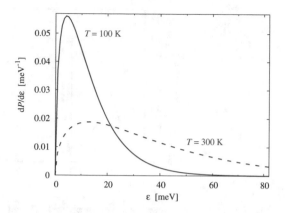

as can be derived from the kinetic-energy expression (4.39).[6] Then, in the spirit of Eq. (4.12), the single-particle kinetic-energy probability distribution is the product of the density of states times the Boltzmann probability that a given state is occupied:

$$\frac{dP(\mathcal{E})}{d\mathcal{E}} = g_{\mathrm{tr}}(\mathcal{E}) \frac{e^{-\beta \mathcal{E}}}{Z_{1\,\mathrm{tr}}} = \frac{M^{3/2} V}{\sqrt{2} \pi^2 \hbar^3} \mathcal{E}^{1/2} \frac{\Lambda^3}{V} e^{-\beta \mathcal{E}} = \frac{2}{\sqrt{\pi}} \beta^{3/2} \mathcal{E}^{1/2} e^{-\beta \mathcal{E}}. \qquad (4.49)$$

This probability distribution, drawn in Fig. 4.4, is remarkably universal: it does not even depend on the mass of the particles, but uniquely on temperature.

Similarly, we derive the translational *velocity distribution*. Every component of $\mathbf{v} = \mathbf{p}/M$ is Gaussian-distributed as

$$\frac{dP(v_u)}{dv_u} = \sqrt{\frac{\beta M}{2\pi}} \exp\left(-\beta \frac{M v_u^2}{2}\right). \qquad (4.50)$$

To obtain the distribution of speed $v = |\mathbf{v}|$, one can simply observe that the kinetic energy $\mathcal{E}$ and speed v are connected by $\mathcal{E} = \frac{M}{2} v^2$, and use the distribution (4.49):

$$\frac{dP(v)}{dv} = \frac{dP(\mathcal{E})}{d\mathcal{E}} \frac{d\mathcal{E}}{dv} = M v \frac{dP(\mathcal{E})}{d\mathcal{E}} \qquad (4.51)$$

$$= M v \frac{2}{\sqrt{\pi}} \beta^{3/2} \left(\frac{M v^2}{2}\right)^{1/2} e^{-\beta \frac{M}{2} v^2} = \sqrt{\frac{2}{\pi}} (M\beta)^{3/2} v^2 e^{-\frac{\beta M}{2} v^2}.$$

---

[6] According to Eq. (4.39), the kinetic energy is proportional to the squared length of the $\mathbf{n}$ vector $\mathcal{E}_{\mathbf{n}} = A |\mathbf{n}|^2$. As the $\mathbf{n}$ points are evenly distributed with unit density, the number of states with energy at most $\mathcal{E}$ equals the volume of the sphere of radius $|\mathbf{n}| = \sqrt{\mathcal{E}/A}$ in $\mathbf{n}$-space. The density of states is the derivative of this number of states with respect to energy: $g_{\mathrm{tr}}(\mathcal{E}) = \frac{d}{d\mathcal{E}} \frac{4\pi}{3} (\mathcal{E}/A)^{3/2} = 2\pi \mathcal{E}^{1/2}/A^{3/2}$. In this expression we substitute the value of $A = (2\pi\hbar)^2/(2ML^2)$ from Eq. (4.39) and $V = L^3$, obtaining Eq. (4.48).

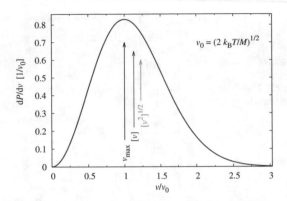

Fig. 4.5 The particle-speed distribution of the high-temperature ideal gas, Eq. (4.51), rescaled by the "thermal speed" $v_0 \equiv [2/(\beta M)]^{1/2} \equiv (2k_B T/M)^{1/2}$. This distribution peaks at $v_{max} = v_0$; the mean velocity is $[v] = \sqrt{4/\pi}\, v_0 \simeq 1.128\, v_0$; and the root mean square velocity is $[v^2]^{1/2} = \sqrt{3/2}\, v_0 \simeq 1.225\, v_0$ (*arrows*)

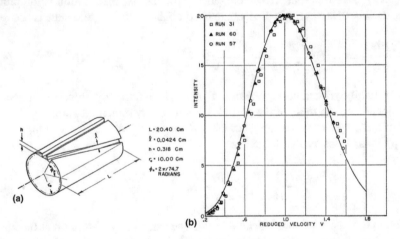

Fig. 4.6 The velocity distribution of the atoms emerging from a oven through a tiny hole was probed by letting them through **a** the spiraling slot of a rotating cylinder, to eventually be counted by a tungsten surface ionization detector. **b** A typical observed distribution of the translational velocities for high-temperature K vapor, compared to the curve of Eq. (4.51), scaled to v_0 as in Fig. 4.5. Reprinted figures with permission from R.C. Miller and P. Kusch, Phys. Rev. **99**, 1314 (1955). Copyright (1955) by the American Physical Society http://link.aps.org/abstract/PR/v99/p1314

The function of Eq. (4.51), the *Maxwell–Boltzmann* equilibrium speed distribution, is drawn in Fig. 4.5. The v^2 factor, with the same "polar" origin discussed for the radial distribution $P(r)$ of one-electron states in Sect. 2.1.3, should not hide the fact that the most likely velocity is $\mathbf{v} = \mathbf{0}$. Measurements of the speed distribution agree well with the statistical analysis, see Fig. 4.6.

4.3.1.2 Internal Degrees of Freedom

Internal degrees of freedom affect neither the equation of state nor any of the translational distributions. However, they do contribute an additive temperature-dependent term to the free energy (4.46) [precisely, the part $F_{int}(T) = N F_{1\,int}(T) = -N k_B T \ln Z_{1\,int}$], to the internal energy

$$U = U_{tr} + U_{int} = \frac{3}{2} N k_B T + N \left[F_{1\,int}(T) - T F'_{1\,int}(T) \right], \qquad (4.52)$$

and therefore also to the heat capacity

$$C_V = \frac{3}{2} N k_B - N T F''_{1\,int}(T). \qquad (4.53)$$

The internal term vanishes for structureless particles (e.g. electrons), while it contributes significantly to thermodynamics whenever the internal degrees of freedom are associated to excitation energies not too far from $k_B T$. This internal contribution is occasionally relevant for atomic gases and crucial for all molecular gases, as illustrated by the following examples.

We summarize here the predictions of statistics for the internal degrees of freedom of diatomic molecules. In the approximation that rotational and vibrational motions are independent (Sect. 3.3), the partition function factorizes $Z_{1\,int} = Z_{1\,vib} Z_{1\,rot}$. Therefore the contributions to all extensive quantities are additive, e.g. $F_{1\,int} = -k_B T \ln Z_{1\,int} = F_{1\,rot} + F_{1\,vib}$.

Statistics of Molecular Rotations. For the *rotational* statistics,

$$Z_{1\,rot} = \sum_{l=0}^{\infty} (2l + 1) \exp\left(-\frac{\Theta_{rot}}{T} l[l + 1] \right) \qquad (4.54)$$

is a function of the dimensionless ratio $\beta \hbar^2/(2I) = \Theta_{rot}/T$, where we introduce the characteristic temperature $\Theta_{rot} = \hbar^2/(2I k_B)$.[7] The series in Eq. (4.54) cannot be evaluated in a closed form. However, the characteristic temperature Θ_{rot} is often small (e.g. 85 K for H_2, 15 K for HCl, 2.9 K for N_2). When $\Theta_{rot}/T \ll 1$ the exponential in Eq. (4.54) varies slowly with l and numerous terms contribute to the sum for $Z_{1\,rot}$: it is a good approximation to replace it with an elementary integral:

$$Z_{1\,rot} \simeq \int_0^{\infty} (2l + 1) \exp\left[-\frac{\Theta_{rot}}{T} l(l + 1) \right] dl = \int_0^{\infty} \exp\left(-\frac{\Theta_{rot}}{T} y \right) dy \quad (4.55)$$

$$= \frac{T}{\Theta_{rot}} \qquad [T \gg \Theta_{rot}].$$

---

[7] Equation (4.54) and its consequences are incorrect for homoatomic molecules [31], where nuclear indistinguishableness and spin play a role, due to ergodicity violations (e.g. the slow ortho–para-hydrogen interconversion). In such cases, one must consider even and odd l separately. However, in the high-T limit the expressions (4.56)–(4.59) remain valid.

The high-temperature rotational contributions to the thermodynamic functions are therefore:

$$F_{1\,\text{rot}} = \frac{F_{\text{rot}}}{N} \simeq -k_B T \ln \frac{T}{\Theta_{\text{rot}}} \tag{4.56}$$

$$U_{1\,\text{rot}} = \frac{U_{\text{rot}}}{N} \simeq k_B T \tag{4.57}$$

$$C_{V\,1\,\text{rot}} = \frac{C_{V\,\text{rot}}}{N} \simeq k_B \tag{4.58}$$

$$S_{1\,\text{rot}} = \frac{S_{\text{rot}}}{N} \simeq k_B \left(1 + \ln \frac{T}{\Theta_{\text{rot}}} \right) \tag{4.59}$$

[see Eqs. (4.22)–(4.25)]. Instead, at lower temperature $T \sim \Theta_{\text{rot}}$, truncating the series (4.54) to a finite number of terms $l \leq l_{\text{max}} \gtrsim 2\sqrt{T/\Theta_{\text{rot}}}$ approximates $Z_{1\,\text{rot}}$ better. Figure 4.7 reports the temperature dependence of the rotational heat capacity and internal energy per molecule. Characteristically, as temperature is raised, the rotational degree of freedom "unfreezes", reaching the *classical equipartition* limit at large temperature $T \gg \Theta_{\text{rot}}$. The equilibrium populations $P_l = (2l + 1)\exp\left(-l[l+1]\Theta_{\text{rot}}/T\right)/Z_{1\,\text{rot}}$ account for the observed relative intensities of the rotational structures in molecular spectra (Figs. 3.9 and 3.10).[8]

Fig. 4.7 The temperature dependence of the rotational contribution to **a** the molecular internal energy $U_{1\,\text{rot}}$ and **b** the molecular heat capacity $C_{V\,1\,\text{rot}}$. The energy scale of the problem is set by $\hbar^2/(2I)$, which translates into a temperature scale $\Theta_{\text{rot}} = \hbar^2/(2Ik_B)$. For $T \gg \Theta_{\text{rot}}$, the curves approach their high-temperature limits Eqs. (4.57) and (4.58)

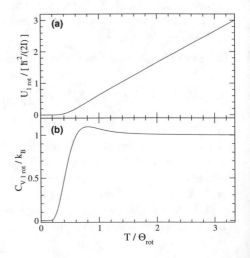

---

[8]The relative intensities of the rotational lines are proportional to the populations P_l provided that the dipole matrix elements averaged over the initial $(2l + 1)$ degenerate states and summed over the final states $l' = l + 1$ (R branch) or $l' = l - 1$ (P branch) are independent of l. The radial part of the matrix element is, but calculation shows that the angular part equals $(l + 1)/(2l + 1)$ for the R branch and $l/(2l + 1)$ for the P branch. For not too small l, these factors are both close to 0.5, and are therefore often assumed to be constant.

Statistics of Molecular Vibrations. In the harmonic approximation, the *vibrational* states form an equally-spaced ladder, Eq. (3.21). The resulting series for $Z_{1\,\text{vib}}$ is evaluated in closed form, thus yielding an *exact* partition function valid at any temperature:

$$Z_{1\,\text{vib}} = \sum_{v=0}^{\infty} \exp\left(-\beta\hbar\omega\left[v + \frac{1}{2}\right]\right) = \exp\left(-\frac{x}{2}\right)\sum_{v=0}^{\infty}\exp(-v\,x) \qquad (4.60)$$

$$= \exp\left(-\frac{x}{2}\right)\frac{1}{1 - \exp(-x)} = \frac{1}{2\sinh(x/2)}, \quad \text{with } x = \beta\hbar\omega = \frac{\Theta_{\text{vib}}}{T}.$$

$Z_{1\,\text{vib}}$ is a function of the dimensionless ratio $x = \Theta_{\text{vib}}/T$, where the characteristic temperature $\Theta_{\text{vib}} = \hbar\omega/k_B$. For the vibrational degree of freedom, the thermodynamic functions of Eqs. (4.22)–(4.25) are:

$$F_{1\,\text{vib}} = \frac{F_{\text{vib}}}{N} = \frac{\hbar\omega}{2} + k_B T \ln\left(1 - e^{-x}\right) \qquad (4.61)$$

$$U_{1\,\text{vib}} = \frac{U_{\text{vib}}}{N} = \frac{\hbar\omega}{2} + \frac{\hbar\omega}{e^x - 1} \qquad (4.62)$$

$$C_{V\,1\,\text{vib}} = \frac{C_{V\,\text{vib}}}{N} = k_B\frac{x^2 e^x}{(e^x - 1)^2} = k_B\left[\frac{x/2}{\sinh(x/2)}\right]^2 \qquad (4.63)$$

$$S_{1\,\text{vib}} = \frac{S_{\text{vib}}}{N} = k_B\frac{x}{e^x - 1} - k_B \ln\left(1 - e^{-x}\right). \qquad (4.64)$$

Figure 4.8 depicts the temperature dependence of the vibrational heat capacity and internal energy per molecule. Similar to rotations, as temperature is raised, the vibrational degree of freedom "unfreezes", reaching the classical limit at $T \gg \Theta_{\text{vib}}$. Note that the high-$T$ limit for *one* oscillator provides a contribution $k_B T$ to U_1 (and thus k_B to $C_{V\,1}$) equal to that of the *two* rotational degrees of freedom. The reason is that one harmonic oscillator is associated to two quadratic terms (kinetic and potential) in the Hamiltonian, the same number as the two kinetic quadratic terms for the two rotational degrees of freedom.

Examples of characteristic vibrational temperatures of a few diatomic molecules are: 6300 K for H_2, 4300 K for $H^{35}Cl$, 3400 K for N_2, 403 K for $K^{35}Cl$. With such values of $\Theta_{\text{vib}} \gg \Theta_{\text{rot}}$, contrary to the rotational unfreezing, the vibrational transition from the quantum-frozen to the classical regime is usually accessible to heat-capacity measurements, which find good accord with Eq. (4.63).

The translational rotational and vibrational contributions to the molecular heat capacity combine additively, Fig. 4.9. In a real gas three physical mechanisms cause deviations from this idealized sketch: (i) Inter-molecular interactions lead to liquid and solid phases for $T \lesssim T_{\text{boil}}$, and usually $\Theta_{\text{rot}} \lesssim T_{\text{boil}} < \Theta_{\text{vib}}$ making it hard to observe the rotational unfreezing; (ii) Even if intermolecular interactions could be neglected, quantum statistical effects (addressed in Sect. 4.3.2 below) would make the low-temperature translational heat capacity deviate from $3/2 k_B$; (iii) at high tem-

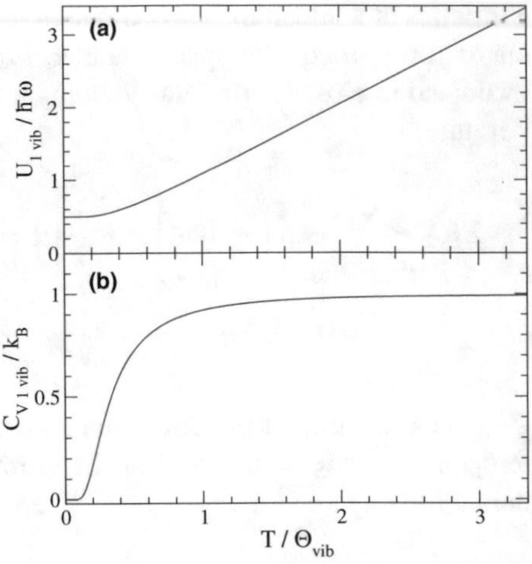

Fig. 4.8 The temperature dependence of **a** the vibrational contribution to the molecular internal energy $U_{1\,\text{vib}}$, Eq. (4.62), and **b** the molecular heat capacity $C_{V\,1\,\text{vib}}$, Eq. (4.63). The energy scale of the problem is set by $\hbar\omega$, which translates into a temperature scale $\Theta_{\text{vib}} = \hbar\omega/k_B$

Fig. 4.9 A broad-range temperature dependence of the total heat capacity of a hypothetical harmonic diatomic molecule (loosely inspired by HCl), compared to that of a monoatomic gas (*dotted line*)

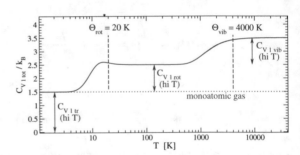

perature $T \gtrsim 10^4$ K anharmonic and molecular-dissociation effects start to affect the vibrational and rotational heat-capacity contributions.

4.3.1.3 Finite Degrees of Freedom (e.g. Spin)

The Boltzmann formalism applies also to the statistics of any set of equal degrees of freedom characterized by a (small) finite number of quantum states. For example, one may wish to evaluate the contributions to thermodynamics of the low-lying electronic and/or magnetic excitations of atomic or molecular gases, accounted for by the internal partition function $Z_{1\,\text{int}}$ introduced in Eq. (4.37). One also applies this method to magnetic "impurities" in solids or liquids, as long as they are so dilute that their interactions can be neglected. For any such system involving a finite number of states, the single-atom/molecule/impurity partition function $Z_{1\,\text{int}}$ (and therefore its derivatives too) is a sum involving a finite number of terms.

Here we examine the simple but important case of non-interacting magnetic moments in an external field. For brevity, we refer to "spins": in reality each magnetic moment is generally proportional to some *total* angular momentum. An angular momentum J coupled to a magnetic field $\mathbf{B} = B\hat{\mathbf{z}}$ as in Eq. (2.24) spans a $(2J + 1)$-dimensional space of states, a basis of which is labeled by the J_z projection quantum number M_J. The "*spin*" partition function is the sum of $(2J + 1)$ terms corresponding to the levels of Eq. (2.59):

$$Z_{1\,\text{spin}} = \sum_{M_J=-J}^{J} \exp\left(-x\, M_J\right), \quad \text{with } x = \beta\, g\, \mu_B B = \Theta_B/T , \tag{4.65}$$

where we introduced the temperature scale $\Theta_B = g\, \mu_B B/k_B$ reflecting the ladder of $\mathcal{E}_{M_J} = g\, \mu_B B\, M_J$ energy levels. g is the relevant g-factor, e.g. the Landé g_J. For any given spin J, evaluating $Z_{1\,\text{spin}}$ is a straightforward exercise.

For the reader's convenience, defining $x = \Theta_B/T$, we summarize the results for the simplest case, *spin* $J = 1/2$:

$$Z_{1\,\text{spin}} = 2 \cosh\left(\frac{\Theta_B}{2T}\right) = 2 \cosh\frac{x}{2} \qquad \left[\text{spin } 1/2\right]. \tag{4.66}$$

Following Eqs. (4.22)–(4.25), the spin-$1/2$ thermodynamic functions are:

$$F_{1\,\text{spin}} = \frac{F_{\text{spin}}}{N} = -k_B T \ln\left(2\cosh\frac{x}{2}\right) \tag{4.67}$$

$$U_{1\,\text{spin}} = \frac{U_{\text{spin}}}{N} = -\frac{g\mu_B B}{2}\tanh\frac{x}{2} \tag{4.68}$$

$$C_{V\,1\,\text{spin}} = \frac{C_{V\,\text{spin}}}{N} = \frac{k_B\, x^2}{2\cosh(x) + 2} \tag{4.69}$$

$$S_{1\,\text{spin}} = \frac{S_{\text{spin}}}{N} = k_B\left[\ln\left(2\cosh\frac{x}{2}\right) - \frac{x}{2}\tanh\frac{x}{2}\right]. \tag{4.70}$$

Figure 4.10 reports the temperature dependence of the internal energy and heat capacity per spin. As temperature is raised from absolute zero, like molecular rotations and vibrations, the spin degree of freedom "unfreezes", with an increase in heat capacity. However, when T is raised further, due to the finite spectrum, the internal energy cannot increase indefinitely: $U_{1\,\text{spin}}$ flattens out. As a result, the spin heat capacity decays to zero at high temperature $T \gg \Theta_B$.

The magnetization density of N uniformly distributed magnetic moments

$$\mathbf{M} = \frac{N}{V}\,[\boldsymbol{\mu}_1] \tag{4.71}$$

is a relevant quantity that can be directly measured in the lab. In such an ideal system $\mathbf{M}$ is necessarily oriented parallel to the magnetic field, $\mathbf{M} = M\hat{\mathbf{z}}$. We have

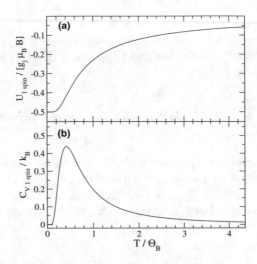

Fig. 4.10 The temperature dependence of **a** the internal energy per spin $U_{1\,\text{spin}}$, Eq. (4.68), and **b** the heat capacity per spin $C_{V\,1\,\text{spin}}$, Eq. (4.69). These quantities characterize the thermodynamics of a magnetic moment in a uniform field B, which sets the energy scale of the problem to $g\mu_B B$, and the corresponding temperature scale $\Theta_B = g\mu_B B/k_B$. Note that, due to the spectrum being bound from above, at high temperature $U_{1\,\text{spin}}$ does not grow indefinitely, but rather approaches a constant (zero)

$$[\mu_{1z}] = \text{Tr}(\hat{\rho}_1 \mu_{1z}) = \sum_{M_J=-J}^{J} -g\mu_B\, M_J\, \frac{\exp(-xM_J)}{Z_{1\,\text{spin}}}$$

$$= -\frac{1}{B} \sum_{M_J=-J}^{J} \mathcal{E}_{M_J} P_{M_J} = -\frac{U_{1\,\text{spin}}}{B} \tag{4.72}$$

$$M_z = \frac{N}{V}[\mu_{1z}] = -\frac{N}{VB} U_{1\,\text{spin}} = \frac{g\mu_B N}{2V} \tanh\frac{x}{2} \qquad \left[\text{spin-}1/2\right]. \tag{4.73}$$

The average magnetization changes with temperature following the same functional dependence as the internal energy, except for a constant factor $-N/(VB)$, thus Fig. 4.10a can also be read as magnetization as a function of temperature for the $J = 1/2$ system. The hyperbolic tangent can be expanded to lowest order in x, obtaining the linear response of the localized spins to a weak magnetic field B:

$$M_z \simeq \frac{g\mu_B N}{4V} x = \chi_B\, B\,, \quad \text{with} \quad \chi_B = \frac{N}{V}\left(\frac{g\mu_B}{2}\right)^2 \frac{1}{k_B T} \qquad \left[\text{spin-}1/2\right]. \tag{4.74}$$

χ_B represents the paramagnetic susceptibility of the spins to the (weak) total magnetic field **B**. The characteristic inverse-T dependence of the *Curie susceptibility* of the noninteracting spins reflects the disordering effect of temperature.

In practice it is more common to measure the susceptibility χ_m relative to the *external* applied field $\mathbf{H} = \epsilon_0 c^2 \mathbf{B}_{ext}$. The relation between χ_B and χ_m derives from

$$\mathbf{M} = \chi_B \mathbf{B} = \chi_B [\mathbf{B}_{ext} + \mathbf{B}_{int}] = \chi_B \left[(\epsilon_0 c^2)^{-1} \mathbf{H} + (\epsilon_0 c^2)^{-1} \mathbf{M} \right], \quad (4.75)$$

where we use the relation $\mathbf{B}_{int} = (\epsilon_0 c^2)^{-1} \mathbf{M}$ for the magnetic field generated by a uniformly magnetized material. We solve Eq. (4.75) for $\mathbf{M}$, obtaining

$$\mathbf{M} = \frac{(\epsilon_0 c^2)^{-1} \chi_B}{1 - (\epsilon_0 c^2)^{-1} \chi_B} \mathbf{H}. \quad (4.76)$$

Noting that the units of $\mathbf{H}$ and those of $\mathbf{M}$ are the same, namely A/m, the desired expression for the dimensionless $\mathbf{H}$-susceptibility χ_m is

$$\chi_m = \frac{(\epsilon_0 c^2)^{-1} \chi_B}{1 - (\epsilon_0 c^2)^{-1} \chi_B}. \quad (4.77)$$

In practice the denominator is always very close to unity, so that $\chi_m \simeq (\epsilon_0 c^2)^{-1} \chi_B$. The reader should not worry about a vanishing denominator in Eq. (4.77), indicating a self-sustained nonzero magnetization for $\mathbf{H} = \mathbf{0}$: even at such a low temperature as 1 K, for $g = 2$ the diverging-χ_m condition $(\epsilon_0 c^2)^{-1} \chi_B = 1$ would require a density $N/V \approx 1.3 \times 10^{29}$ m^{-3}, far too large to describe dilute spin-carrying impurities. Actual ferromagnetic states are never associated to a self-sustained magnetization due to a divergence in Eq. (4.77), but they originate from large exchange interactions similar to those responsible for Hund's rules in open-shell atoms, see Sect. 2.2.9.2.

4.3.2 Low-Temperature Fermi and Bose Gases

In Sect. 4.3.1 we discovered that, at high temperature, non-interacting identical fermions or bosons behave as an ideal classical gas, with the occasional additive contributions of internal degrees of freedom. At low temperature however, spectacular fermion/boson differences show up, as a consequence of the radically different constraints over the occupations of the single-particle states.

We re-examine the evaluation of the partition function Z (4.34) for arbitrary T. We abandon the focus on individual particles (labeled by i). We now focus on single-particle states (labeled by α). We replace the sums over the N single-particle quantum numbers α_i with sums over the occupation numbers n_α of the single-particle states. These n_α are 0 or 1 for fermions, and 0, 1, 2, 3,... for bosons. With this target, we rewrite the total N-particle energy $\sum_i^N \mathcal{E}_{\alpha_i}$ as $\sum_\alpha n_\alpha \mathcal{E}_\alpha$, in terms of the n_α. In this form, the Boltzmann exponential factorizes as $\exp\left(-\beta \sum_\alpha n_\alpha \mathcal{E}_\alpha\right) = \prod_\alpha \exp(-\beta n_\alpha \mathcal{E}_\alpha)$. The exact partition function

$$Z = \sum_{\alpha_1 \alpha_2 \dots \alpha_N} \frac{n_0! \, n_1! \, n_2! \, \cdots}{N!} \, e^{-\beta \sum_i^N \mathcal{E}_{\alpha_i}}$$

$$= \sum_{\substack{\{n_\alpha\} \\ \sum_\alpha n_\alpha = N}} e^{-\beta \sum_\alpha n_\alpha \mathcal{E}_\alpha} = \sum_{\substack{\{n_\alpha\} \\ \sum_\alpha n_\alpha = N}} \prod_\alpha e^{-\beta \mathcal{E}_\alpha n_\alpha} = \sum_{\substack{\{n_\alpha\} \\ \sum_\alpha n_\alpha = N}} \prod_\alpha \left(e^{-\beta \mathcal{E}_\alpha} \right)^{n_\alpha}. \tag{4.78}$$

The binomial coefficient in the first line corrects for overcounting the N-particle states by the $\alpha_1 \dots \alpha_N$ sum. Each N-particle ket is identified uniquely by its set of occupation numbers $\{n_\alpha\}$. Therefore the binomial coefficient must be removed in the n_α-sum, which has no repetitions. The reader is urged to write down Eq. (4.78) explicitly for $N = 2$ bosons or fermions, with three single-particle states available $|\alpha\rangle = |1\rangle$, $|2\rangle$, $|3\rangle$ at energies $\mathcal{E}_1$, $\mathcal{E}_2$, and $\mathcal{E}_3$.

Unfortunately, the constraint fixing the total number of particles to N makes the evaluation of the sum over the occupations in Eq. (4.78) extremely difficult. If we could remove this constraint, we could exchange the product and sums, and carry out the sums. To get rid of the fixed-N constraint we use a trick: we replace the canonical ensemble, where the number of particles is fixed, with the *grand canonical ensemble*, where N is allowed to vary. The latter ensemble describes a thermodynamic system which is weakly exchanging not just energy but also particles with the rest of the universe. In such condition, to maintain, on average, the desired number of particles $[N]$ it is necessary to subtract μN to the energy eigenvalue in the expression (4.11) of the partition function, see Fig. 4.11. For an arbitrary system of equal particles, summing the canonical Z over all N yields the *grand partition function*

$$Q = \sum_{N=0}^{\infty} \sum_{m(N)} e^{-\beta(E_{m(N)} - \mu N)} = \tilde{\mathrm{Tr}} \left(e^{-\beta(H - \mu \hat{N})} \right), \tag{4.79}$$

where $\hat{N}$ is the operator counting the number of particles, and $\tilde{\mathrm{Tr}}$ indicates summing over all N-particle states $|m(N)\rangle$ for all particle numbers N. In the grand canonical ensemble, Q plays the same role as Z in the canonical ensemble. A detailed analysis,

Fig. 4.11 At fixed temperature, the internal energy U is generally a convex function of the number of particles N, which is minimum at some arbitrary N, e.g. $N = 0$. The addition of $-\mu N$ to U shifts the equilibrium point to a tunable average particle number $[N]$, which is an increasing function of the parameter μ. We identify μ with the chemical potential

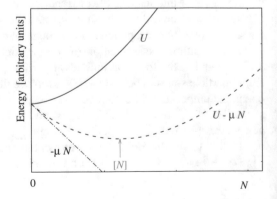

similar to that of Sect. 4.2.1, yields the relation between the grand partition function and thermodynamics. In particular, β identifies with the inverse temperature and μ identifies with the *chemical potential*. We have:

$$J(T, V, \mu) = V P(T, \mu) = k_B T \ln Q(T, V, \mu) \tag{4.80}$$

$$F = \mu [\hat{N}] - J \tag{4.81}$$

$$U = \mu [\hat{N}] - J + TS \tag{4.82}$$

$$G = J + F = \mu [\hat{N}] \tag{4.83}$$

$$[\hat{N}] = \left. \frac{\partial J}{\partial \mu} \right|_{T,V} = V \left. \frac{\partial P}{\partial \mu} \right|_T \tag{4.84}$$

$$\mu = \left. \frac{\partial F}{\partial [\hat{N}]} \right|_{T,V} = \left. \frac{\partial G}{\partial [\hat{N}]} \right|_{T,P} \tag{4.85}$$

$$S = \left. \frac{\partial J}{\partial T} \right|_{\mu,V} = V \left. \frac{\partial P}{\partial T} \right|_{\mu}. \tag{4.86}$$

Equation (4.80) is the grand-canonical analogous of the canonical Eq. (4.22). Relations (4.80)–(4.86) allow us to express all thermodynamic quantities for a system at equilibrium with a reservoir of energy *and* of particles in terms of the microscopic statistical mechanics contained in Q [30].

Armed with this new tool, we proceed to compute the grand partition function Q for a system of noninteracting identical bosons or fermions. The relation $N = \sum_\alpha n_\alpha$ helps us to rearrange the exponent of Eq. (4.78). The new grand canonical summation over N allows us to get rid of the constraint:

$$Q = \sum_{N=0}^{\infty} \sum_{\substack{\{n_\alpha\} \\ \sum_\alpha n_\alpha = N}} e^{-\beta[(\sum_\alpha n_\alpha \mathcal{E}_\alpha) - \mu N]} = \sum_{N=0}^{\infty} \sum_{\substack{\{n_\alpha\} \\ \sum_\alpha n_\alpha = N}} e^{-\beta[(\sum_\alpha n_\alpha \mathcal{E}_\alpha) - \mu \sum_\alpha n_\alpha]}$$

$$= \sum_{\{n_\alpha\}} e^{\sum_\alpha \beta(\mu - \mathcal{E}_\alpha) n_\alpha} = \sum_{\{n_\alpha\}} \prod_\alpha e^{\beta(\mu - \mathcal{E}_\alpha) n_\alpha}$$

$$= \prod_\alpha \sum_{n_\alpha} e^{\beta(\mu - \mathcal{E}_\alpha) n_\alpha} = \prod_\alpha \sum_{n_\alpha} \left(e^{\beta(\mu - \mathcal{E}_\alpha)} \right)^{n_\alpha}.$$

Next, observe that the occupation numbers n_α are mute summation indexes. There is no real reason to tell them apart: we can just call them all n. Accordingly, we write the grand partition function as

$$Q = \prod_\alpha \left[\sum_n \left(e^{\beta(\mu - \mathcal{E}_\alpha)} \right)^n \right]. \tag{4.87}$$

The sum in square brackets can be carried out explicitly: for fermions ($n = 0, 1$) it equals $1 + e^{\beta(\mu - \mathcal{E}_\alpha)}$; for bosons ($n = 0, 1, 2, 3, \ldots$) this sum is a geometric series, whose summation[9] gives $(1 - e^{\beta(\mu - \mathcal{E}_\alpha)})^{-1}$. By introducing a quantity $\theta = +1$ for bosons and $\theta = -1$ for fermions, we can write Q in a form which is valid for both bosons and fermions and factorized over the single-particle states:

$$Q = \prod_\alpha \left(1 - \theta e^{\beta(\mu - \mathcal{E}_\alpha)}\right)^{-\theta} . \tag{4.88}$$

The grand potential (4.80) provides the connection to thermodynamics:

$$J = PV = k_B T \ln Q = -\theta k_B T \sum_\alpha \ln \left(1 - \theta e^{\beta(\mu - \mathcal{E}_\alpha)}\right) . \tag{4.89}$$

This exact expression, together with Eqs. (4.81)–(4.86), allows us to derive the entire thermodynamics of the gas of noninteracting bosons/fermions.

As discussed in Sect. 4.3.1, the single-particle states $|\alpha\rangle$ of noninteracting free particles are labeled by the single-particle momentum $\mathbf{p}$, plus possibly internal degrees of freedom. Focusing to low temperature, any nontrivial internal dynamics is usually "frozen". There remain accessible only g_s degenerate states labeled by a spin variable m_s (representing for example the $\hat{z}$ projection of the total angular momentum of the particle). In practice, like in Eq. (4.40), the α-sum represents a sum over n_x, n_y, n_z, and (if needed) m_s. As the translational levels are very dense, the n-sum can be replaced by an integration over energy $\mathcal{E}$, weighted [in the spirit of Eq. (4.12)] by the density of translational states $g_{\text{tr}}(\mathcal{E})$ of Eq. (4.48):

$$\sum_\alpha \rightarrow \sum_{m_s} \sum_{n_x n_y n_z} \rightarrow g_s \int_0^\infty d\mathcal{E} \, g_{\text{tr}}(\mathcal{E}) \rightarrow \int_0^\infty d\mathcal{E} \, g(\mathcal{E}) . \tag{4.90}$$

Here we introduce the total density of states $g(\mathcal{E}) = g_s \, g_{\text{tr}}(\mathcal{E})$. We obtain the following equation of state:

$$
\begin{aligned}
P = \frac{J}{V} &= -\theta k_B T \int_0^\infty d\mathcal{E} \, \frac{g(\mathcal{E})}{V} \ln \left(1 - \theta \, e^{\beta(\mu - \mathcal{E})}\right) \\
&= -\theta k_B T \, g_s \frac{(2M)^{3/2}}{4\pi^2 \hbar^3} \int_0^\infty d\mathcal{E} \, \sqrt{\mathcal{E}} \ln \left(1 - \theta \, e^{\beta(\mu - \mathcal{E})}\right) .
\end{aligned}
\tag{4.91}
$$

This integral can be rewritten in terms of the dimensionless variable $x = \mathcal{E}\beta$, and integrated by parts to obtain:

---

[9]This series converges only if the quantity $e^{\beta(\mu - \mathcal{E}_\alpha)} < 1$, i.e. only if μ is smaller than the smallest single-particle energy $\mathcal{E}_\alpha$, which is usually 0. Therefore μ must be negative for bosons.

$$P = -\theta (k_{\mathrm{B}} T)^{5/2} g_s \frac{(2M)^{3/2}}{4\pi^2 \hbar^3} \int_0^\infty dx \, \sqrt{x} \, \ln\left(1 - \theta e^{\beta\mu - x}\right) \tag{4.92}$$

$$= \frac{2}{3} \frac{2}{\sqrt{\pi}} k_{\mathrm{B}} T \, g_s \Lambda^{-3} \int_0^\infty dx \, \frac{x^{3/2}}{e^{x - \beta\mu} - \theta} \,.$$

The thermal length Λ was defined in Eq. (4.43). This equation of state expresses the pressure of a gas of ideal particles in terms of its temperature T and chemical potential μ. This form is unpractical, since in experiments μ is more difficult to control than the density $[\hat{N}]/V$: it would be preferable to express P in terms of T and $[\hat{N}]/V$, like in the high-temperature expression (4.47). Alas, we have no simple analytic expression of μ as a function of the density, thus no convenient explicit equation of state either. More interestingly, by computing the internal energy U through Eq. (4.82), it may be verified that the high-temperature ideal-gas relation

$$P = \frac{2}{3} \frac{U}{V} \tag{4.93}$$

applies at any temperature,[10] for both bosons and fermions, even though regarding the equation of state (4.92) at low temperature ideal bosons and fermions deviate radically from an "ideal gas" in the thermodynamic sense.

It is interesting to examine the leading *high-temperature* correction [30] to the ideal-gas behavior Eq. (4.47) due to quantum statistics:

$$P = \frac{[\hat{N}]}{V} k_{\mathrm{B}} T \left[1 - \theta \, 2^{-5/2} \delta + O(\delta)^2\right], \quad \text{for } \delta = \frac{\Lambda^3 [\hat{N}]}{g_s V} = \frac{(2\pi)^{3/2} \hbar^3}{(Mk_{\mathrm{B}} T)^{3/2}} \frac{[\hat{N}]}{g_s V} \ll 1.$$

$$\tag{4.94}$$

The explicit form of the *"degeneracy" parameter* δ makes it clear that Eq. (4.94) is a high-temperature and low-density expansion. The sign of the leading correction indicates the opposite tendencies of boson statistics to decrease pressure, and of fermion statistics to increment it. This is a high-temperature hint at the "better social character" of bosons compared to fermions. The experimental verification of these corrections in real gases (e.g. of atoms) is very difficult, because inter-particle interactions introduce extra corrections to pressure of the same order as or even larger than those associated to indistinguishability.

At low temperature, the behavior of ideal bosons and fermions becomes radically different. The average occupation number $[n_\alpha]$ of single-particle states is a clear indicator of these differences. This quantity is obtained as

---

[10]We focused the present analysis on the low-temperature regime, the so-called "degenerate" gas. In fact, the only assumption about temperature is that all internal degrees should be either frozen or included in g_s. As long as no other internal degree of freedom plays any relevant role, as e.g. in atomic helium, Eqs. (4.89), (4.92), and (4.93) hold for *arbitrary* temperature.

$$[n_\alpha] = \frac{\left(\sum_n n e^{\beta(\mu - \mathcal{E}_\alpha)n}\right) \cdot \prod_{\alpha' \neq \alpha} \sum_n e^{\beta(\mu - \mathcal{E}_{\alpha'})n}}{\prod_{\alpha''} \sum_n e^{\beta(\mu - \mathcal{E}_{\alpha''})n}} = \frac{\sum_n n e^{\beta(\mu - \mathcal{E}_\alpha)n}}{\sum_n e^{\beta(\mu - \mathcal{E}_\alpha)n}}.$$

This fraction is recognized as $\partial/\partial(\beta\mu) \ln(\sum_n e^{\beta(\mu - \mathcal{E}_\alpha)n})$. For evaluating Q, from Eqs. (4.87) to (4.88), we calculated the sum inside the logarithm, obtaining $[1 - \theta e^{\beta(\mu - \mathcal{E}_\alpha)}]^{-\theta}$. Therefore, we have

$$[n_\alpha] = \frac{\partial}{\partial(\beta\mu)} \ln\left(1 - \theta e^{\beta(\mu - \mathcal{E}_\alpha)}\right)^{-\theta} = -\theta \frac{\partial}{\partial(\beta\mu)} \ln\left(1 - \theta e^{\beta\mu - \beta\mathcal{E}_\alpha}\right)$$

$$= -\theta \frac{-\theta e^{\beta(\mu - \mathcal{E}_\alpha)}}{1 - \theta e^{\beta(\mu - \mathcal{E}_\alpha)}},$$

which simplifies to

$$[n_\alpha] = \frac{1}{e^{\beta(\mathcal{E}_\alpha - \mu)} - \theta}. \tag{4.95}$$

In a single expression, Eq. (4.95) collects the celebrated *Bose–Einstein distribution* of boson occupations

$$[n_\alpha]_B = \frac{1}{e^{\beta(\mathcal{E}_\alpha - \mu)} - 1} \tag{4.96}$$

and *Fermi-Dirac distribution* of fermion occupations

$$[n_\alpha]_F = \frac{1}{e^{\beta(\mathcal{E}_\alpha - \mu)} + 1}. \tag{4.97}$$

Given T and μ, the average occupation of each single-particle state is a function uniquely of the energy of the state itself.[11] The presence of all other particles affects each single-particle occupation distribution through the chemical potential μ, which is a function of the total particle density $[\hat{N}]/V$ and temperature T.

4.3.2.1   Fermions

The statistics of a cold gas of ideal fermions is of fundamental interest for the physics of matter, because conduction electrons in many metals can be approximately described as a gas of free non-interacting spin-$1/2$ fermions. At room temperature the thermal length of electrons is $\Lambda \simeq 4$ nm, corresponding to a thermal volume $\Lambda^3 \approx 80$ (nm)^3. With such a large Λ^3, multiplied by the typical electron

---

[11] And of no other property. For example, in the absence of any applied magnetic field, energy, and therefore occupation, is independent of m_s. As a consequence, a gas of ideal bosons or fermions remains in a spin-unpolarized nonmagnetic state at any temperature.

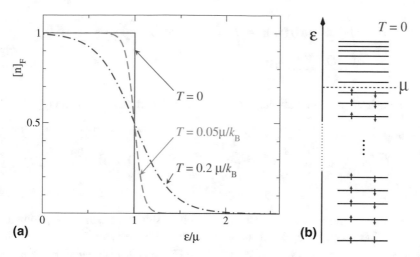

Fig. 4.12 **a** The average filling of the single-particle levels of the ideal Fermi gas as a function of energy $\mathcal{E}$ (scaled by the chemical potential μ), for three values of temperature. For the density of electrons in simple metals, the temperature $0.05\,\mu/k_B$ corresponds to several thousand K. **b** The filling of the single-particle levels of non-interacting fermions (here $g_s = 2$) at $T = 0$

density $[\hat{N}]/V \approx 10^{29}$ m$^{-3} = 100$ (nm)$^{-3}$ of a metal (roughly the inverse cube of a typical interatomic separation), Eq. (4.94) yields a huge degeneracy parameter $\delta \approx 4000$. Electrons in metals are therefore fully outside the range of validity of the high-temperature expansion (4.94). Many of their properties can rather be understood in terms of the ideal Fermi-gas model in the opposite low-temperature limit.

The $T = 0$ properties of a Fermi gas are those of the ground state of N free noninteracting fermions:[12] QM prescribes that this state should be the permutation-antisymmetric state obtained by filling the N lowest-energy single-particle levels (i.e. the N/g_s shortest-$|\mathbf{k}|$ plane-wave states), up to some maximum single-particle energy ϵ_F called *Fermi energy*, and leaving all the states above empty, as illustrated in Fig. 4.12. Indeed, in the $\beta \to \infty$ limit, Eq. (4.97) predicts precisely that the average occupation turns into a step function of energy $\mathcal{E}_\alpha$:

$$\lim_{T \to 0} [n_\alpha]_F = \begin{cases} 1, & \mathcal{E}_\alpha < \mu \\ 1/2, & \mathcal{E}_\alpha = \mu \ , \\ 0, & \mathcal{E}_\alpha > \mu \end{cases} \tag{4.98}$$

which thus identifies the chemical potential at $T = 0$ with the Fermi energy ϵ_F. Then, by requiring that

---

[12]Hence, for brevity, we adopt the symbol N for the average number of particles $[\hat{N}]$.

$$N = \int_0^\infty d\mathcal{E}\, g(\mathcal{E})\, [n_\alpha]_F = \int_0^\mu d\mathcal{E}\, g(\mathcal{E}) = g_s \frac{(2M)^{3/2} V}{4\pi^2 \hbar^3} \int_0^\mu d\mathcal{E}\, \mathcal{E}^{1/2}$$

$$= \frac{(2M)^{3/2} V\, g_s}{4\pi^2 \hbar^3} \frac{2}{3} \mu^{3/2}, \tag{4.99}$$

we establish the relation between the particle density and the Fermi energy:

$$\epsilon_F \equiv \mu(T = 0) = \frac{\hbar^2}{2M} \left(\frac{6\pi^2}{g_s} \frac{N}{V} \right)^{2/3}. \tag{4.100}$$

In the space of momentum $\mathbf{p}$, each state within a sphere of radius p_F [defined by $\epsilon_F = p_F^2/(2M)$] is filled by g_s fermions, while those outside are empty. To this maximum momentum $p_F = \hbar k_F = \hbar \left(6\pi^2 g_s^{-1} N/V\right)^{1/3}$ (the *Fermi momentum*), there corresponds a maximum velocity $v_F = p_F/M$, the *Fermi velocity*. Similarly, the Fermi energy is often expressed in terms of a *Fermi temperature* $T_F = \epsilon_F/k_B$.

In simple metals, typical densities of conduction electrons ($M = m_e$, $g_s = 2$) in the order $N/V \approx 10^{28} - 10^{29}$ m^{-3} yield $\epsilon_F \approx 2 - 10$ eV, i.e. $T_F \approx 20,000 - 100,000$ K. This corresponds to typical $k_F \approx 10^{10}$ m^{-1}, $p_F \approx 10^{-24}$ kg m s^{-1}, and $v_F \approx 10^6$ m s^{-1}.

At $T = 0$ it is also straightforward to evaluate the internal energy:

$$U = \int_0^\infty d\mathcal{E}\, \mathcal{E}\, g(\mathcal{E})\, [n_\alpha]_F = \int_0^{\epsilon_F} d\mathcal{E}\, \mathcal{E}\, g(\mathcal{E})$$

$$= \frac{(2M)^{3/2} V\, g_s}{4\pi^2 \hbar^3} \int_0^{\epsilon_F} d\mathcal{E}\, \mathcal{E}^{3/2} = \frac{(2M)^{3/2} V\, g_s}{4\pi^2 \hbar^3} \frac{2}{5} \epsilon_F^{5/2}. \tag{4.101}$$

U can be expressed in terms of the density N/V by substituting Eq. (4.100) in Eq. (4.101). If we substitute just a factor $\epsilon_F^{3/2}$, we obtain the easier-to-remember relation:

$$U(T = 0) = \frac{3}{5} N \epsilon_F. \tag{4.102}$$

As a special case of Eq. (4.93), at $T = 0$ the pressure

$$P(T = 0) = \frac{2}{5} \frac{N}{V} \epsilon_F. \tag{4.103}$$

P can turn out quite large, due to Pauli's principle forcing the fermions to occupy mutually orthogonal plane-wave states up to a high kinetic energy ϵ_F. In the density conditions of a typical metal, one discovers that the pressure exerted by the $T = 0$ free-electron gas is as large as $P \approx 1 - 10$ GPa! A sharp potential-energy step at the metal surface, generated by the attractive atomic nuclei, maintains this pressure by preventing the conduction electrons from escaping the solid, see Fig. 4.13.

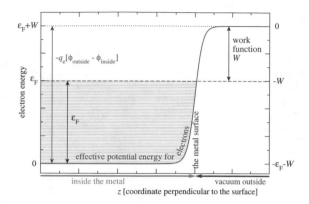

Fig. 4.13 The level scheme of the $T = 0$ Fermi-gas model applied to a metal. Either energy scale, at the left or at the right, can be adopted. Energy levels below the chemical potential (*dashed line*) represent occupied states inside the metal. Electrons promoted to levels above the dotted line by acquiring an energy $\geq W$ (e.g. from electromagnetic radiation) are free to leave the metal

The $T = 0$ free-fermion gas model accounts for several properties of the conduction electrons in simple metals, even at room temperature.[13] The level scheme of Fig. 4.13 illustrates the application of the Fermi-gas model to metals. Note the role played by the work function W, namely the minimum energy required to extract an electron from the solid.

We need to extend our analysis to the finite-T regime, in order to determine those thermodynamic quantities that involve temperature explicitly, such as the heat capacity. In the $T \ll T_F$ limit, a power expansion of the equation of state (4.92) and the relation (4.85) connecting μ to N/V provides the thermal properties at small but finite T/T_F. The mathematical details of this procedure [10] (called Sommerfeld expansion) are slightly intricate, but the qualitative trends are straightforward. For example, one can estimate the leading temperature dependence of μ and U by observing that when a small temperature is turned on, the average occupation $[n_\alpha]_F$ changes slightly as sketched in Fig. 4.12a from the $T = 0$ step function: a weak probability that states above μ are populated arises at the expense of population below μ. As the density of states $g(\mathcal{E}) \propto \mathcal{E}^{1/2}$ is slightly larger above than below ϵ_F, to conserve the fermion number $N = \int_0^\infty g(\mathcal{E})\,[n_\alpha]_F\,d\mathcal{E}$ the chemical potential decreases slowly as T increases.[14] The internal energy U increases due to the few electrons moving up from states $\approx k_B T$ below ϵ_F into states $\approx k_B T$ above (Fig. 4.14): the energy of each excited electron increases by $\approx k_B T$. The number of excited electrons is in the order

---

[13]The accord of the free-electron model with experimental data of many simple metals is surprisingly good despite neglecting the Coulomb interactions between electrons. The reason is that the electron gas screens the long-range Coulomb forces efficiently. An experimentally observed phenomenon directly related to electron-electron Coulomb repulsion is that of *plasmon* collective excitations, which however occur at rather high energy (few eV), and are therefore of little importance to thermodynamics at ordinary temperature.

[14]The low-T dependence of the chemical potential is $\mu = \epsilon_F\left[1 - \pi^2/12\ (T/T_F)^2 + \cdots\right]$.

Fig. 4.14 Thermal excitations/deexcitations across the Fermi sphere involve mainly states within a skin region with thickness $\delta k \propto k_B T$ across the Fermi sphere. Here this skin thickness is greatly exaggerated, compared to the occupation smearing induced by a realistic temperature $T \ll T_F$ of electrons in ordinary metals at room conditions

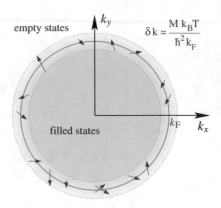

of the density of states $g(\epsilon_F)$ times the energy interval $k_B T$ where excitations occurs. We conclude that thermal excitations raise the total internal energy by approximately $g(\epsilon_F)(k_B T)^2$. The detailed expansion yields

$$U = U(T = 0) + \frac{\pi^2}{6} g(\epsilon_F)(k_B T)^2 + \cdots$$
$$= \frac{3}{5} N \epsilon_F \left[1 + \frac{5\pi^2}{12} \left(\frac{T}{T_F} \right)^2 + \cdots \right] \qquad [T \ll T_F], \qquad (4.104)$$

where the second equality relies on this useful relation valid for free fermions:

$$g(\epsilon_F) = \frac{3N}{2\epsilon_F}. \qquad (4.105)$$

The T-derivative of Eq. (4.104) provides the heat capacity of the ideal Fermi gas:

$$C_V = N k_B \frac{\pi^2}{2} \frac{T}{T_F} + \cdots \qquad [T \ll T_F]. \qquad (4.106)$$

Note that, compared to the high-temperature value $3/2 N k_B$, the $T \ll T_F$ heat capacity of the ideal gas is suppressed by a factor $\propto T/T_F$. The reason is that only few fermions with energy very close to the chemical potential can be involved in thermal excitations, the large majority remaining "frozen" in deeper states. Experimentally, a T-linear contribution to the total heat capacity is observed in solid metals at low temperature (Fig. 4.15), where lattice-vibration contributions are small, see Sect. 5.3 below. The electron gas is the responsible for this T-linear contribution. E.g., the observed T-linear coefficient for potassium $\simeq 2.06$ mJ mol^{-1}K^{-2} (Fig. 4.15) agrees fairly with the free-electron estimate $\pi^2 N_A k_B/(2T_F) \simeq 1.7$mJ mol^{-1}K^{-2},

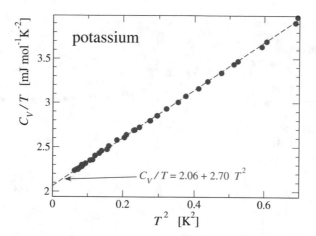

Fig. 4.15 The measured molar heat capacity of metallic potassium divided by T, as a function of T^2, in the sub-K temperature region. For $C_V \simeq a\,T^1 + b\,T^3$, the finite intercept at $T^2 = 0$ of the C_V/T curve measures the coefficient a of the T^1 (electronic) contribution; the slope of the graph measures the coefficient b of T^3, which gives the contribution of lattice vibrations, see Sect. 5.3.2 below (Data from Table 1 of Ref. [32]; the point at $T = 0.4231$ K was omitted)

consistent with the Fermi energy $\epsilon_F \simeq 2.1$ eV, obtained through Eq. (4.100) for the experimental density $N/V \simeq 1.3 \times 10^{28}$ m^{-3} of conduction electrons in potassium.

The Fermi gas is nonmagnetic, as both spin states are equally occupied. The application of a weak external magnetic field strength $\mathbf{H} = H_z\hat{\mathbf{z}}$ produces a total magnetic field $\mathbf{B} = B_z\hat{\mathbf{z}}$. We assume $\mathbf{B}$ to couple only to the spin of the fermions: the degeneracy of the g_s spin components is lifted, and the Fermi gas acquires a magnetization. $\mathbf{M} = M_z\hat{\mathbf{z}}$ denotes the volume density of magnetic moment, like in Eq. (4.71). For spin-1/2 electrons ($g_s = 2$), the magnetization is $M_z = -\mu_B[N_\uparrow - N_\downarrow]/V$. By computing the linear response to the external field, one obtains the magnetic behavior of the ideal Fermi gas. At high $T \gg T_F$, according to Eq. (4.74), independent spins produce a Curie susceptibility $\chi_B \propto N/T$. In contrast, at small $T \ll T_F$ Pauli's principle freezes out the spins of most electrons, the paired ones deep inside the Fermi sphere (Fig. 4.14). Only the approximately $g(\epsilon_F)\,k_B T$ electrons near the Fermi surface do spin-polarize, thus producing a characteristically T-independent magnetization. Detailed calculation yields:

$$M_z = \mu_B^2\,\frac{g(\epsilon_F)}{V}\,B_z = \frac{3}{2}\frac{\mu_B^2}{\epsilon_F}\frac{N}{V}\,B_z = \frac{3^{1/3}}{4\pi^{4/3}}\frac{q_e^2}{m_e}\left(\frac{N}{V}\right)^{1/3} B_z \qquad [T \ll T_F], \quad (4.107)$$

where the last expression provides the explicit density dependence of χ_B. Accordingly, the spin susceptibility of the low-temperature ideal spin-1/2 Fermi gas is given by Eq. (4.77), with

Fig. 4.16 The measured magnetic susceptibility of $Zr_2V_6Sb_9$. Conduction electrons are responsible for the T-independent Pauli susceptibility χ_m of this metal. The surge of χ_m at very low temperature is a Curie-type additional contribution, Eq. (4.74), due to magnetic impurities contaminating this sample (Data from Ref. [33])

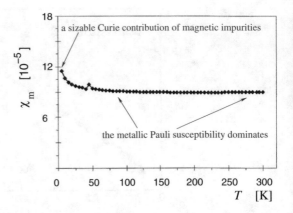

$$(\epsilon_0 c^2)^{-1} \chi_B = \frac{e^2}{m_e c^2} \left(\frac{3N}{\pi V}\right)^{1/3} = \alpha^2 a_0 \frac{1}{\pi} \left(3\pi^2 \frac{N}{V}\right)^{1/3} = \frac{1}{\pi} \alpha^2 a_0 k_F. \qquad (4.108)$$

With an average spacing between electrons $(V/N)^{1/3}$ of a few times the Bohr radius a_0, Eq. (4.108) predicts a very weak T-independent paramagnetic susceptibility $\chi_m \simeq (\epsilon_0 c^2)^{-1} \chi_B \approx 10^{-5}$. This T-independent *Pauli susceptibility*, Eq. (4.108), is a characteristic signature of metals in the experimental study of the magnetic response of materials, see, e.g., Fig. 4.16. This metallic response contrasts sharply with the Curie susceptibility of isolated spins, Eq. (4.74), which is inversely proportional to T and linear in N/V, instead.

4.3.2.2 Bosons

Proper Bosonic Particles. The ground state of N non-interacting bosons is far simpler than that of N fermions: they all occupy the lowest-energy state $|\mathbf{k} = \mathbf{0}\rangle$, with energy $\mathcal{E} = 0$. If a spin degeneracy g_s is present, each spin projection is populated by an N/g_s bosons, on average. Expression (4.96) reflects this $T \to 0$ limit for $[n_\alpha]_B$: the occupancies of all positive-energy states vanish; meanwhile $\mu \simeq -k_B T g_s / N \to 0^-$, in such a way to ensure that the occupation $[n_0]_B$ of each of the g_s ground states at $\mathcal{E} = 0$ approaches N/g_s. Unfortunately, the density of states (4.48) vanishes at $\mathcal{E} = 0$: this indicates that the conversion (4.90) of the discrete sum over the one-particle states into an energy integral is missing completely the $|\mathbf{k} = \mathbf{0}\rangle$ state, whose population becomes dominant at low temperature. A correct treatment, including explicitly the population $[n_0]_B$ of the $\mathcal{E} = 0$ state, predicts a phase transition at a finite temperature

$$T_{BE} = \frac{6.625}{k_B} \frac{\hbar^2}{2M} \left(\frac{N}{g_s V}\right)^{2/3}, \qquad (4.109)$$

signaled by the macroscopic filling of the $\mathcal{E} = 0$ level at $T \leq T_{BE}$.[15] This low-temperature collective state is called a *"Bose–Einstein condensate"*.

Even though many atoms and molecules form boson gases, upon cooling inter-atomic/intermolecular interactions make them all turn into solids at a temperatures far higher than the T_{BE} appropriate for their standard densities. The only (partial) exception in ordinary matter is ^{4}He, and indeed a *superfluid* transition similar to the Bose–Einstein condensation of an ideal boson gas is observed in ^{4}He near 2 K at ordinary pressure. However, ^{4}He at low temperature is a liquid rather than a gas, indicating that inter-particle interactions play an important role. Accordingly the helium fluid can hardly be regarded as an ideal system: to fully understand the actual nature of the superfluid transition of ^{4}He, more sophisticated tools are required.

Since the mid 1990's, ultracold droplets of atoms are being produced and kept in a metastable gaseous state inside electromagnetic traps. Progress in cooling techniques, down to the sub-μK range, permits to cross routinely T_{BE} with droplets of boson atoms [34]. In these droplets, atoms are far more dilute than in a liquid, thus the atom-atom interactions play an almost irrelevant role. Therefore these droplets provide concrete experimental realizations of the ideal Bose–Einstein condensate.

Quanta of Harmonic Oscillators. In addition to proper material bosons, the Bose–Einstein distribution describes the thermodynamic properties of fictitious particles related to harmonic oscillators, too. Equations (3.21) and (C.65) yield the ladder of eigenvalues of the harmonic oscillator. The step quantum number v counts the nodes of the corresponding wavefunction. Wherever several harmonic oscillators with frequencies ω_α are present, as e.g. in a polyatomic molecule, a vibrational state is labeled by all the v_α quantum numbers, and the associated total energy is

$$E_{\text{vib}}(v_1, v_2, \ldots) = \sum_\alpha \left(v_\alpha + \frac{1}{2} \right) \hbar \omega_\alpha. \tag{4.110}$$

Compare Eq. (4.110) to the expression

$$E(n_1, n_2, \ldots) = \sum_\alpha n_\alpha \mathcal{E}_\alpha \tag{4.111}$$

used, e.g. in Eq. (4.78), for the energy of noninteracting particles in terms of the occupation numbers n_α of the single-particle states with energy $\mathcal{E}_\alpha$. Apart from an irrelevant constant zero-point shift $\frac{1}{2} \sum_\alpha \hbar \omega_\alpha$, expressions (4.110) and (4.111) are identical, provided that the following identifications are made:

---

[15]The average occupation $[n_\alpha]_B$ of each state $|\alpha\rangle$ follows Eq. (4.95), which is remarkably independent of the system size. If we, e.g., double the system size (both N and V), then the average occupation of $|\alpha\rangle$ does not change: the doubling of the density of states $g_{\text{tr}}(\mathcal{E}_\alpha)$ takes care of the extra particles, see Eq. (4.48). The $\mathcal{E} = 0$ state marks an exception. Below T_{BE}, its occupation $[n_0]_B$ becomes a finite fraction of N: thus, if the system size doubles, its occupation also doubles.

label of individual oscillator $\alpha \longleftrightarrow \alpha$ label of single-particle state

oscillator quantum number $v_\alpha \longleftrightarrow n_\alpha$ occupation number of state α

oscillator energy quantum $\hbar\omega_\alpha \longleftrightarrow \mathcal{E}_\alpha$ single-particle energy of state α

oscillator eigenvalue $v_\alpha \hbar\omega_\alpha \longleftrightarrow n_\alpha \mathcal{E}_\alpha$ energy of n_α particles in state α .

As $v_\alpha = 0, 1, 2, 3, \ldots$, this identification makes sense for boson occupations. The equality of the spectrum produces a completely equivalent statistical behavior. In detail, consider the average vibrational energy of one oscillator, Eq. (4.62) and remove the zero-point term $\hbar\omega_\alpha/2$. Individual energy levels $v_\alpha \hbar\omega_\alpha$ are proportional to the quantum number v_α. Thus the average energy $\hbar\omega_\alpha/(\exp x - 1)$, reflects an average value of v_α, which is $[v_\alpha] = 1/(\exp x - 1)$, where $x = \hbar\omega_\alpha/(k_B T)$. This expression coincides with the average occupation $[n_\alpha]_B$, Eq. (4.96), of a single-boson state $|\alpha\rangle$, provided that $\mathcal{E}_\alpha = \hbar\omega_\alpha$ and $\mu = 0$. One is then led to think of each step in the harmonic ladder as a fictitious boson-type particle, called "*phonon*" (sound particle) or "*photon*" (light particle), depending on the nature of the involved oscillator. Accordingly, an oscillator in its $|v = 4\rangle$ state is said to carry 4 photons, while an oscillator in its ground state $|v = 0\rangle$ holds no photons. In this way, the Bose–Einstein distribution can be profitably employed to describe the thermodynamics of a set of harmonic oscillators, by replacing the single-particle energy $\mathcal{E}_\alpha$ with the relevant $\epsilon = \hbar\omega_\alpha$. The lack of chemical potential μ indicates that the average number N of bosons is impossible to control. Contrary to material particles enclosed in a vessel, phonons/photons are not conserved: N varies widely as a function of temperature.

We can now reformulate the statistical properties of a set of independent oscillators in terms of Bose statistics. We summarize here the main results for the "photon gas", i.e. the thermodynamics of the normal modes of the electromagnetic fields at thermal equilibrium inside an isothermal cavity. This system is also known as *blackbody radiation*.

The components of the electromagnetic fields in vacuum obey a (Laplace) stationary equation formally identical to the Eq. (C.52) of Schrödinger particles, but with c^2 in place of $\hbar^2/(2m)$ and ω^2 in place of $\mathcal{E}$. The position-dependent equation being mathematically the same, solutions are also plane waves. Under the same periodic boundary conditions, the allowed values of wave vector $\mathbf{k}$ are connected to the box size by the same Eq. (C.55). The resulting *dispersion relation* $\epsilon(\mathbf{p}) = \hbar\omega(\mathbf{p}/\hbar)$ is

$$\epsilon(\mathbf{p}) = c|\mathbf{p}| \qquad \text{or} \qquad \omega(\mathbf{k}) = c|\mathbf{k}|, \qquad (4.112)$$

where c is the speed of light. Compare this relation with the quite different one

$$\mathcal{E}(\mathbf{p}) = \frac{|\mathbf{p}|^2}{2m} \qquad \text{or} \qquad \omega(\mathbf{k}) = \frac{\hbar|\mathbf{k}|^2}{2m}, \qquad (4.113)$$

appropriate for free material particles with mass m. By counting the states within a **k**-sphere one obtains the total density of oscillator energies

$$g_s \, g_{\text{ph}}(\epsilon) = \frac{V}{\pi^2 \hbar^3 c^3} \epsilon^2, \tag{4.114}$$

to be compared with the analogous distribution for Schrödinger particles, Eq. (4.48). The "spin" degeneracy of photons is $g_s = 2$, corresponding to the two orthogonal transverse (i.e. perpendicular to **k**) polarizations.

By executing the sum over the independent harmonic oscillators, i.e. integrating over their energy $\epsilon = \hbar \omega$, one obtains the following thermodynamic relations:

$$U = V \int_0^\infty u(\epsilon, T) \, d\epsilon = V \frac{\pi^2 (k_B T)^4}{15 \, \hbar^3 c^3},$$

$$\text{with } u(\epsilon, T) = \frac{1}{V} \, g_s \, g_{\text{ph}}(\epsilon) \, [n_\epsilon]_B \, \epsilon = \frac{1}{\pi^2 \hbar^3 c^3} \frac{\epsilon^3}{e^{\epsilon/k_B T} - 1}, \tag{4.115}$$

$$[\hat{N}] = \int_0^\infty g_s \, g_{\text{ph}}(\epsilon)[n_\epsilon]_B \, d\epsilon = V \frac{1}{\pi^2 \hbar^3 c^3} (k_B T)^3 \int_0^\infty \frac{y^2}{e^y - 1} \, dy$$

$$= V \frac{2 \xi(3)}{\pi^2 \hbar^3 c^3} (k_B T)^3, \tag{4.116}$$

where ξ is the Riemann function [$\xi(3) \simeq 1.20206$]. These results were first derived by M.K.E.L. Planck to interpret the experimental data of thermal-radiation spectra. Indeed, Eq. (4.115) makes an important prediction for the spectral density $u(\epsilon, T)$ (energy per unit volume and energy interval) of radiation at equilibrium.[16]

Experimentally, the spectral irradiance $R(\epsilon, T)$ (radiated power per unit surface and spectral energy interval) of radiation is more straightforward to probe than its energy density. As it turns out, R and u are proportional:[17]

$$R(\epsilon, T) = \frac{c}{4} u(\epsilon, T). \tag{4.117}$$

---

[16]The same Planck distribution is occasionally quoted in terms of frequency ν or wavelength λ, rather than photon energy $\epsilon = \hbar \omega = 2\pi \hbar \nu = 2\pi \hbar c / \lambda$. Explicitly, because $\tilde{g}(\nu, T) = 8\pi V \nu^2/c^3$, $U = V \int_0^\infty \tilde{u}(\nu, T) \, d\nu$, with $\tilde{u}(\nu, T) = 16\pi^2 \hbar c^{-3} \nu^3 / \{\exp[(2\pi \hbar \nu)/(k_B T)] - 1\}$. And, likewise, $U = V \int_0^\infty \tilde{u}(\lambda, T) \, d\lambda$, with $\tilde{u}(\lambda, T) = 16\pi^2 \hbar c \lambda^{-5} / \{\exp[(2\pi \hbar c)/(\lambda k_B T)] - 1\}$. Despite expressing the same property, the physical dimensions of these spectral densities are of course different: $[u] = $ Length^{-3}, $[\tilde{u}] = $ Energy$\times$Length$^{-3}\times$Time, and $[\tilde{u}] = $ Energy$\times$Length^{-4}.

[17]The proportionality factor $c/4$ in Eq. (4.117) originates from the fact that each photon carries its energy at the speed of light c. Moreover, only a fraction $(4\pi)^{-1} \int_0^1 \cos\theta \, d\cos\theta \int_0^{2\pi} d\varphi = 1/4$ of the photons in the surroundings of a given infinitesimal surface crosses it in a given direction.

Fig. 4.17 The spectral irradiance $R(\epsilon, T) = \frac{c}{4}u(\epsilon, T)$ of the equilibrium (blackbody) electromagnetic fields at three temperatures. Note that as T increases, the area under the curve, i.e. the total radiated power per unit surface—Eq. (4.118), increases very rapidly. Also, the curve maximum shifts to higher energy (shorter wavelength)

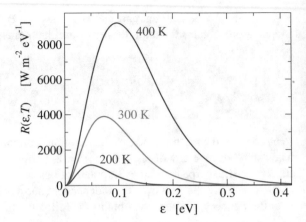

Figure 4.17 reports the spectral irradiance $R(\epsilon, T)$ as a function of the photon energy. The area under each curve equals the total radiated power per unit surface

$$\int_0^\infty R(\epsilon, T)\, d\epsilon = \frac{\pi^2 (k_B T)^4}{60\, \hbar^3 c^2}. \tag{4.118}$$

This result is know as the *Stefan–Boltzmann law*. According to Eq. (4.118), e.g., at 300 K one square meter of blackbody surface[18] radiates a total 459 W. The photon energy where the irradiance $R(\epsilon, T)$ is maximum shifts linearly with T: $\epsilon_{\max} \simeq 2.8214\, k_B T \simeq 0.2431$ meV/K $\times\, T$ (*Wien's displacement law*).

The spectral distribution of energy density (and, equivalently, radiative power) of electromagnetic fields at equilibrium is a universal function of temperature (called *blackbody spectrum*), and does not depend on the precise way this equilibrium is established (e.g. on the properties of the material of the cavity enclosing the fields). The distribution (4.117) agrees perfectly with the spectral analysis of radiation escaping from an isothermal cavity through a tiny hole. This agreement is not surprising, because a description of the electromagnetic fields in terms of harmonic oscillators is basically exact. Blackbody spectra are usually associated to radiation emitted by hot surfaces, e.g. those of Appendix B.1. A spectacular realization of the photon gas is the cosmic microwave background, Fig. 4.18, a "fossil" relic of an early stage of the universe when it was all at thermal equilibrium.

In Sect. 5.3.2, we are going to apply the same statistics, with an appropriately different spectral density of oscillators, to describe the thermal properties of the vibrations of solids in terms of a "gas of phonons".

---

[18]The name "blackbody" to indicate the spectrum of equilibrium radiation originates from the fact that a perfectly absorbing (0% reflectivity over the entire spectral range of Fig. 1.2) surface irradiates precisely the spectrum $R(\epsilon, T)$, where T is the temperature of the surface itself. The reason is the detailed energy balance of the fluxes of incoming and outgoing radiation between the surface and the fields, which would be necessarily established if the surface was to be a part of the inside surface of an isothermal cavity containing the equilibrium electromagnetic fields.

Fig. 4.18 The observed cosmic microwave background frequency spectrum, compared to a 2.73 K blackbody spectrum derived from Eq. (4.117). Note the $\propto \epsilon^2$ increase at small frequency, and the rapid (exponential) decay past the maximum (From G.F. Smoot and D. Scott, Cosmic Background Radiation, http://www.astro.ubc.ca/people/scott/cbr_aph_00.ps)

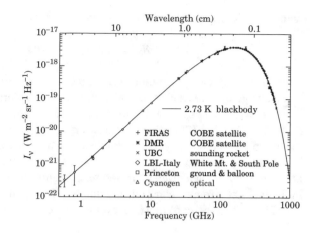

4.4 Matter-Radiation Interaction

In Sect. 2.1.9 we have sketched the basic result for the rate—Eq. (2.45)—of spontaneous decay of an atom (or a molecule) from an excited state, as occurs in emission experiments. In the absence of any external stimulation, the material system decays to a lower-energy state transferring its excess energy to the electromagnetic radiation field in an average time γ^{-1}. In this section we will use the results of statistics to explore how an external *stimulating radiation*, as e.g. in an absorption experiment, affects this transition. First off, this radiation needs to resonate with the energy of a transition between two eigenstates of the system. For simplicity, we shall ignore off-resonance transitions, which occur at much smaller rates.

Consider an ensemble of identical noninteracting quantum "particles" (e.g. atoms or molecules in gas phase or diluted impurities in a solid) and, in the single-particle spectrum, focus on two states: $|1\rangle$ and $|2\rangle$, with energies $\mathcal{E}_1 < \mathcal{E}_2$. We use n_1 and n_2 to indicate the instantaneous populations of these states. Resonant photons with energy $\epsilon = \hbar\omega = \mathcal{E}_2 - \mathcal{E}_1$ can induce transitions between these levels. In particular, the excitation $|1\rangle \rightarrow |2\rangle$ is driven by the presence of radiation. Accordingly, the probability per unit time (i.e. the rate) that an atom initially in state $|1\rangle$ gets excited to the upper state $|2\rangle$ is proportional to the spectral energy density (per unit volume and spectral interval) $\rho(\epsilon)$ of the electromagnetic field at the resonant energy:

$$R_{1\rightarrow 2} = B_{12}\,\rho(\epsilon)\,, \tag{4.119}$$

where B_{12} is a suitable proportionality constant, depending on the microscopic characteristics of the system and on its coupling to the field. Here we neglect nonlinear effects $O(\rho^2)$. In the de-excitation direction, an atom initially in state $|2\rangle$ has a probability per unit time A_{21} to decay spontaneously to state $|1\rangle$—in the dipole approximation, A_{21} equals the γ_{21} of Eq. (2.45). In addition, we need to consider the probability of downward transitions stimulated by the presence of resonant radiation, whose rate is proportional to the same spectral density. In total:

$$R_{2\to1} = A_{21} + B_{21}\,\rho(\epsilon)\,, \tag{4.120}$$

where B_{21} is another yet-to-be-determined constant of proportionality.

At any given time, the total number of particles undergoing the $|1\rangle \to |2\rangle$ transition is $n_1 R_{1\to2}$, and the total number of particles going $|2\rangle \to |1\rangle$ is $n_2 R_{2\to1}$. Relations (4.119) and (4.120) hold under arbitrary radiation conditions, for example when the ensemble is probed by a radiation beam in an absorption experiment (Fig. 1.3). In particular, these relations hold also when the ensemble and the radiation field are *at equilibrium* at a given temperature. At equilibrium the average populations of individual states remain constant, and this implies that the total number of $|1\rangle \to |2\rangle$ and $|2\rangle \to |1\rangle$ transitions must, on average, be equal:

$$[n_1]\,R_{1\to2} = [n_2]\,R_{2\to1} \quad \text{[at equilibrium]}\,. \tag{4.121}$$

We substitute Eqs. (4.119) and (4.120) in the balance equation (4.121)

$$[n_1]\,B_{12}\,\rho(\epsilon) = [n_2]\,(A_{21} + B_{21}\,\rho(\epsilon))\,. \tag{4.122}$$

We isolate the energy density $\rho(\epsilon)$, obtaining

$$\frac{\frac{A_{21}}{B_{21}}}{\frac{[n_1]}{[n_2]}\frac{B_{12}}{B_{21}} - 1} = \rho(\epsilon)\,. \tag{4.123}$$

The ratio of the populations equals the ratio of the equilibrium probabilities which, according to Boltzmann statistics, is simply $[n_1]/[n_2] = P_1/P_2 = \exp(\beta[\mathcal{E}_2 - \mathcal{E}_1]) = \exp(\beta\epsilon)$. At equilibrium, the spectral density of the radiation field follows the universal energy dependence described in Sect. 4.3.2.2, in particular by Eq. (4.115): $\rho(\epsilon) = u(\epsilon, T)$. Accordingly,

$$\frac{A_{21}}{B_{21}}\frac{1}{\exp\left(\frac{\epsilon}{k_{\mathrm{B}}T}\right)\frac{B_{12}}{B_{21}} - 1} = \rho(\epsilon) = u(\epsilon, T) = \frac{\epsilon^3}{\pi^2\hbar^3c^3}\frac{1}{\exp\left(\frac{\epsilon}{k_{\mathrm{B}}T}\right) - 1}\,. \tag{4.124}$$

The comparison of the two explicit expressions, which must coincide for *any* temperature, requires the following identities for the coefficients:

$$\frac{B_{12}}{B_{21}} = 1\,, \qquad \frac{A_{21}}{B_{21}} = \frac{\epsilon^3}{\pi^2\hbar^3c^3}\,, \tag{4.125}$$

known as *Einstein relations*. Relations (4.125), which can be reformulated as

$$B_{12} = B_{21} = \frac{\pi^2 \hbar^3 c^3}{\epsilon^3} A_{21}, \qquad (4.126)$$

were derived under the assumption of equilibrium. Since the coefficients B_{12} B_{21} A_{21} depend only on electromechanical properties of the particles, these relations hold unchanged for arbitrary field conditions. Once one of these coefficients is evaluated, either experimentally or, e.g., by $A_{21} = \gamma_{21}$ of Eq. (2.45), the two others are fixed by Eq. (4.126).

The first equality expresses the symmetric role of the initial and final states in QM, implying that a radiation field induces equal rates of excitation $|1\rangle \to |2\rangle$ and of stimulated emission $|2\rangle \to |1\rangle$. For a strong radiation intensity, such that $B_{21} \rho(\epsilon) \gg A_{21}$, the spontaneous emission rate becomes negligible, and $B_{12} = B_{21}$ implies in particular that $R_{1\to2} \simeq R_{2\to1}$, thus rapidly also $n_1 \simeq n_2$ (*saturated transition*). This result clarifies the role of the intense pump beam in the experiment of Fig. 2.11: it saturates the individual components of the $|n = 2\rangle \longleftrightarrow |n = 3\rangle$ transition of hydrogen.

The second relation implies that, for a given spectral energy density $\rho(\epsilon)$ (independent of ϵ), the ratio of spontaneous emission to stimulated emission varies with ϵ^3. Accordingly, in a low-energy (microwave, IR) transition, stimulated emission tends to prevail, while at higher energy (UV, X-rays) spontaneous emission usually dominates. Even when equilibrium radiation is considered, the ratio of spontaneous to stimulated emission

$$\frac{A_{21}}{B_{21} u(\epsilon, T)} = \exp\left(\frac{\epsilon}{k_B T}\right) - 1 \qquad (4.127)$$

indicates, unsurprisingly, that thermal radiation is effective in stimulating emission for high temperature $k_B T \gtrsim \epsilon$ only. Finally, in the context of spectroscopy, the proportionality between the three coefficients $B_{12} = B_{21} \propto A_{21}$ proves that an emission-forbidden transitions is also absorption-forbidden and that, *vice versa*, a fast allowed transition in emission is intense in absorption, too.

4.4.1 The Laser

Ordinary media attenuate traversing electromagnetic waves. Absorption spectroscopies (Fig. 1.3) measure precisely this attenuation. Consider instead an optical medium composed of non-interacting quantum systems which *amplifies*—rather than attenuates—light. For this to occur, the total emission rate needs to exceed absorption: $n_2 R_{2\to1} > n_1 R_{1\to2}$. Equivalently the ratio

$$\frac{\text{rate of emission}}{\text{rate of absorption}} = \frac{n_2 [A_{21} + B_{21}\rho(\epsilon)]}{n_1 B_{12}\rho(\epsilon)} = \frac{n_2}{n_1}\left[1 + \frac{A_{21}}{B_{21}\rho(\epsilon)}\right] \qquad (4.128)$$

needs to exceed unity. Since the term in brackets approaches unity as soon as a sufficient radiation intensity builds up, this equation tells us that light amplification (emission overcoming absorption) requires a single condition: a population ratio $n_2/n_1 > 1$, i.e. inverted compared to a regular equilibrium population $[n_2]/[n_1] < 1$. An inverted population is a radical deviation from equilibrium, and is thus highly unstable: precisely the prevalence of emission over absorption leads an ensemble of quantum systems prepared in a population-inverted state to relax spontaneously toward the regular Boltzmann population $[n_2] < [n_1]$. In practice, some kind of electronic or optical *pumping* is needed to generate and sustain a population inversion for an extended period of time, and replace the radiated energy at the expense of an external power source, as, e.g., in the schemes of Fig. 4.19.

Once an optical *active medium* which amplifies light is realized, light must be channeled through it in a precise direction, by building a resonating one-dimensional (1D) optical cavity around it. The ruby laser of Fig. 4.20 illustrates the working principle of an historically relevant type of device. A crucial feature of stimulated emission is coherence: the emitted photon is not emitted at random as in spontaneous emission, but prevalently in the same direction and with the same phase and polarization as the stimulating photon, which thus is "cloned".

The use of a long cavity lets photons emitted at odd directions escape basically unamplified, while photons directed along the cavity axis bounce back and forth several times through the active medium, thus getting strongly amplified. As a result, a powerful highly coherent beam of radiation builds up for as long as the population inversion is maintained. A device such as described here, producing a coherent beam of photons by means of Light Amplification by the Stimulated Emission of Radiation is named *laser*.

Diverse commercial applications of such coherent beams extend from the industrial to the consumer side, including telecommunications, optical data storage, telemetry, cutting, welding, surgery, etc. Lasers have also become irreplaceable

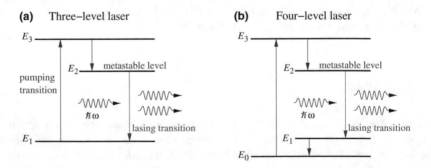

Fig. 4.19 The principle of operation of **a** three- and **b** four-level lasers. These level schemes illustrate two tricks to realize optically a population inversion, namely a population of level $|2\rangle$ exceeding that of level $|1\rangle$, as needed to obtain an "active medium". The spontaneous decay $|2\rangle \rightarrow |1\rangle$ must occur at a far slower rate than all other indicated downward transitions. For $|3\rangle \rightarrow |2\rangle$ and $|1\rangle \rightarrow |0\rangle$ one usually picks rapid electric-dipole–allowed transitions

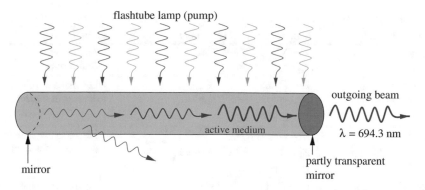

Fig. 4.20 The ruby laser is a three-level solid-state laser. Chrome impurities in an (otherwise transparent) Al_2O_3 crystal rod act as noninteracting quantum systems carrying the levels involved in the population inversion, as in Fig. 4.19a. A white flashtube lamp pumps this "active medium" optically, raising many Cr impurities from their ground state $|1\rangle$ to a band of broad short-lived states $|3\rangle$. The impurities decay rapidly to metastable state $|2\rangle$. As further spontaneous decays are very slow, the $|2\rangle \rightarrow |1\rangle$ transition is mainly induced by stimulated emission. Since the energy of state $|2\rangle$ is defined sharply, the emitted radiation is highly monochromatic ($\lambda = 694.3 \pm 0.5$ nm). Off-axis photons arising from spontaneous decay escape from the sides of the rod. Repeatedly-reflected axially-moving photons get amplified and stimulate further coherent emission. The output photon beam escapes through the partly transparent mirror at one end of the rod

tools for research in spectroscopy, photochemistry, ultracold trapped-gas cooling, microscopy, adaptive optics of telescopes, interferometry, etc.

Final Remarks

The present chapter connects thermodynamics with the principles of equilibrium statistics, based on a simple "democratic" assumption of equal probability of all states at a given energy. We focus on applications to ordinary matter, especially on the few systems (mainly gas phases) where ideal noninteracting systems provide semi-quantitative results.

Eventually, we obtain several quite different formulations and results, which are applicable in diverse contexts, with a special focus on ideal gases of noninteracting particles. Table 4.1 collects the main formulas and their range of applicability.

Statistical physics becomes more intricate and exciting when moving beyond ideal systems and addressing applications to real interacting systems [29, 30]. Physicists have devised all sorts of techniques to investigate interacting systems both theoretically (many-body methods, the renormalization group, ...) and experimentally (the definition and measurement of density/momentum/spin correlation functions, high-pressure techniques to investigate phase diagrams in extreme regimes, analysis of fluctuations, size effects, exotic nematic or magnetic phases...).

Table 4.1 A summary of the main results of this chapter, with their appropriate applicability range

Physical context	Main relations $[\beta = (k_B T)^{-1}]$	Applies to
General Boltzmann statistics	$Z = \sum_{m'}^{\text{global states}} e^{-\beta E_{m'}}$ $P_m = e^{-\beta E_m}/Z$	Any system at equilibrium
Noninteracting "distinguishable" particles (any T)	$Z = \prod_i Z_i$ $Z_i = \sum_\alpha^{\text{states of particle } i} e^{-\beta E_\alpha^i}$ $P_\alpha^{(i)} = e^{-\beta E_\alpha^i}/Z_i$	Localized or internal degrees of freedom
Noninteracting identical particles (high T)	$Z = (Z_1)^N/N!$ $Z_1 = Z_{1\,\text{tr}} = V/\Lambda^3, \quad \Lambda = \hbar\sqrt{\frac{2\pi}{M k_B T}}$ $PV = N k_B T = 2/3\, U$	Ideal-gas translational degrees of freedom
Noninteracting identical bosons (any T)	$Q = \prod_\alpha \left[1 - e^{\beta(\mu - \mathcal{E}_\alpha)}\right]^{-1}$ $[n_\alpha]_B = 1/(e^{\beta(\mathcal{E}_\alpha - \mu)} - 1)$ $PV = -k_B T \sum_\alpha \ln\left(1 - e^{\beta(\mu - \mathcal{E}_\alpha)}\right) = 2/3\, U$	Ideal-gas translational degrees of freedom
Noninteracting identical fermions (any T)	$Q = \prod_\alpha \left[1 + e^{\beta(\mu - \mathcal{E}_\alpha)}\right]$ $[n_\alpha]_F = 1/(e^{\beta(\mathcal{E}_\alpha - \mu)} + 1)$ $PV = k_B T \sum_\alpha \ln\left(1 + e^{\beta(\mu - \mathcal{E}_\alpha)}\right) = 2/3\, U$	Ideal-gas translational degrees of freedom

The present introduction to statistical thermodynamics focuses on equilibrium, and omits important *dynamical* out-of-equilibrium quantities. These quantities will become useful to discuss transport (e.g. the electric and thermal conductivities in solids discussed in the next chapter) and collective "hydrodynamic" properties, in particular when relating them to the microscopic interactions.

Problems

A $\star$ marks advanced problems.

4.1* A gas-phase HCl sample is traversed by a microwave field resonating with the rotational transition between the rotational states of angular momentum $l = 1$ and $l = 2$. Evaluate the ratio between the spontaneous and stimulated emission rates, given the equilibrium distance $R_M = 127$ pm between the proton and the ^{35}Cl nucleus, and the microwave spectral density at resonance $\rho_v = 0.250$ J m^{-3} Hz^{-1}.

4.2* The inner density of "white dwarf" stars can reach approximately 10^{11} kg m^{-3}. Assume for simplicity that:

- these stars consist of a gas of non-interacting protons and electrons in equal number and uniform density;
- the electrons are ultra-relativistic, with energy $E = c|\mathbf{p}| = \hbar c|\mathbf{k}|$;
- temperature is 0 K.

Within the above hypothesis, evaluate the pressure of this electron gas.

4.3 To a fair approximation, the valence electrons of metal lithium form a Fermi gas, while positive ions remain essentially immobile at crystalline equilibrium positions. The density of this material is 535 $kg\,m^{-3}$. Evaluate the pressure of the electron gas at temperature 0 K. Instead, if one could generate a weakly-interacting *atomic-^{6}Li* gas with the same density, evaluate the pressure of such Fermi gas at temperature 0 K.

4.4 Consider the dissociated single-atom fraction of a fluorine gas at temperature 5,000 K, thus neglecting all diatomic molecules. Neglect also all atoms in electronic states with a configuration other than $1s^2 2s^2 2p^5$. Evaluate the fraction of atoms in the state with total angular momentum $J = 1/2$, given that its excitation energy is 50.1 meV above the atomic ground state. Compute also the translational and electronic contribution of each atom to the molar heat capacity at the assigned temperature.

4.5 Comparing gas-phase samples of Cl atomic at equilibrium at temperatures 1000 and 2000 K, evaluate the intensity ratio $I\,(2000\,K)/I\,(1000\,K)$ of the least energetic absorption line exhibited by a sample in the $3s^2 3p^5 (^2P) \rightarrow 3s^2 3p^4 4s (^2P)$ transition, due to the different thermal population of the ground and first-excited state of the $3s^2 3p^5 (^2P)$ configuration. These $3s^2 3p^5 (^2P)$ states are 109.4 meV apart.

4.6 The three atomic levels associated to the ground configuration $3d^2 4s^2\,{}^3F$ of titanium are found at energy 0, 0.02109, and 0.04797 eV. Evaluate the contribution of these electronic excitations to the molar heat capacity of a vapor of monoatomic titanium at temperature 1000 K, and their fractional contribution to the total heat capacity of the gas.

4.7 An insulating solid contains non-interacting atomic impurities (at a number-density level $3 \times 10^{21}\,m^{-3}$) characterized by localized levels F (threefold degenerate) and A (nondegenerate), separated by an energy $E_A - E_F = 50$ meV. Evaluate the contribution of these impurities to the heat capacity per kilogram of this solid at $T = 700$ K, given the material's density 2300 $kg\,m^{-3}$.

4.8 Evaluate the ratio of the heat capacity per unit volume of the electromagnetic fields in a blackbody cavity at temperatures 10000 and 4000 K. Evaluate also the ratio between the total radiated power at the two indicated temperatures.

4.9 Evaluate the mean speed $\langle|v|\rangle$ and the pressure of the conduction electrons in the ground state of metallic gold (density $19.3 \times 10^3\,kg\,m^{-3}$, one electron per atom in the conduction band), in the approximation of free non-interacting electrons.

4.10* Approximate the spectrum emitted by a furnace with blackbody radiation. The radiated power (in $W\cdot m^{-2}$) in the wavelength range between 3150 and 3250 nm (infrared) is measured. One observes that when the furnace absolute temperature is doubled, this power increases by a factor 10. Evaluate the final furnace temperature. At such temperature, what is the ratio of the radiated power in the range 3150–3250 nm to the radiated power in the visible range 695–705 nm? [Hint: evaluate integration approximately as a finite sum, taking advantage of the narrowness of the integration intervals.]

4.11 Estimate, (with at least 10% precision) the temperature needed to produce with a 0.1% efficiency, soft X rays of energy ≥ 10 eV using a thermal blackbody source. [Hints: (a) when $x \gg 1$, the error in the approximation $(e^x \pm 1)^{-1} \simeq e^{-x}$ is negligible; (b) $\int_0^\infty x^3/(e^x - 1)dx = \pi^4/15$; (c) approximate solutions of non-algebraic equations can be obtained by numeric bisection.]

4.12 Approximate a human body to a blackbody at temperature $37\,°C$. Evaluate the total power that it irradiates, assuming a surface area 1.8 m^2. Taking also into account the power that reaches the human body when inside a blackbody cavity, at what temperature should this cavity be placed for the net power lost by the human body to be reduced by a factor 10?
[Recall the expression for the Stefan–Boltzmann proportionality constant between T^4 and the total irradiance: $\pi^2 k_B^4/(60\,\hbar^3 c^2)$.]

4.13 Consider a metallic sodium sample (density $\rho = 950$ kg m^{-3}). Assuming the free non-interacting Fermi model for the conduction electrons, evaluate the electronic pressure. Assume that 10% of the sodium atoms is replaced by aluminum atoms, with no change in the crystal structure (even the lattice spacing remains unchanged). Evaluate the variation of the electronic pressure compared to pure sodium due to the presence of 3 (rather than 1) electrons per aluminum atom in the conduction band.

4.14* The maximum phonon frequency of NaCl is $\nu_{max} = 5$ THz. Assume that all vibrational modes whose frequency is smaller than of equal to ν_{max} contribute to electromagnetic radiation absorption, to the extent that NaCl is equivalent to a black body in the $0 - \nu_{max}$ frequency interval. Assume moreover that at frequencies larger than ν_{max} the solid neither absorbs nor emits electromagnetic radiation. Estimate (with less than 10% error) the total radiated power emitted by a NaCl crystal with a 1 cm^2 surface kept at temperature 2000 K.
[Recall the density of states of the electromagnetic fields in a cavity of volume V: $g(\epsilon) = V\epsilon^2/(\pi^2 \hbar^3 c^3)$.]

Chapter 5
Solids

Macroscopic systems realize a thermodynamic equilibrium state in a balance between the tendency of internal energy to decrease and that of entropy to increase. Temperature tunes the balance in favor of the entropic contribution over the energetic one: entropy prevails at high temperature, energy dominates at low temperature. In the solid state, where each atom sits most of the time around a definite position, entropy is usually smaller than in fluid states. Indeed, experience shows that almost all materials solidify when temperature is lowered sufficiently.

The internal energy results from a kinetic plus a potential term. Like entropy, the translationally-invariant kinetic energy T_n tends to favor states characterized by delocalized and uncorrelated positions of individual atoms, On the contrary, the adiabatic potential energy V_{ad} takes advantage of characteristic optimal interatomic spacings (see e.g. Fig. 3.1) and angles, and attempts therefore to impose strong positional localization and correlations. The solid state signals the prevalence of the adiabatic potential energy over the translational kinetic energy in Eq. (3.9). For almost all materials at zero temperature, V_{ad} prevails over T_n, thus leading to solid states. The one remarkable exception is helium, which at ordinary pressure remains fluid (a *quantum fluid*) down to zero temperature, due to its exceptionally weak interatomic attraction associated to a scarce atomic polarizability (see Table 3.1). If atoms are squeezed together by an applied pressure, the He-He repulsion described by V_{ad} starts to prevail over T_n, and even helium manifests low-temperature solid phases. These phases are *quantum solids* because zero-point motion, tunneling, and all sorts of quantum-kinetic effects play a significant role.

The tendency of atoms to stick together and form molecules, driven by the forces which lead to a minimization of V_{ad}, is measured by a "molecular binding energy". For solids, the analogous quantity is the *cohesive energy*, or lattice energy, namely the energy necessary to disaggregate the solid into its *atomic* components. This quantity is often expressed per atom, or per formula unit, or per mole. Like for molecules, in the atomized state it is convenient to assume that $V_{ad} = 0$; and clearly $[T_n] = 0$ in the atomized state at $T = 0$. The solid state, a bound state, has $[V_{ad}] < 0$ and

© The Editor(s) (if applicable) and The Author(s), under exclusive license
to Springer Nature Switzerland AG 2020
N. Manini, *Introduction to the Physics of Matter*, Undergraduate Lecture Notes
in Physics, https://doi.org/10.1007/978-3-030-57243-3_5

$[T_n] > 0$. In ordinary solids, the kinetic term $[T_n]$ is much smaller than $|[V_{ad}]|$, while in quantum solids the two terms are comparable, see Fig. 3.15.

Solid matter is characterized by long-distance rigidity: a force applied to one or a few of the atoms in a solid sample acts through the whole sample, which accelerates maintaining its average shape unchanged. This is due to the ability of solids (as opposed to fluids) to resist shear forces. This macroscopic property, on which our daily experience relies, is far from trivial from the point of view of the microscopic equations governing the dynamics of electrons and nuclei composing a solid. Indeed, our analysis of the adiabatic potential acting directly between two neutral atoms at large distance reveals that V_{ad} usually decays with a rather fast power law R_{12}^{-6}. As the number of atoms at distance R_{12} from a given atom grows as R_{12}^2, the total interaction energy with faraway atoms decays as R_{12}^{-4}. In practice, each atom interacts significantly only with the atoms sitting in its close neighborhood, of a few nm^3 say. Long-distance rigidity therefore is unrelated to long-range forces. This means that, in solids, short-range forces propagate from one atom to the next ones, and from those to farther atoms again and again, across the whole sample.

5.1   The Microscopic Structure of Solids

Many solids exhibit ordered microscopical structures, but highly disordered solids are very frequent as well. Before coming to the experimental evidence for the ordered structure of many solids, we try to understand why regular spatial arrangements of atoms should emerge spontaneously.

In our initial study of many-atoms system (Chap. 3), we analyzed the typical shape of the adiabatic potential (Fig. 3.1) for a diatom. We also characterized the adiabatic potential of many atoms as an explicit function of the relative positions of all of them, including all distances and angles. For exceptionally simple systems, such as the noble-gas elements, the total adiabatic potential of N_n atoms is fairly well approximated by a sum of 2-body terms (e.g. of the Lennard-Jones type)

$$V_{ad}\left(\mathbf{R}_1, \mathbf{R}_2, \ldots \mathbf{R}_{N_n}\right) = \sum_{\alpha < \alpha'}^{N_n} V_2(|\mathbf{R}_\alpha - \mathbf{R}_{\alpha'}|). \tag{5.1}$$

Let us neglect the nuclear kinetic energy: the minimum-energy state of two such atoms is realized by placing them at the equilibrium distance R_M of the potential V_2, with an adiabatic energy equal to the depth $-\varepsilon$ of the potential well.[1] To start, assume that all equal atoms are constrained to move along a line: a third added atom can join the two others on either side, at approximately a distance R_M from the nearest atom, as illustrated in Fig. 5.1. Neglecting the weak attraction of second- third- etc. -neighbors (see Fig. 5.1b), the adiabatic potential energy decreases to approximately -2ε. Likewise, N_n atoms along a chain place themselves at almost perfectly regular

---

[1] For the Lennard-Jones potential—Eq. (3.19)—$R_M = \sqrt[6]{2}\,\sigma$, and $V_2(R_M) = V_{LJ}(R_M) = -\varepsilon$.

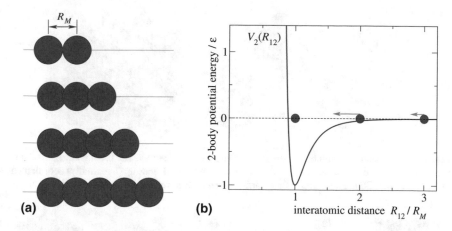

Fig. 5.1 **a** A 1D solid constructed by the successive addition of atoms. After the second one, atoms add at fairly regular distances, slightly smaller than the optimal equilibrium distance R_M of the 2-body adiabatic potential V_2. **b** The reason of the slightly smaller equilibrium separation in the solid than in the diatom: if the separation was R_M, forces acting on second, third… neighbors would all be attractive and uncompensated, thus some energy is gained by contracting the interatomic separation. This contraction is tiny, because V_2 explodes at short distance. It is even tinier at the surface than in the bulk, due to local scarcity of second and farther neighbors

distances,[2] and gain a total cohesive energy $\simeq (N_n - 1)\,\varepsilon$, i.e. essentially ε per atom. Depending on the precise shape of V_2, the actual cohesive energy turns out slightly larger due to the attraction of second and farther neighbors.

In 1D, regularity is trivial and indeed unavoidable for a single species of atoms. In larger dimensions, the greater geometric freedom allows several competing regular and irregular atomic patterns to form. In 2D, a third atom can join a dimer, forming an equilateral triangle; further atoms joining this cluster find a lowest-energy arrangement by progressively extending the triangular pattern, called *hexagonal* or *triangular lattice*, as illustrated in Fig. 5.2. Each atom (except the few ones at the edges) is surrounded by 6 nearest neighbors: it thus forms 6 bonds, each shared by two atoms. Accordingly, the cohesive energy per atom is approximately 3ε. The attraction of farther neighbors adds a small correction to this cohesive energy. The important message here is that two-body interactions favor configurations with a *maximal coordination*, i.e. arrangements where each atom is surrounded by as many nearest neighbors as geometry allows.[3]

---

[2]Slight distortions occur at the surface. Deep inside the bulk, however, each atom is subject to basically the same interactions as its neighboring ones, thus it reaches an equilibrium position relative to the surrounding ones which involves perfectly regular spacings.

[3]By comparison, if the atoms were arranged as a square lattice, each atom would bind to 4 nearest neighbors, rather than 6. The resulting cohesive energy per atom would amount to $\simeq 2\varepsilon$ instead of $\simeq 3\varepsilon$: the macroscopic cohesive energy difference $\Delta U = U^{\text{square}} - U^{\text{triang}} \simeq N_n\,\varepsilon$ makes the square lattice strongly unstable, and ready to deform spontaneously to the triangular lattice shape.

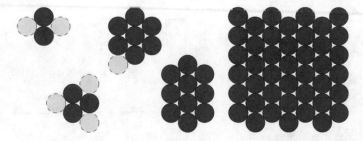

Fig. 5.2   A 2D solid constructed by the successive addition of atoms. The basic pattern, the equilateral triangle, tends to repeat itself, so that each atom in the bulk ends up (maximally) coordinated
to 6 other atoms. Contraction due to 2nd-, 3rd-,...-neighbor attraction is neglected in this figure

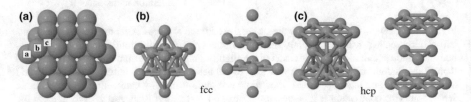

Fig. 5.3   **a** A close packing of spheres (e.g. cannonballs) in 3D is obtained by stacking 2D triangular
arrays. If a third layer is superposed to the "b" layer at sites of type "c", the fcc stacking is initiated.
If instead a third layer is placed vertically above the spheres of layer "a", the hcp crystal is realized.
b The conventional cubic cell of the fcc lattice, built as an abcabc... stacking sequence. **c** The
conventional hexagonal cell of the hcp crystal, built as an ababab... stacking

This principle of maximal coordination affects 3D structures, too.[4] In 3D, 4 atoms
maximize their coordination by placing themselves at the vertexes of a regular tetrahedron. Extra atoms extend this basic unit in space building either one of the following
regular patterns: the *face-centered cubic* (fcc) lattice or the *hexagonal close-packed*
(hcp) structure. As illustrated in Fig. 5.3, both lattices are the result of a regular
stacking of 2D triangular lattices. In both hcp and fcc lattices, the second layer is
stacked above the first one so that the atoms sit on top of the centers of one half of
the triangles of the lower layer. Likewise, the third layer is stacked on top of one
half of the centers of the second-layer triangles: in the hcp, it goes directly above
the atoms of the first layer, while in the fcc above the remaining triangles. In both
patterns, each atom is surrounded by 12 nearest neighbors, thus the cohesive energy
is approximately 6ε per atom. When the attraction of farther neighbors is accounted
for, the cohesive energy is usually marginally more favorable to fcc than to hcp.

At low temperature the noble gases are indeed observed to crystallize in fcc structures. The optimal equilibrium distance (accounting for interactions with all neighbors, not just the nearest ones) for the Lennard-Jones fcc solid equals $0.971\,R_M =$

---

[4]3D structures are visualized better in 3D than with flat projections. The atomic coordinates depicted
in Fig. 5.3 and other 3D structures described below are available for download and visualization at
the web page of Ref. [25].

Table 5.1 The experimental equilibrium nearest-neighbor interatomic distance d and cohesive energy per atom of solid noble-gas fcc crystals [10], compared to the estimations provided by the Lennard-Jones model with the parameters of Table 3.1; discrepancies in the order of few percent are not surprising, especially for the light Ne, in view of the neglect of the kinetic energy of the nuclei; note the correlation between the cohesive energy per atom and the melting temperature T_{melt} of the solid ($|U|/N_n \simeq 12\,k_B T_{melt}$)

Element	d [pm]	$1.09\,\sigma$ [pm]	$|U|/N_n$ [meV]	$8.6\,\varepsilon$ [meV]	T_{melt} [K]
	Exp	Theory	Exp	Theory	Exp
Ne	313	300	20	27	24.6
Ar	375	371	80	90	83.8
Kr	399	401	110	124	115.8
Xe	433	443	170	167	161.4

$1.09\,\sigma$, with a total cohesive energy per atom of $8.6\,\varepsilon$, significantly more strongly bound than the $6\,\varepsilon$ nearest-neighbor estimate. By plugging the parameters of Table 3.1 in this simple model, one obtains the bond lengths and cohesive energies of Table 5.1. Expectedly, the experimental lattice cohesive energy is generally slightly smaller than the prediction of the simple Lennard-Jones model, which neglects the ionic kinetic energy T_n and the associated zero-point motion. The good overall agreement confirms the concept of maximizing the coordination, leading to compact structures, usually fcc, similar to the packing of hard spheres, a concept valid whenever the adiabatic potential can be decomposed into 2-body terms, Eq. (5.1).

The 2-body adiabatic-potential model applies with fair accuracy to other solids, in addition to the noble gases (and mixtures thereof): it can describe the overall structure of many solids formed by "spherically symmetric" close-shell molecules, e.g. methane CH_4 [35]. With suitable adaptations, similar 2-body models can describe other *molecular solids* (e.g. H_2, N_2, Cl_2) where again weak van der Waals interactions provide cohesion, but the asymmetry of the individual molecules can favor different lattice structures.

Like the Ar_2 dimer is a rather exotic example of a molecule, the hitherto discussed molecular solids, where each molecular unit retains many of its molecular properties with only weakly bonds to other units, constitute a marginal, atypical class of solids. In more common *covalent* or *metallic solids*, at variance with molecular solids, electrons of the outer atomic shells modify substantially their quantum state, like the electronic states of H N and O get modified in forming proper covalent molecules such as H_2, N_2, H_2O, as discussed in Sect. 3.2. A covalent or metallic solid can indeed be viewed as a huge molecule, with the electronic states responsible for the molecular bonding extending to the whole solid. Experimentally, cohesive energies per atom $|U|/N_n$ of solids are comparable to the bond energies of covalently bound diatomic molecules, i.e. in the order of several eV. These are of course much larger than those of noble-gas solids (Table 5.1); for example we report the cohesive energies per atom of a few elemental solids: Li 1.65 eV, C (diamond) 7.36 eV, Si 4.64 eV, Fe 4.29 eV, Cs 0.83 eV. *Ionic solids* (e.g. NaCl) exhibit similar cohesive energies, of several eV per ion pair.

To estimate quantitatively the cohesive energy of a covalent or metallic solid, it is necessary to study the dynamics of its electrons in detail. Like for many-electron molecules, in practice, reliable estimates can be computed by means of detailed self-consistent methods. Before returning to the electronic states of solids in Sect. 5.2, observe that the adiabatic potential associated to such nontrivial electronic states is expected to depend strongly on all bond lengths and angles, pretty much as it does in molecules where it enforces relatively rigid equilibrium molecular geometries (Fig. 3.8). Therefore, in covalent and metallic solids, the 2-body approximation Eq. (5.1) is bound to fail completely: therefore different structures, other than fcc, are to be expected, depending on the detailed chemistry and thus on the relevant V_{ad} involved. Indeed, different elements exhibit ordered crystalline structures including, beside fcc: hcp, *body-centered cubic* (bcc), *diamond*, and others. Long-range crystalline order reaches spectacular levels of perfection, for example in industrial-grade Si single crystals, where a sub-nanometer unit cell repeats itself over and over in three dimensions for distances exceeding 1 m (Fig. 5.4). We shall soon discuss these structures within the standard formalism of periodic crystals, based on the infinite periodic repetition in space of a small piece of matter.

A perfect infinitely repeated lattice is an idealization which is never exactly realized in nature. In a real material, the crystalline structure is doomed to contain *defects*. A localized defect, such as a vacancy or an interstitial atom (Fig. 5.5) raises the total energy by an amount ε, in the order of a few times the typical bond energy of an

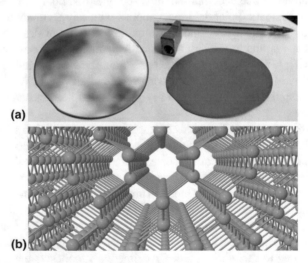

(a)

(b)

Fig. 5.4   a Two 0.3 mm thick wafers sliced from ∼1 m long silicon boules, i.e. single-crystal ingots. The diameters of these wafers are 100 mm (left) and 50 mm (right)—larger ones are available commercially. The surface of the wafer at the left is polished, thus mirror-reflecting, while at the right the unpolished side is shown. (The wafers were kindly provided by A. Podestà and R. Manenti.) **b** A balls-and-sticks view "from the inside" of the silicon crystal lattice structure down the (110) direction. Each atom binds four other nearest-neighbor atoms only. Atoms of diamond (C) and germanium (Ge) follow this same geometric arrangement

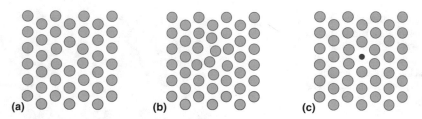

Fig. 5.5 Point defects in an otherwise perfect triangular lattice: **a** a vacancy, **b** an interstitial atom, and **c** an impurity atom

atom to its neighbors. This energy is irrelevant for the total cohesive energy per atom U/N_n, in the large-N_n limit. At equilibrium at low temperature, Boltzmann statistics predicts a small relative concentration (in the order of $\sim e^{-\beta\varepsilon}$) of localized defects to survive. The modest difference in energy between the fcc and hcp lattices makes the generation even of extended defects (e.g. dislocations—Fig. 5.6a, or stacking faults—Fig. 5.7) possible and relatively likely during crystal formation. The energy cost of extended defects is macroscopically large and should therefore suppress them strongly at equilibrium. However defects remain easily "frozen" within the solid, the typical time for a defect to drift out of a macroscopically large sample being often astronomically large. As a result, at low temperature a solid often remains locked in a metastable non-ergodic non-equilibrium state with a finite (often large) concentration of defects depending on preparation. Extended defects of the type of Figs. 5.6, 5.7 and 5.8 are crucial for understanding the plastic deformations of real crystals under strain. Ultimately, even without internal defects, real solids are never ideal because the lattice periodicity must end at the inevitable terminating surface to vacuum, or interface with another material, or grain boundary, see Fig. 5.8.

In many materials (e.g. multiple-component off-stoichiometric compounds, glasses, polymers, alloys…), the cost of the formation of defects is so small that it is highly nontrivial (often impossible) to obtain crystalline samples. These materials form *amorphous solids*, with a microscopic structure often resembling that of a frozen liquid, with no long-range periodic order, see Fig. 5.9. The formalism that we are going to set up for crystals is of little help for amorphous solids: their investigation involves more advanced tools than appropriate for present textbook.

Fig. 5.6 a A dislocation in a simple-cubic lattice. **b** The motion of dislocations decreases substantially the shear resistance of a real crystal relative to a defect-free crystal

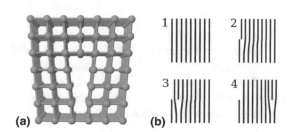

Fig. 5.7 Side view of a
stacking fault in a fcc crystal.
The defective interface
involves an irregular relative
stacking of two successive
(111) hexagonal planes, see
Fig. 5.3

stacking
fault

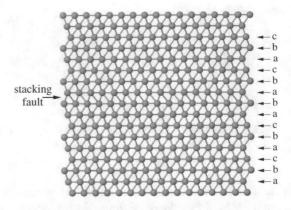

← c
← b
← a
← c
← b
← a
← b
← a
← c
← b
← a
← c
← b
← a

Fig. 5.8 A grain boundary
in a fcc solid, viewed down
the (100) direction

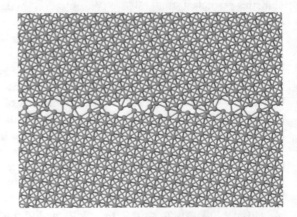

Fig. 5.9 Simulated
atomic-resolution
microscopy image of an
amorphous zirconium (Zr)
alloy. Contrasted to, e.g.,
Fig. 1.1, atoms are arranged
randomly, due to the
noncrystalline glassy
structure of this material

4 nm

The tendency to grow and to break along flat planes (*cleave*) at fixed characteristic
relative angles strongly hints at crystalline order in many solids, e.g. gemstones, see
Fig. 5.10. An even more compelling evidence is provided by the diffraction of X-rays,
neutrons and electrons with wavelengths in the a_0 region. As Fig. 5.11 illustrates, several wave-like probes with wavelength in the correct range interact with crystals and

Fig. 5.10 A few naturally grown minerals, resulting in aggregates of imperfect crystals with typical ~few-mm size. **a** Quartz (SiO_2). **b** Fluorite (CaF_2). Corundum Al_2O_3: **c** a Ti-doped (sapphire) sample from Sri Lanka [Rob Lavinsky, iRocks.com—CC-BY-SA-3.0], and **d** a Cr-doped (ruby) sample from Tanzania [Rob Lavinsky, iRocks.com—CC-BY-SA-3.0]

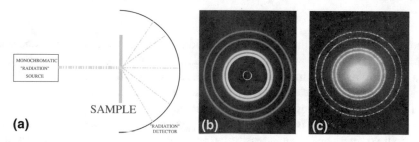

Fig. 5.11 **a** The scheme of a powder or polycrystalline sample diffraction experiment. The radiation beam can consist of any wave field interacting with matter, typically X-rays, neutrons, or electrons. The sample must be thin compared to the attenuation length of that radiation in that material. The diffraction patterns made by a beam of **b** X rays and of **c** electrons passing through the same thin Al foil. The angles characterizing the diffraction rings result from the ratio of the lattice periodicity to the wavelength of the incident radiation

generate diffracted beams, as expected of regular, periodic, arrays of scatterers.[5] We proceed to introduce the basic mathematics describing periodic solids, with the central concepts of direct and reciprocal lattice, and employ the related Fourier analysis to address diffraction experiments.

5.1.1 Lattices and Crystal Structures

The basic property of a crystalline solid is the practical equivalence of many different regions in space. Any Ne atom in its fcc crystalline position "sees" an essentially equivalent surrounding environment, unless it lies close to the crystal surface or to some defect. In a sufficiently perfect crystal, most atoms are far enough, say at least

---

[5]Occasional compounds have a quasicrystalline structure: they are perfectly ordered *non-periodic* materials, which also generate regular diffraction patterns.

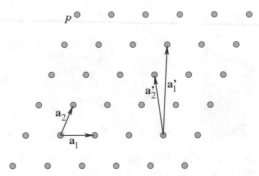

Fig. 5.12   A finite portion of a generic 2D lattice generated by the primitive vectors $\mathbf{a}_1$ and $\mathbf{a}_2$. Other equally good primitive vectors $\mathbf{a}_1'$ and $\mathbf{a}_2'$ are indicated, based on a different origin. Any lattice point $\mathbf{R}$, can be expressed as $n_1\mathbf{a}_1 + n_2\mathbf{a}_2$, for example $P = -1\,\mathbf{a}_1 + 4\,\mathbf{a}_2 = -4\,\mathbf{a}_1' + 8\,\mathbf{a}_2'$

5 neighbors away, from the nearest imperfection. The fields experienced by one such "bulk" atom practically equal those the same atom would experience if it belonged in a perfectly periodic structure. It thus makes sense to address many properties of crystalline solids by modeling them as ideal crystals extending through all space.

The specific feature of a crystal is its *discrete translational symmetry*.[6] Given a point $\mathbf{r}$ in the crystal, perfectly equal physical properties (including electric potential, electric field, mass and charge density, current, etc.) are observed at all other points

$$\mathbf{r}' = \mathbf{r} + \mathbf{R}, \qquad \text{with } \mathbf{R} = n_1\mathbf{a}_1 + n_2\mathbf{a}_2 + n_3\mathbf{a}_3, \qquad (5.2)$$

with n_j arbitrary integers, and $\mathbf{a}_j$ three linearly independent vectors. All translations of the type $\mathbf{R}$ in Eq. (5.2) form an infinite array extending through space, named a *Bravais lattice*. The vectors $\mathbf{a}_j$ are said to *generate the lattice*. They are called *primitive* if, for any $\mathbf{r}$, the points $\mathbf{r}'$ defined by Eq. (5.2) are *all* the points which have equal physical properties as $\mathbf{r}$ (i.e. $\mathbf{a}_j$ are taken "as short as possible", to make the array of $\mathbf{R}$ points as dense as possible).

As a simplest example, Fig. 5.12 illustrates these ideas for a 2D lattice. Figure 5.13 shows a portion of a simple-cubic crystal, viewed from different angles. Primitive vectors can be chosen as $a\,\hat{\mathbf{x}}$, $a\,\hat{\mathbf{y}}$, $a\,\hat{\mathbf{z}}$, i.e. orthogonal and all with the same length a, the side of the smallest cube in the lattice. The drawn portion of the crystal includes 125 points generated by $5 \times 5 \times 5$ consecutive values of the n_j indexes. Polonium

---

[6]In an isolated atom, the rotational symmetry commuting with the effective single-electron Hamiltonian makes the angular momenta l_i of individual electrons good quantum numbers, used to label atomic states such as, e.g., $1s^2 2s^2 2p^4$. Sections 5.1.1 and 5.1.2 address the similar problem to determine and diagonalize the symmetry operators of electrons in a crystal, in order to provide electrons with appropriate quantum numbers.

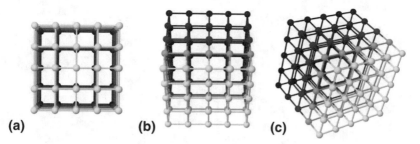

Fig. 5.13 A portion ($5 \times 5 \times 5$ lattice spacings) of a simple-cubic lattice observed from **a** the $\hat{\mathbf{z}}$ direction (001), **b** the $\hat{\mathbf{y}} + \hat{\mathbf{z}}$ direction (011), and **c** the $\hat{\mathbf{x}} + \hat{\mathbf{y}} + \hat{\mathbf{z}}$ direction (111)

represents the only material whose equilibrium structure exhibits one atom at each simple-cubic lattice point.

Other examples of 3D Bravais lattice—already encountered above—are the fcc and bcc lattices. Beside the noble gases, many elements have fcc equilibrium structures, e.g. aluminum, nickel, copper, silver, gold, lead. Others are bcc crystals, e.g. lithium, vanadium, chromium, iron, molybdenum, tungsten. These lattices are built by adding sites to the simple-cubic lattice: in the fcc at the center of each cube face, see Fig. 5.14; in the bcc at the geometric center of each cube, see Fig. 5.15. These added sites are perfectly equivalent to the original sites at the cube corners. The lattice-point density of fcc is four times and that of bcc is twice that of simple cubic of the same cube side a. For fcc and bcc, the same cubic-lattice vectors $a\,\hat{\mathbf{x}}, a\,\hat{\mathbf{y}}, a\,\hat{\mathbf{z}}$ as for the simple-cubic lattice are often conveniently used. However, these are not primitive: Fig. 5.16 depicts a standard choice of primitive lattice vectors.

The primitive vectors can be used as three converging edges which define a parallelepiped with volume $V_c = (\mathbf{a}_1 \times \mathbf{a}_2) \cdot \mathbf{a}_3$. This parallelepiped contains all "different", translationally-inequivalent points $\mathbf{r}$ in space: any $\mathbf{r}'$ lying outside this parallelepiped is equivalent to some other $\mathbf{r}$ inside, to which it can be translated by using Eq. (5.2) with suitable n_j. This parallelepiped is an example of a *primitive cell*, or *unit cell*: a minimal volume which, by applying lattice translations **R**, fills

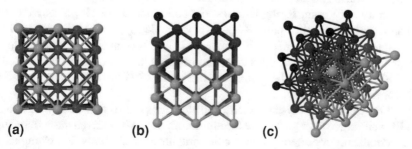

Fig. 5.14 A portion ($2 \times 2 \times 2$ cubic lattice spacings) of a face-centered cubic (fcc) lattice observed from **a** the $\hat{\mathbf{z}}$ direction (001), **b** the $\hat{\mathbf{y}} + \hat{\mathbf{z}}$ direction (011), and **c** the $\hat{\mathbf{x}} + \hat{\mathbf{y}} + \hat{\mathbf{z}}$ direction (111)

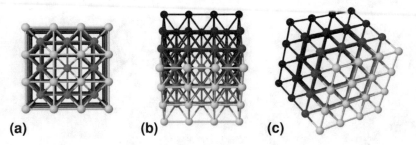

Fig. 5.15 A portion ($3 \times 3 \times 3$ cubic lattice spacings) of a body-centered cubic (bcc) lattice observed from **a** the $\hat{z}$ direction (001), **b** the $\hat{y} + \hat{z}$ direction (011), and **c** the $\hat{x} + \hat{y} + \hat{z}$ direction (111)

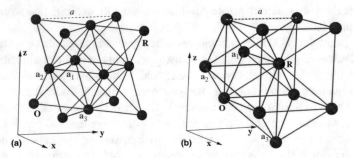

Fig. 5.16 **a** Primitive lattice vectors for the fcc Bravais lattice: $\mathbf{a}_1 = \frac{a}{2} (\hat{y} + \hat{z})$, $\mathbf{a}_2 = \frac{a}{2} (\hat{z} + \hat{x})$, $\mathbf{a}_3 = \frac{a}{2} (\hat{x} + \hat{y})$. **b** Primitive lattice vectors for the bcc lattice: $\mathbf{a}_1 = \frac{a}{2} (\hat{y} + \hat{z} - \hat{x})$, $\mathbf{a}_2 = \frac{a}{2} (\hat{z} + \hat{x} - \hat{y})$, $\mathbf{a}_3 = \frac{a}{2} (\hat{x} + \hat{y} - \hat{z})$. As an example, in both panels the translation marked by $\mathbf{R}$ can be expressed as $\mathbf{R} = \mathbf{a}_1 + \mathbf{a}_2 + \mathbf{a}_3$

up the whole space without overlapping. The primitive cell contains all the relevant information about the entire periodic crystal: what happens outside the primitive cell amounts to essentially boring repetitions of what occurs inside it. According to its definition, a primitive cell needs not be a parallelepiped (in 2D, a parallelogram—see Fig. 5.17). Note that each unit cell contains exactly one Bravais-lattice point. Figure 5.18 illustrates the unit cells of the fcc and bcc lattices.

Given a lattice point $\mathbf{R}$, the *Wigner-Seitz cell* is the set of all points $\mathbf{r}$ closer to $\mathbf{R}$ than to any other lattice point $\mathbf{R}'$ (see for example Fig. 5.19). One can show that the Wigner-Seitz cell is indeed a primitive cell. This special choice of primitive cell retains the full symmetry of the lattice, and resolves the arbitrariness (illustrated by Fig. 5.17) in the choice of the primitive cell. Figure 5.20 illustrates the Wigner-Seitz cells for two common lattices.

In many elementary crystals exactly one atom happens to sit in each primitive cell of a Bravais lattice as, e.g., in solid fcc aluminum or bcc iron. Even more frequently, each periodically repeated primitive cell contains several atoms. For example, the nuclei of solid NaCl (and many similar compounds such as LiCl, NaBr, KI, AgF,

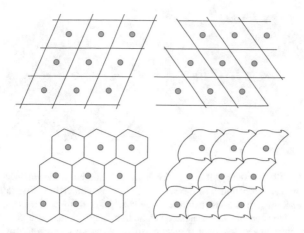

Fig. 5.17 Different primitive cells of the same 2D lattice. Depending on the choice of the primitive vectors, equivalent primitive cells can be parallelograms with different shape, but can take other polygonal or non-polygonal shapes. Contrasted to this "tile"-shape freedom, the lattice symmetry fixes the area of the primitive cell

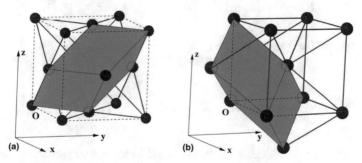

Fig. 5.18 Primitive cells of **a** the fcc and **b** bcc lattices. The parallelepipeds are obtained as $x_1\mathbf{a}_1 + x_2\mathbf{a}_2 + x_3\mathbf{a}_3$, with $0 \leq x_i < 1$, and the $\mathbf{a}_i$ drawn in Fig. 5.16. The primitive cells have the following volumes: fcc $\rightarrow a^3/4$; bcc $\rightarrow a^3/2$

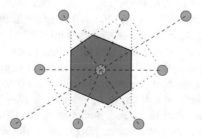

Fig. 5.19 The geometric construction of the Wigner-Seitz primitive cell of a generic 2D lattice. The Wigner-Seitz cells of 2D lattices are hexagons, except when the lattice is rectangular or square

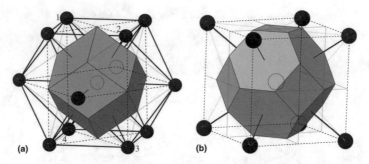

Fig. 5.20 **a** The Wigner-Seitz cell of the fcc lattice, a rhombic dodecahedron. A face of the conventional cubic cell with side a is the square defined by points 1–4. The lattice site around which the Wigner-Seitz cell is constructed (*dashed circle*) sits at the center of this square. **b** The Wigner-Seitz cell of the bcc lattice, a truncated octahedron. Each regular hexagon bisects a segment joining the cube center to a vertex. Each truncation square bisects a segment joining the cube center to the center of one of the 6 adjacent cubes (not drawn). The corner of each truncation square sits midway between the center and one edge of the corresponding cube face

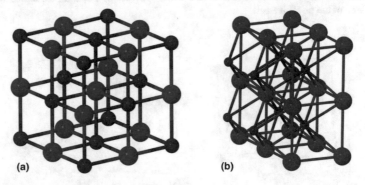

Fig. 5.21 **a** The sodium-chloride structure. The ions of each kind form two interpenetrating fcc lattices. **b** The cesium-chloride structure. The ions of each kind form interpenetrating simple-cubic lattices. Small ball = cation; large ball = anion

CaO, BaSe, …) sit around simple-cubic lattice points, alternately as in Fig. 5.21a, with each Na surrounded by 6 neighboring Cl, and *vice versa*. Similarly, the nuclei of CsCl (and similar compounds CsBr, TlCl, …) sit at the lattice sites of a bcc lattice, alternately as in Fig. 5.21b, with each Cs at the center of a cube with 8 neighbor Cl at its vertexes, and *vice versa*. To describe such structures, it suffices to recognize the true periodicity of the lattice. For example the NaCl structure can be viewed as a fcc Bravais lattice containing two atoms per cell: a Na atom at $\mathbf{0}$ and a Cl atom at the center of the fcc primitive cell $(\mathbf{a}_1 + \mathbf{a}_2 + \mathbf{a}_3)/2 = (\hat{\mathbf{x}} + \hat{\mathbf{y}} + \hat{\mathbf{z}})\, a/2$. This leads us to the necessity of introducing a *basis*, namely a list of atoms with their positions within a primitive cell. A *Bravais lattice plus a basis* define completely a *crystal structure*.

A basis is sometimes needed even when all atoms in the crystal are chemically equal. For example, Fig. 5.22 illustrates the 2D honeycomb net (all sites hosting "equal" atoms), which is not a simple Bravais lattice, because two neighboring atoms are *geometrically inequivalent*.

The honeycomb net is the 2D structure of *graphene*, an atomically thin layer of the *graphite* form of carbon. The structure of graphite, shown in Fig. 5.23, is a "vertical" alternating stack of graphene sheets. The graphite structure is then a hexagonal Bravais lattice with a basis of 4 atoms per primitive cell. The hcp structure displayed

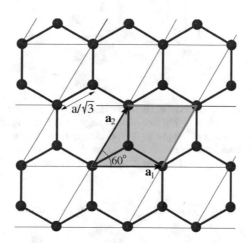

Fig. 5.22 The honeycomb structure is a 2D triangular net to which one third of the points has been removed. It is not a simple Bravais lattice, since the geometric environment of two neighboring atoms is inequivalent. The honeycomb net involves a lattice defined by two primitive vectors $\mathbf{a}_1$ and $\mathbf{a}_2$ of equal length a, separated by a 60° angle, plus a basis consisting of $\mathbf{d}_1 = 0$ and $\mathbf{d}_2 = (\mathbf{a}_1 + \mathbf{a}_2)/3$. Graphene is carbon in a honeycomb structure with $a \simeq 245$ pm

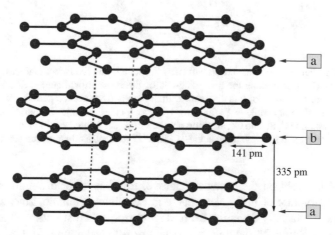

Fig. 5.23 The structure of the graphite form of carbon: an alternating stack of 2D honeycomb nets (graphene sheets, see Fig. 5.22). This crystal structure is a 3D hexagonal lattice with a basis of 4 atoms per primitive cell: 2 atoms in sheet "a" plus 2 atoms in sheet "b"

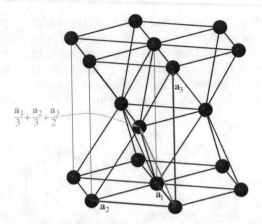

Fig. 5.24 The hcp structure consists of an alternate stacking of 2D hexagonal lattices, see Fig. 5.3. A parallelepiped primitive cell is highlighted. The hexagonal-lattice primitive vectors $\mathbf{a}_1$ and $\mathbf{a}_2$ have equal length a and form a 60° angle; $\mathbf{a}_3$, perpendicular to the $(\mathbf{a}_1, \mathbf{a}_2)$ plane, has a different length c. For the ideal close-packed structure $c = a\sqrt{8/3}$. A basis consisting of 2 atoms, e.g. one at $\mathbf{d}_1 = 0$ and one at $\mathbf{d}_2 = 1/3\,(\mathbf{a}_1 + \mathbf{a}_2) + 1/2\,\mathbf{a}_3$ generates the two geometrically inequivalent planes

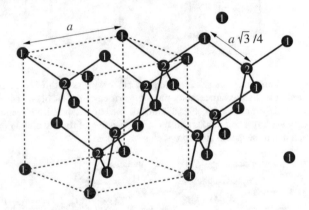

Fig. 5.25 The diamond structure consists of two interpenetrating fcc Bravais lattices, labeled "1" and "2", mutually shifted along the body diagonal of the conventional cubic cell (*dashed lines*) by one quarter of the length of this diagonal. It then results from the fcc periodic repetition of a two-atoms basis: $\mathbf{d}_1 = 0$ and $\mathbf{d}_2 = 1/4\,(\mathbf{a}_1 + \mathbf{a}_2 + \mathbf{a}_3) = 1/4\,(\hat{\mathbf{x}} + \hat{\mathbf{y}} + \hat{\mathbf{z}})a$. In fourfold-coordinated elementary crystalline C, Si, and Ge, all atoms are the same. The related zincblende (ZnS) structure has chemically different species sitting at the "1" and "2" geometrically nonequivalent sites

in Fig. 5.24 is based on the hexagonal lattice, too, but it has a 2-atom basis. The equilibrium structure of many elements, e.g. beryllium, titanium, zinc, zirconium, is hcp, although usually deviating from ideal close-packed structure, characterized by $c = a\sqrt{8/3}$. Similarly, the diamond structure (of C-diamond, Si, Ge, and α-Sn) drawn in Fig. 5.25 is a fcc lattice with a 2-atom basis. Each atom binds to four nearest neighbors placed at the vertexes of a regular tetrahedron, see also Fig. 5.4.

Given the possibility that several atoms belong to each cell, one often finds it more convenient to adopt a *conventional cell* containing several *equal* atoms, rather than the primitive unit cell of the Bravais lattice. For example, fcc and bcc are often conveniently described in terms of the bigger nonprimitive simple-cubic cell with the same a. The fcc lattice is then viewed as a simple-cubic Bravais lattice with a 4-atom basis $[\mathbf{0}, (\hat{\mathbf{y}} + \hat{\mathbf{z}})\,a/2, (\hat{\mathbf{z}} + \hat{\mathbf{x}})\,a/2$, and $(\hat{\mathbf{x}} + \hat{\mathbf{y}})\,a/2]$. The bcc lattice is viewed as a simple-cubic lattice with a 2-points basis $[\mathbf{0}$ and $(\hat{\mathbf{x}} + \hat{\mathbf{y}} + \hat{\mathbf{z}})\,a/2]$. When this conventional "lattice with basis" formalism is adopted in place of the natural description in terms of the appropriate Bravais lattice, all results should coincide.

In this Section we introduced the Bravais lattice as an infinite array of discrete geometric points. In fact, the precise mathematical meaning of a Bravais lattice is a *group of translations*, which transform the corresponding array of points back into itself.[7] We leave the details of the classification of space and point groups of 3D crystals[8] to specific solid-state courses, and only note that extra symmetries often induce extra degeneracies of the electronic and vibrational states of the crystal.

---

[7]Any vector $\mathbf{R}$ can be also viewed as a translation operator $T_{\mathbf{R}}$ such that $T_{\mathbf{R}}\,\mathbf{r} = \mathbf{R} + \mathbf{r}$, for any point $\mathbf{r}$. It is easily verified that the set of all lattice translations $\{T_{\mathbf{R}}\}$ of a Bravais lattice is closed for composition ($T_{\mathbf{R}}\,T_{\mathbf{R}'} = T_{\mathbf{R}+\mathbf{R}'}$), it contains a neutral element (T_0), and for each $T_{\mathbf{R}}$ there exists an inverse element ($T_{-\mathbf{R}}$) such that the composition of $T_{\mathbf{R}}$ with its inverse yields the neutral element. This means that all lattice translations $\{T_{\mathbf{R}}\}$ form a *group of geometric transformations*. As $\mathbf{R} + \mathbf{R}' = \mathbf{R}' + \mathbf{R}$, this group is *Abelian*, i.e. commutative. When the Hamiltonian has the entire symmetry of this group of discrete translations, its energy eigenstates are simultaneous eigenstates of all group operations, i.e. they are labeled by the group *irreducible representations*, see Sect. C.4. The math of the following Sect. 5.1.2 is useful to identify these irreducible representations: we shall see that their structure is very general and not especially complicated.

[8]In addition to the discrete translations, most lattices and crystal structures have *extra symmetries*. For example, a simple-cubic lattice transforms back into itself when rotated around a lattice point by 90° around the $\hat{\mathbf{x}}$ axis, or by 120° around a body diagonal direction such as $\hat{\mathbf{x}} + \hat{\mathbf{y}} + \hat{\mathbf{z}}$. These extra transformations extend the group of discrete translations $\{T_{\mathbf{R}}\}$ outlined above. The full symmetry group of the crystal (*space group*) is a proper combination of the *point group* (a finite group of rigid rotations and reflections about one point) and the lattice group of discrete translations $\{T_{\mathbf{R}}\}$. Back in the first half of the 19th century it was recognized that only *7 inequivalent point groups* can occur for 3D crystals. Two groups are equivalent if they contain the same type of symmetry operations (e.g., rotations by a certain angle, discrete translations up to suitable scaling factors). In particular, there do not exist point groups including fivefold axes (rotations by $2\pi/5 = 72°$), or sevenfold or higher-order axes, as they cannot replicate infinitely in space.

These 7 point groups combine differently with the translations to form *14 inequivalent Bravais lattices*. The introduction of a basis into the Bravais lattices can reduce the symmetry of the replicated objects in the primitive cell, and thus the overall space group. When combined with all possible symmetries of the basis, the different point groups become 32, rather than 7, and the space groups 230, rather than 14.

5.1.2   The Reciprocal Lattice

The *Fourier transform of a periodic function involves only discrete "frequencies"*:
this key statement is the reason for the introduction of the reciprocal lattice. We start
illustrating this concept in 1D. By definition, a periodic function $f(x)$ with period
a (the lattice spacing) satisfies $f(x) = f(x - na) = f(x - R)$ [$R = na$ is a lattice
translation, like in Eq. (5.2)]. Then, in the Fourier expansion

$$f(x) = F^{-1}[\tilde{f}](x) = \frac{1}{\sqrt{2\pi}} \int\limits_{-\infty}^{\infty} e^{ikx} \tilde{f}(k)\, dk, \tag{5.3}$$

all Fourier components $\tilde{f}(k)$ vanish except those whose $\exp(ikx)$ has the same periodicity as $f(x)$. The corresponding k must satisfy $\exp(ikx) = \exp[ik(x - a)]$, i.e.
$\exp(-ika) = 1$, i.e. $ka = 2\pi l$, i.e. $k = l \cdot 2\pi/a$, for any integer $l = 0, \pm 1, \pm 2, \ldots$.
We indicate those special k-values compatible with the lattice periodicity with the
notation

$$G = G(l) = l \cdot \frac{2\pi}{a}. \tag{5.4}$$

At any value of $k \neq G$ which does not respect the lattice periodicity the Fourier
component $\tilde{f}(k)$ vanishes. Accordingly, the Fourier expansion (5.3) can be written
as a discrete Fourier series

$$f(x) = \sum_{G} e^{iGx} \tilde{f}(G), \tag{5.5}$$

with coefficients

$$\tilde{f}(G) = \frac{1}{a} \int\limits_{0}^{a} \exp(-iGx) f(x) dx. \tag{5.6}$$

When periodic functions with period a are represented in Fourier space, the G points
acquire therefore a special role among all k's. According to Eq. (5.4), in k space
the G points form a regular lattice, with unit vector $2\pi/a$. Apart from the physical
dimension of inverse length rather than length, the k space is similar to the x space,
thus the lattice of G points holds all the properties of a Bravais lattice on its own: the
lattice of G points of Eq. (5.4) is called *reciprocal lattice*. By definition, *direct-lattice*
points R and reciprocal-lattice points G satisfy

$$e^{iRG} = e^{i\,na\,l2\pi/a} = e^{i\,2\pi\,nl} = 1. \tag{5.7}$$

The simple 1D example discussed above can be generalized to real-life 3D crys-
tals. A function $f(\mathbf{r})$ has the periodicity of a Bravais lattice if $f(\mathbf{r}) = f(\mathbf{r} - \mathbf{R})$
for any lattice vector $\mathbf{R} = n_1 \mathbf{a}_1 + n_2 \mathbf{a}_2 + n_3 \mathbf{a}_3$, with integer n_j as in Eq. (5.2).
Then, the only nonzero components in its Fourier expansion satisfy $\exp(i\mathbf{G} \cdot \mathbf{r}) =$

$\exp[i\mathbf{G} \cdot (\mathbf{r} - \mathbf{R})]$ for all $\mathbf{R}$ in the direct lattice. This relation is satisfied for all $\mathbf{G}$ such that

$$e^{i\mathbf{R} \cdot \mathbf{G}} = 1. \qquad (5.8)$$

In words, *the vectors* $\mathbf{G}$ *of the reciprocal lattice are all the* $\mathbf{k}$ *vectors whose associated plane wave has the periodicity of the direct Bravais lattice*. In particular, one finds that the $\mathbf{G}$ vectors are all the vectors of the type

$$\mathbf{G} = \mathbf{G}(l_1, l_2, l_3) = l_1\mathbf{b}_1 + l_2\mathbf{b}_2 + l_3\mathbf{b}_3 \quad (l_i = 0, \pm1, \pm2, \dots), \qquad (5.9)$$

with

$$\mathbf{b}_1 = \frac{2\pi}{V_c} \mathbf{a}_2 \times \mathbf{a}_3, \qquad \mathbf{b}_2 = \frac{2\pi}{V_c} \mathbf{a}_3 \times \mathbf{a}_1, \qquad \mathbf{b}_3 = \frac{2\pi}{V_c} \mathbf{a}_1 \times \mathbf{a}_2, \qquad (5.10)$$

where $V_c = (\mathbf{a}_1 \times \mathbf{a}_2) \cdot \mathbf{a}_3$ is the volume of the unit cell in terms of the real-space primitive vectors $\mathbf{a}_j$. The $\mathbf{G}$ vectors form a Bravais lattice, like in 1D. $\{\mathbf{b}_i\}$, Eq. (5.10), are a set of primitive vectors. Note that as $\mathbf{b}_i \cdot \mathbf{a}_j = 2\pi\delta_{ij}$, the dot product in Eq. (5.8) $\mathbf{R} \cdot \mathbf{G} = 2\pi (n_1l_1 + n_2l_2 + n_3l_3)$. Any other vector $\mathbf{k} \neq \mathbf{G}$ is associated to a plane wave which does not respect the lattice periodicity, thus the corresponding Fourier component $\tilde{f}(\mathbf{k})$ vanishes. Like in 1D, the Fourier expansion of the periodic function is a discrete Fourier summation over the reciprocal lattice

$$f(\mathbf{r}) = \sum_{\mathbf{G}} e^{i\mathbf{G} \cdot \mathbf{r}} \tilde{f}(\mathbf{G}), \qquad (5.11)$$

with coefficients

$$\tilde{f}(\mathbf{G}) = \frac{1}{V_c} \int_{V_c} \exp(-i\mathbf{G} \cdot \mathbf{r}) f(\mathbf{r}) \, d^3r, \qquad (5.12)$$

where the integration is carried out over a single unit cell.

The volume of the reciprocal-lattice primitive cell is $\mathbf{b}_1 \times \mathbf{b}_2 \cdot \mathbf{b}_3 = (2\pi)^3/V_c$. The Wigner-Seitz cell of the reciprocal lattice is called *first Brillouin zone* (BZ).

By applying the transformations (5.10), it is easy to verify that the reciprocal lattice of a simple-cubic lattice with side a is another simple-cubic lattice with side $2\pi/a$. The fcc lattice with cube side a has as reciprocal lattice a bcc lattice with cube side $4\pi/a$. Conversely, the bcc lattice with side a has a fcc reciprocal lattice with cube side $4\pi/a$.[9] As a consequence, the first BZ of the fcc lattice has the shape of the bcc Wigner-Seitz cell (Fig. 5.20b), and that of the bcc lattice has the shape of the

---

[9]This is a consequence of the general fact that the reciprocal of the reciprocal lattice is the original lattice. This can be verified by applying twice the transformations (5.10), or even more simply by observing that the roles of $\mathbf{R}$ and $\mathbf{G}$ in Eq. (5.8) can be exchanged.

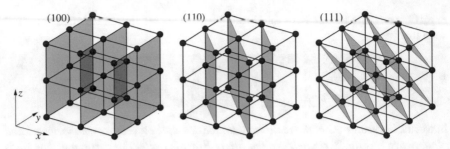

Fig. 5.26 A **G** vector identifies a well-defined family of parallel lattice planes, of constant phase **G** · **r**. of the plane wave $e^{i\mathbf{G}\cdot\mathbf{r}}$. These planes are perpendicular to **G** and separated by a distance $\frac{2\pi}{|\mathbf{G}|}$. The labels report the standard notation for the Miller indexes $(l_1\, l_2\, l_3)$ identifying families of lattice planes in cubic lattices: $(1\,0\,0)$, $(1\,1\,0)$, and $(1\,1\,1)$

fcc Wigner-Seitz cell (Fig. 5.20a). It is a useful exercise to determine the reciprocal lattice of the hexagonal lattice, starting from its standard primitive vectors, Fig. 5.24.

Each **G** vector in the reciprocal lattice defines a plane wave $e^{i\mathbf{G}\cdot\mathbf{r}}$ with the direct-lattice periodicity. Consider the constant-wave surfaces of a given plane wave, namely its "wave fronts", for example those fixed by $e^{i\mathbf{G}\cdot\mathbf{r}} = 1$: these surfaces form a family of parallel planes, perpendicular to **G** and separated by one wavelength $\lambda = 2\pi/|\mathbf{G}|$. Certain of these planes pass through some lattice points. In particular all these planes pass through lattice points if the integer indexes l_1, l_2, and l_3 have no common nontrivial divisors. Otherwise, for **G** given by nl_1, nl_2, nl_3, for integer $n > 1$, one plane out of n goes through lattice points. The integer indexes l_1, l_2, and l_3 are called *Miller indexes* of the family of lattice planes. These indexes are inversely proportional to the intercepts of these planes with the crystal primitive directions. The standard notation to indicate planes and directions in **k** space is $(l_1\, l_2\, l_3)$—no commas, with minuses conventionally represented by overbars; e.g. $(2-1\,0)$ is noted as $(2\,\bar{1}\,0)$. Figure 5.26 illustrates the relation of a few popular plane families with the real-space cubic cell. Traditionally, for all cubic lattices, including fcc and bcc, lattice planes are labeled relative to the conventional cubic directions $\hat{\mathbf{x}}$, $\hat{\mathbf{y}}$, $\hat{\mathbf{z}}$, not relative to the primitive vectors of Fig. 5.16.[10]

---

[10]The reciprocal lattice helps us to identify all different irreducible representations (see Sect. C.4) of the group of the discrete direct-lattice translations. These representations, labeled by an arbitrary **k** vector, are 1-dimensional and the corresponding character of a group operation $T_{\mathbf{R}}$ is simply $\exp(-i\mathbf{k}\cdot\mathbf{R})$. Two irreducible representation labeled by **k** and **k**′ have all equal characters (thus are the same representation) whenever $\mathbf{k} - \mathbf{k}' = \mathbf{G}$ for some **G** in the reciprocal lattice. Explicitly: $\exp(-i\mathbf{k}\cdot\mathbf{R}) = \exp(-i[\mathbf{k}' + \mathbf{G}]\cdot\mathbf{R}) = \exp(-i\mathbf{k}'\cdot\mathbf{R})\exp(-i\mathbf{G}\cdot\mathbf{R}) = \exp(-i\mathbf{k}'\cdot\mathbf{R})$, using Eq. (5.8). Accordingly, one primitive cell of the reciprocal lattice (e.g. the first BZ) contains the **k** points labeling all different irreducible representations of the discrete translational group.

5.1.3 Diffraction Experiments

Diffraction of "wave probes" is the main quantitative source of structural data about solids. The wavelength of the following wave-like objects match the typical unit-cell size of ordinary crystals, namely a few interatomic distances, $0.1-1$ nm: electrons with kinetic energy in the $1.5-150$ eV range; electromagnetic radiation in the $1-10$ keV range—X rays, see Fig. 1.2; and neutrons in the $1-100$ meV range. The periodic density modulation of crystals diffracts these three kinds of waves. Electrons interact very strongly with matter, and are thus sensitive to few topmost surface layers only. Avoiding energies resonant with any of the core excitations of the atoms in the sample (see Fig. 2.22), the penetration depth of X rays in solids easily exceeds thousands of unit-cell lengths, sufficient to generate sharp bulk diffraction patterns. Slow neutrons are even better fit for structural diffraction studies, as they interact weakly and mostly with the nuclei, boasting penetration depths exceeding the centimeter scale. A sufficiently small sample guarantees that the total probability of probe-sample interaction is small, so that most probing radiation goes unscattered through the sample, a small fraction scatters once, and almost none scatters twice or more. Under these conditions, understanding scattering is straightforward.

As sketched in Fig. 5.27a the incoming beam is assumed to be produced by a monochromatic source placed, for simplicity, at a large distance from the sample, so that it is characterized by a well-defined wave vector $\mathbf{k}$ (i.e. momentum $\hbar\mathbf{k}$). The detector is also very far from the sample, so that it detects outgoing radiation scattered elastically to another well-defined wave vector $\mathbf{k}'$. As analyzed quantitatively in the theory of scattering, every infinitesimal volume $d^3\mathbf{r}$ of a continuous distribution of matter scatters radiation in proportion to the number of scatterers $n(\mathbf{r})\, d^3\mathbf{r}$ present locally at $\mathbf{r}$. As sketched in Fig. 5.27b, the complex amplitude of the radiation scattered by this element of matter is proportional to $e^{-i\mathbf{k}'\cdot\mathbf{r}}\, e^{i\mathbf{k}\cdot\mathbf{r}}\, n(\mathbf{r})\, d^3\mathbf{r} = e^{-i\mathbf{q}\cdot\mathbf{r}}\, n(\mathbf{r})\, d^3\mathbf{r}$, introducing the *transferred wave vector* $\mathbf{q} = \mathbf{k}' - \mathbf{k}$. Summing these scattered amplitudes at the detector, the total probability

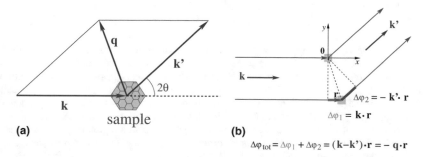

(a)   **(b)**

Fig. 5.27 **a** In an elastic-scattering experiment ($|\mathbf{k}'| = |\mathbf{k}|$), the transferred wave vector $\mathbf{q} = \mathbf{k}' - \mathbf{k}$ holds information about the scattering angle 2θ and radiation wavelength $\lambda = 2\pi/|\mathbf{k}|$. **b** The "far-field" wave scattered by a point-like element of matter at $\mathbf{r}$ is dephased by $-\mathbf{q} \cdot \mathbf{r}$ compared to the one at the origin $\mathbf{0}$, due to the extra path length $\hat{\mathbf{q}} \cdot \mathbf{r}$ emphasized by bold segments. The evaluation of the total interfering wave from a continuous distribution leads to a Fourier-transform "summation"

for the probing radiation to scatter $\mathbf{k} \to \mathbf{k}'$ is proportional to the square modulus of $\int e^{-i\mathbf{k}'\cdot\mathbf{r}} n(\mathbf{r}) e^{i\mathbf{k}\cdot\mathbf{r}} d^3\mathbf{r} = \int e^{-i(\mathbf{q})\cdot\mathbf{r}} n(\mathbf{r}) d^3\mathbf{r} \propto F[n](\mathbf{q}) = \tilde{n}(\mathbf{q})$. Here $\tilde{n}(\mathbf{q})$ indicates the 3D Fourier transform of the number density of scatterers $n(\mathbf{r})$. We conclude that the scattering rate of the $\mathbf{k}$-incoming wave probe into the $\mathbf{k}' = \mathbf{k} + \mathbf{q}$ direction is *proportional to the square modulus of the Fourier transform* $\tilde{n}(\mathbf{q})$ *of the matter density* $n(\mathbf{r})$, see Fig. 5.27. For neutrons the relevant density is the nuclear-matter density $n_{\text{nuc}}(\mathbf{r})$ while for X rays it is the density $n_{\text{el}}(\mathbf{r})$ of the electrons:

$$I_{\substack{\text{neutr} \\ \text{X-ray}}}(\mathbf{q}) \propto \left| F[n_{\substack{\text{nuc} \\ \text{el}}}](\mathbf{q}) \right|^2 \equiv \left| \tilde{n}_{\substack{\text{nuc} \\ \text{el}}}(\mathbf{q}) \right|^2 . \tag{5.13}$$

At atomic resolution, a nucleus is essentially point-like. Its density distribution resembles a Dirac δ. Accordingly, the Fourier transform of the density distribution of a nucleus is flat: one atomic nucleus scatters neutrons essentially independently of $\mathbf{q}$, see Fig. 5.28a. When *two* atoms diffuse neutrons, the scattered matter waves *interfere*: interference leads to intensity reinforcement and reduction in alternating directions. Constructive interference occurs whenever the two source-atom-detector path lengths (Fig. 5.27b) differ by an integer multiple of the radiation wavelength λ. Quantitatively, the square modulus of the Fourier transform $\tilde{n}_{\text{nuc}}(\mathbf{q})$ of two equal point-like objects at $\mathbf{R}_1$ and $\mathbf{R}_2$

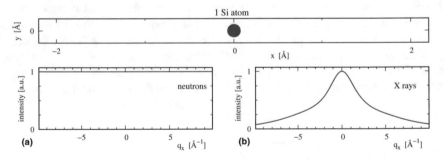

Fig. 5.28 Comparison of **a** neutron and **b** X-ray elastic scattering by a single Si atom. q_x represents the x-component of the change in wave vector $\mathbf{k}' - \mathbf{k}$ of the scattered radiation. Intensity is proportional to the square modulus of the Fourier transform of the appropriate density. Neutrons probe the nuclear-matter density, which is concentrated in lumps of typical size ~ 1 fm, practically similar to a Dirac $\delta(\mathbf{r})$ distribution: its Fourier transform is a constant, essentially independent of $\mathbf{q}$ (until a huge $|q| \gtrsim 10^4$ Å^{-1}, irrelevant for diffraction experiments of crystals). X rays probe the Fourier transform of the electronic charge density, which lumps in a region of ~ 1 Å size: as a result, the intensity, proportional to the atomic form factor, varies significantly over a q_x range of a few Å^{-1} (Generated using the software of Ref. [36])

$$|\tilde{n}_{2\,\text{nuclei}}(\mathbf{q})|^2 \propto |\exp(-i\mathbf{R}_1 \cdot \mathbf{q}) + \exp(-i\mathbf{R}_2 \cdot \mathbf{q})|^2$$

$$= \left|\exp\left(-i\,\frac{\mathbf{R}_1 + \mathbf{R}_2}{2} \cdot \mathbf{q}\right) 2\cos\left(\frac{\mathbf{R}_1 - \mathbf{R}_2}{2} \cdot \mathbf{q}\right)\right|^2$$

$$= 2\left[1 + \cos((\mathbf{R}_1 - \mathbf{R}_2) \cdot \mathbf{q})\right] = 2\left[1 + \cos(a\,q_x)\right] \quad (5.14)$$

shows characteristic oscillations (Fig. 5.29a, c). Assume that $\mathbf{R}_1 - \mathbf{R}_2 = a\hat{\mathbf{x}}$: constructive interference yields maximum scattered intensity in directions characterized by a projection q_x along the line joining the two nuclei such that $(\mathbf{R}_1 - \mathbf{R}_2) \cdot \mathbf{q} \equiv a\,q_x = 2\pi \times$ an integer l_1. The role of the interatomic separation a on the interference pattern of the scattered wave is illustrated by the comparison of Figs. 5.29a, c: an increased separation in real space produces closer interference maxima in $\mathbf{q}$ space. The scattered intensity is independent of the q_y, q_z components orthogonal to the line joining the two atoms (see also Fig. 5.32a for a 2D view).

Neutron scattering from a crystal arises from the interference of the waves scattered coherently by a regular array of point scatterers. For a 1D chain, the Fourier transform in Eq. (5.13) equals $\sum_{n_1} e^{-iq_x n_1 a}$, recalling the Fraunhofer theory of light

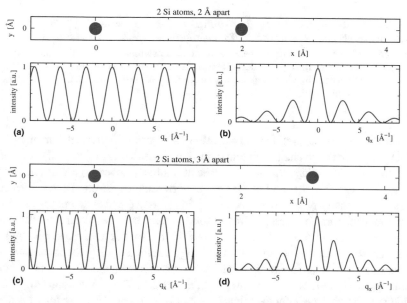

Fig. 5.29 **a, c** Intensity patterns of neutrons and **b, d** X-rays scattered by two Si atoms. The patterns show the characteristic interference periodicity $2\pi/a$, where $a = |\mathbf{R}_1 - \mathbf{R}_2|$ is the distance between the two nuclei, (**a, b**) $a = 2$ Å and (**c, d**) $a = 3$ Å. According to Eq. (5.18), the X-ray pattern is the product of the neutron pattern (uniquely determined by the atomic positions) times the atomic form factor, carrying information about the charge distribution of one atom. Note that, due to the enlarged interatomic separation a, the interference fringes in (**c, d**) are closer than in (**a, b**), while the enveloping atomic form factor is the same

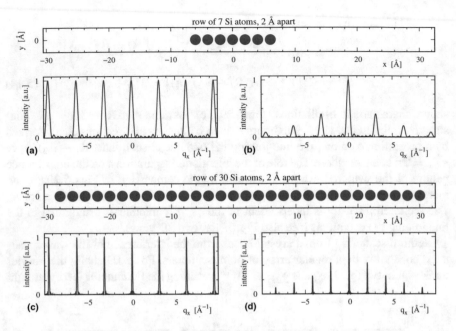

Fig. 5.30 Neutron (**a**, **c**) and X-ray (**b**, **d**) diffraction patterns generated by 7 (**a**, **b**) and by 30 (**c**, **d**) Si atoms rather than 2 as in Fig. 5.29. The main diffraction peaks at distance $\frac{2\pi}{2.0}$ Å^{-1} reflect the separation $a = 2.0$ Å. Their width decreases as the number of atoms increases. The relative intensity of the secondary intermediate peaks decreases quickly as the number of atoms is increased, and vanishes in the limit of an infinitely large crystal. In the X-ray pattern, the diffraction-peak intensities are modulated by the atomic form factor

diffracted by an optical grating with narrow slits. Figure 5.30 shows the intensity scattered by short periodic chains of atoms: despite the smallness of such 1D "crystals", sharp intense *diffraction structures* emerge at regularly-spaced **q** directions characterized by $e^{-iq_x n_1 a} = 1$ for all integer n_1. In between these strong Bragg-diffraction peaks, the intensity of the weak secondary interference peaks decreases rapidly as the number of atoms increases. Already for 30 atoms these weak structures become almost invisible (Fig. 5.30c, d), and disappear in the limit of a macroscopically large perfect crystal.[11] The values of q_x for the Bragg maxima of diffracted intensity, with all phases $q_x n_1 a =$ an integer multiple of 2π are precisely the 1D reciprocal lattice vectors $q_x = G_x = l_1 \times 2\pi/a$, with $l_1 = 0, \pm 1, \pm 2, \ldots$ The same occurs also for 2D and 3D periodicity: we conclude that Bragg *diffraction peaks occur for transferred wave vector equaling reciprocal-lattice points* **q** = **G**.

---

[11]The modest regularity sufficient to produce diffracted peaks grants the possibility to generate detectable diffraction patterns even in the event that the coherence length of the probing radiation available in the lab extends to several lattice cells spacings only, much shorter than the whole sample size.

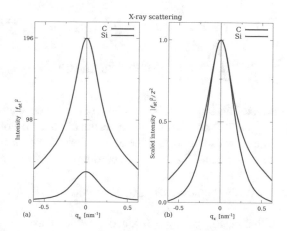

Fig. 5.31   The square moduli of the X-ray atomic form factors $|f_{at}(\mathbf{q})|^2$ of carbon ($Z = 6$) and of silicon ($Z = 14$). **a** Actual values, indicating that the total X-ray scattering increases $\propto Z^2$. **b** Intensities scaled to the $\mathbf{q} = 0$ value, showing that heavier atoms have broader form factors, due to the sharper charge localization of its inner-core electrons Spherical symmetry makes f_{at} independent of the direction of $\mathbf{q}$: $f_{at}(\mathbf{q}) = f_{at}(|\mathbf{q}|)$

The patterns of Figs. 5.29 and 5.30 show that the X-ray diffractograms differ from neutrons diffractograms by an amplitude modulation. This is easily understood in steps:

- X rays scattered by a single atom probe its smooth electronic density distribution $n_{at}(\mathbf{r})$. Following Eq. (5.13), one atom scatters X rays elastically with the nontrivial $\mathbf{q}$ dependence given by Fourier transform of its electronic density, called *atomic form factor* $f_{at}(\mathbf{q}) \equiv \tilde{n}_{at}(\mathbf{q})$, as drawn for Si in Fig. 5.28b. Figure 5.31 reports two examples of X-ray atomic form factors: these are characteristic bell-shaped functions whose value at $\mathbf{q} = 0$ equals the squared number of electrons.

- For atoms with $Z \gg 1$, core electrons scatter X rays more than the less numerous valence electrons. Core electrons retain an atomic-like charge distribution. Accordingly, it is a fair approximation to assume that the electronic distribution of a collection of atoms (as in a molecule, or in a solid) equals *the sum of the individual atomic electronic distributions*. For example an elementary crystal consisting of many equal atoms sitting one per cell at the points $\mathbf{R}$ of a Bravais lattice has

$$n_{el}(\mathbf{r}) = \sum_{\mathbf{R}} n_{at}(\mathbf{r} - \mathbf{R}). \tag{5.15}$$

As discussed in Sect. 3.2.1 and 3.2.2, chemical bonding does modify the electronic states of a molecule or a solid, so that the electronic charge density deviates from the sum of Eq. (5.15). However only the comparably few valence electrons deform significantly due to involvement in bonding.

- Substitute Eq. (5.15) in the calculation of the Fourier transform:

$$\tilde{n}_{el}(\mathbf{q}) = \sum_{\mathbf{R}} \int n_{at}(\mathbf{r} - \mathbf{R})e^{-i\mathbf{q}\cdot\mathbf{r}}d^3\mathbf{r} = \sum_{\mathbf{R}} \int n_{at}(\mathbf{r}')e^{-i\mathbf{q}\cdot(\mathbf{r}'+\mathbf{R})}d^3\mathbf{r}' \quad (5.16)$$

$$= \sum_{\mathbf{R}} e^{-i\mathbf{q}\cdot\mathbf{R}} \int n_{at}(\mathbf{r})e^{-i\mathbf{q}\cdot\mathbf{r}}d^3\mathbf{r} = \tilde{n}_{Bravais}(\mathbf{q})\, f_{at}(\mathbf{q}).$$

Observe that the resulting amplitude is the product of two independent functions of $\mathbf{q}$: $\tilde{n}_{Bravais}(\mathbf{q}) = \sum_{\mathbf{R}} e^{-i\mathbf{q}\cdot\mathbf{R}}$ describing the scattering of point-like objects arranged at the nodes of the Bravais lattice; and the atomic form factor accounting for the scattering from a single atom, Fig. 5.28b. Accordingly, the scattered intensity

$$I_{X\text{-ray}}(\mathbf{q}) \propto |\tilde{n}_{el}(\mathbf{q})|^2 = |\tilde{n}_{Bravais}(\mathbf{q})\, f_{at}(\mathbf{q})|^2 = |\tilde{n}_{Bravais}(\mathbf{q})|^2 \, |f_{at}(\mathbf{q})|^2. \quad (5.17)$$

- Finally, observe that $n_{nuc}(\mathbf{r}) \propto n_{Bravais}(\mathbf{r})$, and conclude that

$$I_{X\text{-ray}}(\mathbf{q}) \propto |f_{at}(\mathbf{q})|^2 \, |\tilde{n}_{nuc}(\mathbf{q})|^2 \propto |f_{at}(\mathbf{q})|^2 \, I_{neutr}(\mathbf{q}). \quad (5.18)$$

This relation clarifies the role of the atomic form factor $f_{at}(\mathbf{q})$ in X-ray diffraction from a monoatomic crystal: X rays scatter in the *same* $\mathbf{q}$ *directions as neutrons* of the same wavelength, but the peak *intensities are modulated multiplicatively by the squared atomic form factor*. This observation accounts for the compared diffractograms of Figs. 5.28, 5.29 and 5.30.

The 1D patterns of Figs. 5.28, 5.29 and 5.30 generalize to 2D and 3D. Figure 5.32 shows the 2D scattering patterns produced by 2, 3, 4, and 5 equal atoms located at the vertexes of regular polygons. The fact that the interference pattern of 3 and 4 atoms recalls a Bravais lattice, while that of the pentagonal arrangement is nonperiodic, suggests that this symmetry is incompatible with repetition in space. Figure 5.33 illustrates the typical effect of lattice geometric deformations on the diffraction pattern, as described mathematically by Eq. (5.10). Figures 5.34, 5.35, 5.36 and 5.37 display neutron diffraction patterns for square, rectangular, oblique, and triangular 2D crystals.

The 3D patterns follow similar rules: diffracted beams come out in the directions where the transferred $\mathbf{q}$ matches a $\mathbf{G}$-vector in the 3D reciprocal lattice. In practice, when shining a monochromatic neutron or X-ray beam on a single crystal, one generally obtains no diffracted beams, since, for that given $\mathbf{k}$, all possible $\mathbf{k}' = \mathbf{k} + \mathbf{G}$ have mismatching $|\mathbf{k}'| \neq |\mathbf{k}|$, and would then correspond to *inelastic* scattering. Diffracted beams are retrieved by carefully orienting the crystal until, for some $\mathbf{G}$, the incident and scattered wave numbers match:

$$|\mathbf{k}| = |\mathbf{k}'| = |\mathbf{k} + \mathbf{G}|. \quad (5.19)$$

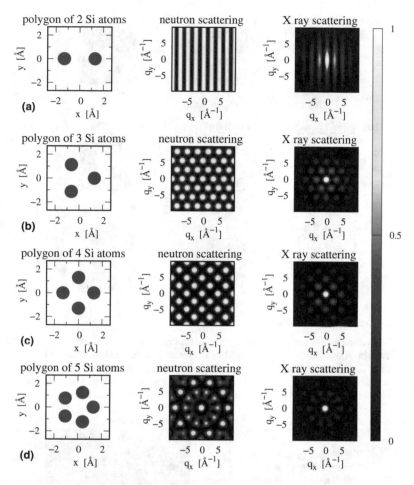

Fig. 5.32 Color-intensity 2D neutron and X-ray scattering patterns generated by regular polygons of 2 to 5 Si atoms. For 2 atoms (**a**), as indicated by Eq. (5.14), scattering is independent of the **q** component perpendicular to the line joining them. Note that the regular triangle (**b**) and square (**c**) generate Bravais lattices as "diffraction" patterns, while the pentagon (**d**) does not

This geometric condition is represented by the Ewald construction of Fig. 5.38a. The scattering angle, i.e. the angle between **k** and **k′** is related to |**k**| and **G** by squaring Eq. (5.19), obtaining

$$\mathbf{k} \cdot \hat{\mathbf{G}} = -\frac{|\mathbf{G}|}{2}. \tag{5.20}$$

Fig. 5.38b illustrates the geometric relation $\mathbf{k} \cdot \hat{\mathbf{G}} = -|\mathbf{k}| \sin \theta$ between θ, **k**, and **G**. According to Eq. (5.20) this projection must also equal $-1/2\,|\mathbf{G}|$. Thus, scattering is observed at angles 2θ from the **k** direction, related to |**k**| and |**G**| by

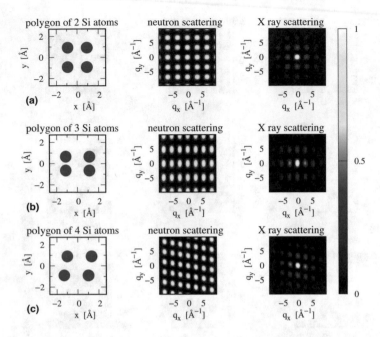

Fig. 5.33 Neutron and X-ray scattering patterns generated by 4 atoms. These patterns hint at the general features of diffraction from the square (**a**), rectangular (**b**), and oblique (**c**) nets. A large crystal, rather than 4 atoms, would produce proper sharp diffracted spots at the locations of these intensity maxima, namely the reciprocal-lattice points. Observe the inverse proportionality and orthogonality relation (5.10) of the reciprocal-lattice and direct-lattice vectors

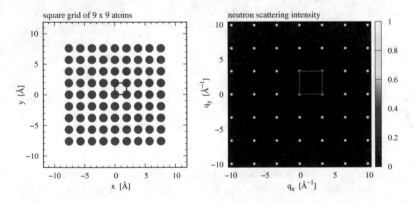

Fig. 5.34 The 2D neutron diffraction pattern (*right*) generated by a 9×9 square-lattice crystal (*left*). A red square highlights the real-space primitive cell; a blue square marks the reciprocal-space primitive cell. Note that the peaks are much sharper than in the 2×2 example of Fig. 5.33a

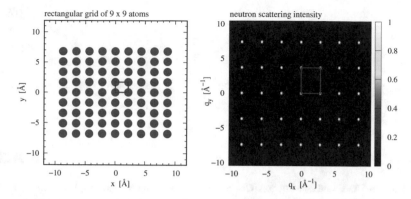

Fig. 5.35 The neutron diffraction pattern generated by a rectangular-lattice crystal

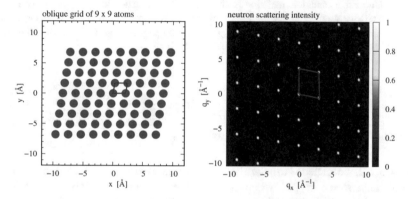

Fig. 5.36 The neutron diffraction pattern generated by an oblique-lattice crystal

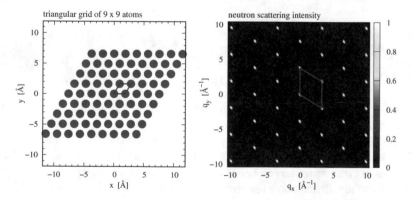

Fig. 5.37 The neutron diffraction pattern generated by a triangular-lattice crystal

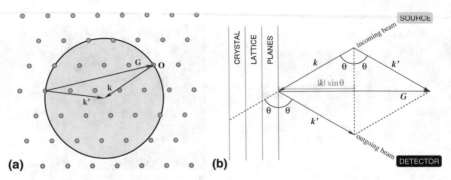

Fig. 5.38 **a** The Ewald construction. Given the incident vector **k**, draw a sphere with radius |**k**| about the point **k** in reciprocal space. Diffraction peaks corresponding to reciprocal lattice vectors **G** occur only if −**G** happens to lie on the surface of the Ewald sphere (*shaded*), as in the drawn example. Under this condition, radiation is diffracted to the direction **k**' = **k** + **G**. **b** The relation between the scattering angle 2θ and the lengths of **k** and **G** vectors. The angle θ between the incident beam and the family of Bragg planes fixed by **G** yields the projection $\mathbf{k} \cdot \hat{\mathbf{G}} = -|\mathbf{k}| \sin\theta$, which is compatible with elastic scattering when it equals $-1/2 |\mathbf{G}|$

$$\sin\theta = \frac{|\mathbf{G}|}{2|\mathbf{k}|}. \tag{5.21}$$

In Sect. 5.1.2, we related **G** to a family of lattice planes (drawn in Fig. 5.38b) separated by a distance $d = n2\pi/|\mathbf{G}|$, where n is the greatest common divisor of the Miller indexes defining **G**. On the other hand, |**k**| is connected to the radiation wavelength λ by $|\mathbf{k}| = 2\pi/\lambda$: substitution in Eq. (5.21) yields the celebrated *Bragg condition for diffraction*

$$2d\sin\theta = n\lambda. \tag{5.22}$$

According to these relations and to Fig. 5.38a, *no diffraction occurs for* $2|\mathbf{k}| < |\mathbf{G}|$, or equivalently for $\lambda > 2d$.

In practice, to generate diffracted beams off a single crystal, one must move the reciprocal lattice relative to the Ewald sphere until some **G** point touches the Ewald sphere, as in Fig. 5.38a. One can either vary the radiation wavelength, thus changing the Ewald-sphere diameter, or else rotate the crystal sample, because its reciprocal lattice rotates by the same amount, see Eq. (5.10).

In the lab, it is common to characterize the structure of *powder samples*, i.e. collections of microcrystals rotated randomly in space. This uniform distribution of orientations is equivalent to averaging over all possible rotations of the reciprocal

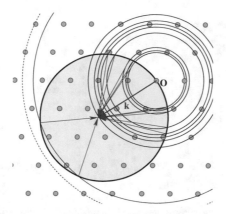

Fig. 5.39 The Ewald construction for a powder sample. The wave vector **k** of the incoming radiation defines the (*shaded*) Ewald 2D circle (3D sphere). Crystal grains are rotated in all possible ways: the entire reciprocal lattice (solid points) must then be "averaged" over all rotations: every **G** point generates a circle in 2D (a sphere in 3D). In 2D, each one of these circles intersects the fixed Ewald circle at a pair of points **k**′, both forming an angle 2θ with **k**. In 3D, each sphere intersects the Ewald sphere at a circle of points **k**′, forming the same angle 2θ with **k**: we conclude that a crystal powder scatters radiation to coaxial *cones* of $\hat{\mathbf{k}}'$ directions, axially symmetric around the $\hat{\mathbf{k}}$ direction. **G** vectors longer than $2|\mathbf{k}|$ generate circles/spheres such as the *dashed* one, which do not intersect the Ewald sphere, thus producing no diffraction

lattice. As illustrated by Fig. 5.39, the powder scatters radiation at fixed angles 2θ [given by Eq. (5.21)] away from the incident **k**. This means that diffracted radiation forms *cones* whose axis is the incident direction $\hat{\mathbf{k}}$. The rings of Fig. 5.40a illustrate an example of such a pattern recorded on film. Also Fig. 5.11 shows an example of such a pattern, generated by an Al foil (not a proper powder): the sharp cones of radiation diffracted at characteristic angles prove that Al is microcrystalline. Microcrystalline structures of this kind (tightly bound collections of randomly oriented microscopic individual crystals separated by grain boundaries) are responsible for the plastic deformable character of most solid metals, as opposed to the rigidity of those solids which crystallize in the form of macroscopic single crystals (e.g. Si, NaCl).

Structural data about powder or microcrystalline samples are conveniently collected in plots of the diffracted intensity as a function of the angle 2θ, as in the patterns of Fig. 5.40. In this kind of diffractograms, the vertical axis reports the total scattered intensity, integrated along circles at fixed 2θ.

For polyatomic crystals, with n_d atoms in each periodically repeated cell, the derivation of Eq. (5.17) is readily generalized by assuming that the electron density is $n_{el}(\mathbf{r}) = \sum_{\mathbf{R}} \sum_{j=1}^{n_d} n_{at\,j}(\mathbf{r} - \mathbf{d}_j - \mathbf{R})$. The result is that the atomic form factor in Eqs. (5.16–5.18) must be replaced by a *structure factor* $|S(\mathbf{q})|^2$, with

$$S(\mathbf{q}) = \sum_{j=1}^{n_d} e^{-i\mathbf{q}\cdot\mathbf{d}_j} f_{at\,j}(\mathbf{q}) \qquad (5.23)$$

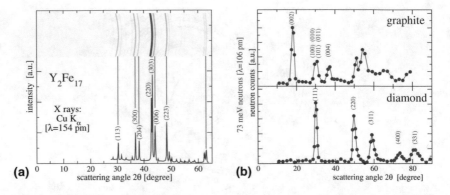

Fig. 5.40 Powder diffraction patterns: intensity as a function of the scattering angle 2θ. **a** Scattering of the copper K_α X-ray radiation by a Y_2Fe_{17} powder sample, with the intensity plot compared to the rings on the film recording. (Data from Ref. [37].) **b** Scattering of low-energy neutrons by powder samples of elementary carbon (Data from Ref. [38])

representing the Fourier transform of the matter distribution of the n_d atoms sitting at positions $\mathbf{d}_j$ in the repeated cell. Expression (5.23) holds for X rays, $f_{\text{at }j}(\mathbf{q})$ being the atomic form factors of the individual atoms in a cell. For neutrons, the same expression applies, provided that the $f_{\text{at }j}(\mathbf{q})$ are replaced by $\mathbf{q}$-independent neutron scattering amplitudes of the individual nuclei in the cell. The decomposition $I(\mathbf{q}) \propto |\tilde{n}_{\text{Bravais}}(\mathbf{q})|^2 |S(\mathbf{q})|^2$ implies a fundamental result: a given Bravais lattice yields the same characteristic diffraction pattern irrespective of the number, kinds and positions of the atoms populating its unit cell. These details affect the structure factor $|S(\mathbf{q})|^2$, which in turn applies a multiplicative intensity modulation to the *same* peaks.

The possibility of describing a same crystal structure in terms of different lattices with different bases might make us worry of some ambiguity. For example, a square lattice of side a could also be viewed as a square lattice with side $2a$, and $n_d = 4$ atoms per cell: the reciprocal-lattice $\mathbf{G}$ points of the $2a$ lattice are twice as dense in each direction, but the diffracted pattern must remain the same, because this is the same crystal described in a formally different way. Indeed, the structure factor (shown in Fig. 5.33a) vanishes for all $\mathbf{G}$ points of the denser reciprocal lattice which do not belong to the reciprocal lattice of the true a-side square. However, the $2a$ lattice may become the correct minimal description of the actual structure, e.g. as a consequence of a structural deformation—one or several atoms in each 2×2 square moving away from their perfect-square position. With such a deformation, some of the $2a$ denser $\mathbf{G}$ peaks acquire nonzero intensity. Likewise, a fcc crystal with cube side a, as a Bravais lattice, generates diffraction spots at the $\mathbf{q} = \mathbf{G}$ points forming a bcc reciprocal lattice of conventional side $4\pi/a$. However, the same fcc structure can be viewed as a simple-cubic lattice of side a with 4 atoms/cell. The reciprocal lattice of this simple-cubic lattice has denser $\mathbf{G}$ points (a simple-cubic lattice of side $2\pi/a$) than that of the fcc. However, the diffraction pattern is the same, regardless of the adopted formalism, because the structure factor computed following Eq. (5.23) for 4 equal atoms [same $f_{\text{at}\,j}(\mathbf{q})$] sitting at positions $\mathbf{0}$, $(\hat{\mathbf{x}} + \hat{\mathbf{y}})a/2$, $(\hat{\mathbf{x}} + \hat{\mathbf{z}})a/2$, and $(\hat{\mathbf{y}} + \hat{\mathbf{z}})a/2$ gets rid of peaks at $\mathbf{G}$ points of the simple-cubic lattice (side $2\pi/a$) not belonging to the actual bcc reciprocal lattice (side $4\pi/a$).

Defects and the finite crystal size displace some scattered intensity from the sharp Bragg peaks into a diffuse continuous background (see e.g. Fig. 5.30). If disorder increases, the intensity of this continuous background grows until, for amorphous or liquid samples, neutron or X-ray scattering does not exhibit the sharp Bragg peaks characteristic of lattice periodicity any more: the scattered intensity is then a smooth function of the angle 2θ, which provides useful information about the statistical structural properties of the sample, retrievable by numerical Fourier analysis.

5.2 Electrons in Crystals

Within the adiabatic framework (Sect. 3.1), electrons move in a solid according to the electronic equation (3.2). The total electronic energy obtained by solving Eq. (3.2), added to the inter-nuclear repulsion, yields the adiabatic potential (3.10) which, in turn, determines the dynamics of the nuclei through Eq. (3.9). For a crystalline solid at low enough temperature, V_{ad} keeps the atomic configuration close to its minimum, characterized by a regular arrangement of the nuclei, basically a large chunk of an ideal crystal of the kind described in Sect. 5.1. In the adiabatic spirit, we omit the nuclear kinetic energy, assuming that all ions sit at the ideal crystal-structure positions, and turn our attention to the motion of electrons. Within this idealized scheme, we investigate the solutions of the electronic equation (3.2), starting from their general properties resulting from symmetry. We defer the study of the ionic motions about their equilibrium configurations to Sect. 5.3.

The many-body equation (3.2) is plagued by the same difficulties discussed for atoms and molecules. The Schrödinger problem of many electrons in a crystal can usually be approached in a one-electron mean-field scheme, e.g. of the Hartree-Fock type, like that of atoms and molecules. As discussed in Sect. 2.2.5, this type of approximation maps the N-electron equation to a single-electron self-consistent equation for the motion of one electron in the field of the nuclei and the charge distribution of the $N - 1$ other electrons. The mean-field effective potential $V_{\text{eff}}(\mathbf{r})$ has the same symmetry as the potential created by the bare nuclei, i.e. the full crystal symmetry[12] (Fig. 5.41). Similarly to atoms, where the single-electron HF orbitals carry spherical-symmetry labels l and m, the single-electron states of crystals carry

---

[12]This assumption is better grounded than the approximation of a spherically-symmetric mean field in atoms. Indeed, in contrast with atoms where non-s states have anisotropic angular distributions, we shall shortly see that the density distribution (wavefunction square modulus) of crystal states is identical in all cells. Another example of symmetry-preserving solutions similar to electrons in crystals concerns the reflection across the mid-plane separating the nuclei of homoatomic dimers: the symmetric and antisymmetric wavefunctions of H_2^+, Eq. (3.11) and Fig. 3.3, have both a perfectly symmetric square modulus under this reflection.

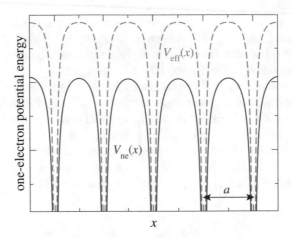

Fig. 5.41 The bare nuclear potential energy $V_{ne}(x)$ (*solid*) and the screened effective one-electron potential energy $V_{eff}(x)$ (*dashed*) experienced by an electron moving in a crystal along a line through a direct-lattice primitive direction. The screened potential energy is less attractive than that produced by the bare nuclei, but it exhibits the same lattice translational symmetry: $V_{eff}(\mathbf{r}) = V_{eff}(\mathbf{r} + \mathbf{R})$

Bravais-lattice group representations, labeled by $\mathbf{k}$ vectors chosen within a primitive cell of the reciprocal lattice, often the first BZ. Specifically, the $\mathbf{k}$ quantum number carries information about the way a wavefunction changes under the action of lattice translation $T_{\mathbf{R}}$, i.e. in moving from one lattice cell to the next. Like in atoms, additional non-symmetry-related quantum numbers identify states with different nodal structure (within each primitive cell) and spin projection.

The determination of the eigenfunction and eigenenergy for an electron moving according to $H_1 = [T_{e\,1} + V_{eff}(\mathbf{r})]$ requires a detailed calculation, which is usually carried out numerically. In this context, symmetry plays a twofold role:

- simplify greatly the solution of the HF-Schrödinger equation in the crystal;
- understand general features of its solutions.

Bloch's theorem expresses the consequences of the lattice symmetry on the crystal electronic eigenstates. This theorem states that in a lattice-periodic potential all Schrödinger eigenfunctions can be chosen in the factorized form

$$\psi_j(\mathbf{r}) = e^{i\mathbf{k}\cdot\mathbf{r}}\, u_{\mathbf{k}\,j}(\mathbf{r}), \tag{5.24}$$

where the function $u_{\mathbf{k}\,j}(\mathbf{r})$ has the same periodicity as the potential $[u_{\mathbf{k}\,j}(\mathbf{r}) = u_{\mathbf{k}\,j}(\mathbf{r} + \mathbf{R})]$, and $\mathbf{k}$ is an arbitrary wave vector. This fundamental result can be interpreted in two alternative, but equally instructive ways:

- In a periodic context all electronic eigenfunctions display a nontrivial spatial dependence within one primitive unit cell only: in any other cell displaced by $T_{\mathbf{R}}$, the wavefunction is equal to the one in the original cell, apart from a constant phase factor $e^{i\mathbf{k}\cdot\mathbf{R}}$ which leaves the probability distribution $|\psi_j|^2$ unaffected. This observation provides also a demonstration of Bloch's theorem.[13]
- In a periodic potential, the Schrödinger eigenstates are essentially free-electron plane-wave–like states, except for a periodic (thus trivial) amplitude modulation.

Observe that we can as well restrict $\mathbf{k}$ to one primitive cell of the reciprocal lattice, e.g. a primitive parallelepiped, or the first BZ[14]—see Fig. 5.17. Indeed, if $\mathbf{k}$ was outside this primitive cell, we could always find a reciprocal lattice vector $\mathbf{G}$ such that $\mathbf{k}' = \mathbf{k} + \mathbf{G}$ is in the primitive cell of our choice. But then $\psi_j(\mathbf{r}) = e^{i\mathbf{k}\cdot\mathbf{r}} u_{\mathbf{k}j}(\mathbf{r}) = e^{i(\mathbf{k}'-\mathbf{G})\cdot\mathbf{r}} u_{\mathbf{k}j}(\mathbf{r}) = e^{i\mathbf{k}'\cdot\mathbf{r}} e^{-i\mathbf{G}\cdot\mathbf{r}} u_{\mathbf{k}j}(\mathbf{r})$, and the function $u'_{\mathbf{k}'j}(\mathbf{r}) = e^{-i\mathbf{G}\cdot\mathbf{r}} u_{\mathbf{k}j}(\mathbf{r})$ is lattice-periodic, thus $e^{i\mathbf{k}'\cdot\mathbf{r}} u'_{\mathbf{k}'j}(\mathbf{r})$ is a valid Bloch function.

It is instructive to obtain the equation satisfied by the functions $u_{\mathbf{k}j}(\mathbf{r})$ of Eq. (5.24). To this purpose we substitute the decomposition (5.24) in the stationary Schrödinger equation for the electrons in the periodic potential:

$$\left[-\frac{\hbar^2}{2m_e}\nabla^2 + V_{\text{eff}}(\mathbf{r})\right] e^{i\mathbf{k}\cdot\mathbf{r}} u_{\mathbf{k}j}(\mathbf{r}) = \mathcal{E} e^{i\mathbf{k}\cdot\mathbf{r}} u_{\mathbf{k}j}(\mathbf{r}). \tag{5.25}$$

To deal with the kinetic term, observe that

$$\begin{aligned}
\nabla^2 e^{i\mathbf{k}\cdot\mathbf{r}} u_{\mathbf{k}j}(\mathbf{r}) &= \nabla \cdot \left[e^{i\mathbf{k}\cdot\mathbf{r}}\nabla u_{\mathbf{k}j}(\mathbf{r}) + i\mathbf{k}e^{i\mathbf{k}\cdot\mathbf{r}} u_{\mathbf{k}j}(\mathbf{r})\right] \\
&= e^{i\mathbf{k}\cdot\mathbf{r}}\nabla^2 u_{\mathbf{k}j}(\mathbf{r}) + 2e^{i\mathbf{k}\cdot\mathbf{r}} i\mathbf{k}\cdot\nabla u_{\mathbf{k}j}(\mathbf{r}) - |\mathbf{k}|^2 e^{i\mathbf{k}\cdot\mathbf{r}} u_{\mathbf{k}j}(\mathbf{r}) \\
&= e^{i\mathbf{k}\cdot\mathbf{r}}\left(\nabla + i\mathbf{k}\right)^2 u_{\mathbf{k}j}(\mathbf{r}).
\end{aligned}$$

By substituting this decomposition into the Schrödinger equation (5.25), and dividing by the common factor $e^{i\mathbf{k}\cdot\mathbf{r}}$, we obtain

$$\left[-\frac{\hbar^2}{2m_e}\left(\nabla + i\mathbf{k}\right)^2 + V_{\text{eff}}(\mathbf{r})\right] u_{\mathbf{k}j}(\mathbf{r}) = \mathcal{E}_{\mathbf{k}j}\, u_{\mathbf{k}j}(\mathbf{r}). \tag{5.26}$$

This fundamental equation allows us to compute the stationary states of every electron in a crystal, given its wave number $\mathbf{k}$. Thanks to the periodicity of $u_{\mathbf{k}j}$ established by

---

[13]The single-electron effective Hamiltonian $T_e + V_{\text{eff}}(\mathbf{r})$ commutes with all lattice translations $T_{\mathbf{R}}$. Accordingly, its eigenfunctions may be chosen as simultaneous eigenfunctions of all discrete translation operators. But then $T_{\mathbf{R}}\psi_j(\mathbf{r}) = e^{-i\mathbf{k}'\cdot\mathbf{R}}\psi_j(\mathbf{r})$ for some $\mathbf{k}'$ in the first BZ. Accordingly, one can call $\mathbf{k} = -\mathbf{k}'$ and define $u_{\mathbf{k}j}(\mathbf{r}) = \psi_j(\mathbf{r}) e^{i\mathbf{k}\cdot\mathbf{r}}$, which is periodic, as can be readily verified by the substitution of $\mathbf{r} - \mathbf{R}$ in place of $\mathbf{r}$.

[14]In 1D, any k-interval of size $2\pi/a$. For example the first BZ, i.e. the interval $-\pi/a < k \leq \pi/a$.

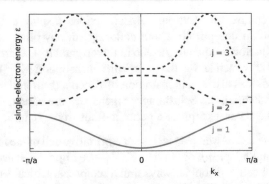

Fig. 5.42 For each given **k** in the first BZ, the solution of Eq. (5.26) produces discrete electronic energies $\mathcal{E}_{\mathbf{k}\,j}$, for $j = 1, 2, \ldots$, three of which are sketched here. These energies depend parametrically on **k**. The continuous functional dependence $\mathcal{E}_j(\mathbf{k}) = \mathcal{E}_{\mathbf{k}\,j}$ is an energy band of the crystal

Bloch's theorem, Eq. (5.26) must be solved within a *single cell* of the direct lattice (with applied periodic boundary conditions) rather than over the macroscopically large volume of the *whole crystal*. In practice, the need to solve Eq. (5.26) in a microscopically small unit cell represents a substantial technical advantage.[15] Once $u_{\mathbf{k}\,j}(\mathbf{r})$ is evaluated by exact or approximate techniques, the full electronic wavefunction $\psi_j(\mathbf{r})$ can then be extended to the entire crystal by means of Eq. (5.24).

Before attempting the solution of Eq. (5.26), we discuss the qualitative properties of the single-electron eigenenergies and eigenstates in a crystal. The second-order differential equation (5.26) must be solved in the unit-cell volume V_c, with standard periodic boundary conditions. Apart for the wave-vector shift by **k**, Eq. (5.26) is like a stationary Schrödinger equation. As such, for any fixed **k**, its solutions should be analogous to those of a standard Schrödinger equation (C.30) in a microscopically small volume, namely: a ladder of discrete eigenenergies $\mathcal{E}_{\mathbf{k}\,j}$ associated to eigenfunctions $u_{\mathbf{k}\,j}(\mathbf{r})$, characterized by an increasing number of nodal planes for increasing energy. The index $j = 1, 2, 3, \ldots$ labels precisely these eigensolutions, for increasing energy, see Fig. 5.42.

As **k** varies in the first BZ, the stationary-states equation (5.26) changes analytically, thus one should expect that its solutions depend smoothly on **k**. This dependence justifies attaching the label **k** to the resulting eigenenergies $\mathcal{E}_{\mathbf{k}\,j}$ and eigenfunctions $u_{\mathbf{k}\,j}(\mathbf{r})$. Indeed, for fixed j, the eigenenergies $\mathcal{E}_{\mathbf{k}\,j}$ depend on **k** as continuous functions called *energy bands*, or simply "bands" (Fig. 5.42). For each j, as **k** spans the first BZ, $\mathcal{E}_{\mathbf{k}\,j}$ spans a continuous interval of available energies (the range of the $\mathcal{E}_{\mathbf{k}\,j}$ function, sometimes itself called a "band"). The ranges of two successive bands $\mathcal{E}_{\mathbf{k}\,j}$ and $\mathcal{E}_{\mathbf{k}\,j+1}$ can either overlap (like bands 2 and 3 in Fig. 5.42) or not overlap (like bands 1 and 2 in Fig. 5.42). Both possibilities are compatible with Eq. (5.26), and do occur in actual solids. The main consequence of Bloch's theorem is then a

---

[15] No such simplification is possible in the absence of lattice periodicity. The motion of electrons, e.g., in non-periodic solids is therefore far harder to investigate.

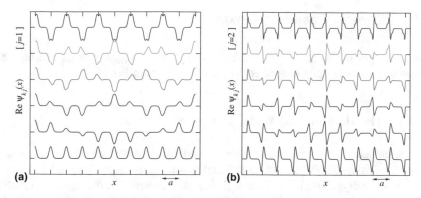

Fig. 5.43   A sketch of the real part of a few Bloch wavefunction $\psi_{\mathbf{k}\,j}(x) = e^{i\mathbf{k}\cdot\mathbf{r}}\,u_{\mathbf{k}\,j}(x)$ as a function of the x component of $\mathbf{r}$. The $\mathbf{k}$ components are $k_y = k_z = 0$, and (bottom to top): $k_x = 0$, $k_x = 0.1 \times 2\pi/a, k_x = 0.2 \times 2\pi/a, k_x = 0.3 \times 2\pi/a, k_x = 0.4 \times 2\pi/a, k_x = \pi/a$, for the states belonging to bands **a** $j = 1$ and **b** $j = 2$. As k_x increases from 0 to the first-BZ boundary, the phase difference of the wavefunction at neighboring cells increases. At the zone boundary $k_x = \pi/a$, this phase difference is maximum, and equals π, corresponding to the sign alternation of the topmost curves. Negative k_x values yield wavefunctions with the same real part as drawn here, but opposite $\mathrm{Im}\,\psi_{\mathbf{k}\,j}$ (not drawn)

spectrum of electronic energies involving intervals of allowed energies, often separated by ranges of forbidden energies called *band gaps*. The energy spectrum of electrons in a crystal is somewhat intermediate between that of a free particle (all positive energies) and that of an atom (isolated eigenvalues separated by gaps).

Figure 5.43 sketches the real part of the band eigenfunctions along a line in a monoatomic crystal. The imaginary part is similar, although with different phases, so that $|\psi_{\mathbf{k}\,j}(x)|^2$ is periodic, i.e. it repeats identically in all cells. The primary role of $\mathbf{k}$ is tuning the wavefunction phase change in moving from one cell to the next. The wavelength associated to these *Bloch waves* is $2\pi/|\mathbf{k}|$. For the selected values of $\mathbf{k}$, the waves of Fig. 5.43 are commensurate to the drawn 10-cells region. Other intermediate choices of $\mathbf{k}$ would produce waves not periodic in that region. In addition, $\mathbf{k}$ affects also the local "shape" of the wavefunction $\psi_{\mathbf{k}\,j}(x)$ within each cell, through the explicit dependence of Eq. (5.26), and thus of $u_{\mathbf{k}\,j}(x)$. This $\mathbf{k}$-dependence is smooth and often relatively mild, while one usually observes more significant wavefunction differences when comparing different bands.

The quantitative prediction of the band energies $\mathcal{E}_{\mathbf{k}\,j}$ and wavefunctions $u_{\mathbf{k}\,j}(x)$ of an actual solid is typically carried out by means of a self-consistent evaluation of the mean-field potential $V_{\mathrm{eff}}(\mathbf{r})$, associated to the numerical solution of Eq. (5.26) in a direct-lattice primitive unit cell. Like in Sect. 4.3.1, periodic boundary conditions are applied to a macroscopic box, chosen as a parallelepiped repetition of the unit cell, in order to preserve the lattice symmetry, and to generate a sufficiently dense sampling of the $\mathbf{k}$ points. One usually refrains from applying more realistic open-end boundary conditions, which would break the periodicity, and would involve the subtle properties of *crystal surfaces*, themselves an entire branch of science.

5.2.1 Models of Bands in Crystals

Useful insight in the physics of the band states and energies can be obtained by considering substantially simplified "model" solutions of Eq. (5.26).

5.2.1.1 The Tight-Binding Model

In a region near each atomic nucleus, the crystal effective potential (Fig. 5.41) acting on an electron resembles that of an isolated atom. Accordingly, for the band wavefunctions of a crystal it makes sense to follow the construction of molecular orbitals in terms of atomic ones, as sketched in Chap. 3 for H_2^+ and H_2. In this scheme, the Bloch functions are constructed as the natural generalization of bonding and antibonding molecular orbitals of a huge molecule. Models for extended electronic states expressed as *linear combinations of atomic orbitals* are often referred to as "tight binding".

The 2-atoms calculation of Sect. 3.2.1 is the simplest example of tight-binding model. It addresses the 1s "band" of a "crystal" consisting of $N_n = 2$ H atoms only. $|1s\,L\rangle$ represents the state at the left atom. $|1s\,R\rangle$, represents that at the right atom, obtained from $|1s\,L\rangle$ by a "lattice" translation. Fictitious periodic boundary conditions bring $|1s\,R\rangle$ back again to $|1s\,L\rangle$ upon a further translation (Fig. 5.44). The symmetry-adapted states of Eq. (3.11) can be viewed as bonding and antibonding combinations

$$|S\rangle = N_S(|1s\,L\rangle + |1s\,R\rangle) = N_S \sum_{p=0,1} |p\rangle = N_S \sum_{p=0}^{N_n-1} e^{ikpa}|p\rangle, \qquad k=0$$

$$|A\rangle = N_A(|1s\,L\rangle - |1s\,R\rangle) = N_A \sum_{p=0,1} (-1)^p |p\rangle = N_A \sum_{p=0}^{N_n-1} e^{ikpa}|p\rangle, \quad k=\frac{\pi}{a},$$

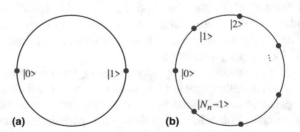

Fig. 5.44 **a** A diatomic molecule with its bond duplicated seen as a tiny "crystal" made of $N_n = 2$ equal atoms. **b** A 1D crystal of N_n equal atoms with periodic boundary conditions exhibits the connectivity of a ring

with $|0\rangle = |1s\,L\rangle$, $|1\rangle = |1s\,R\rangle$, and $\mathcal{N}_{S/A}$ are suitable normalization constants. The only symmetry-allowed phase relations in these two-site combinations are "in phase" ($|S\rangle$, or $k = 0$) and "out of phase" ($|A\rangle$, or $k = \pi/a$).

As suggested in Fig. 5.44, these linear combinations can be generalized to a chain of $N_n \geq 2$ atoms, forming a 1D "crystal". To construct kets compatible with the global "ring" periodic boundary condition $\langle r + N_n a|\phi\rangle = \langle r|\phi\rangle$, we adopt the phase factors $\exp(ikpa)$ generated by

$$k = \left(-\frac{1}{2} + \frac{1}{N_n}\right)\frac{2\pi}{a}, \; \left(-\frac{1}{2} + \frac{2}{N_n}\right)\frac{2\pi}{a}, \; \left(-\frac{1}{2} + \frac{3}{N_n}\right)\frac{2\pi}{a}, \; \dots, \; \frac{1}{2}\frac{2\pi}{a}, \tag{5.27}$$

the 1D version of Eq. (C.55) with $L = N_n a$. We construct N_n combinations of the 1s orbitals by using these phase factors in the linear combination:

$$|\phi_k\rangle = \mathcal{N}_k \sum_{p=0}^{N_n-1} e^{ikpa}|p\rangle. \tag{5.28}$$

In the limit of macroscopically large N_n, the energies of the $|\phi_k\rangle$ states form a continuum in the first BZ $-\pi/a < k \leq \pi/a$. As k spans the BZ, the band energy spans continuously the range from the bonding value (band bottom) to the antibonding value (band top), Fig. 5.45.

By extending the derivation of the molecular Eq. (3.13), we can work out a method to compute the bands for a general crystal, which also provides approximate analytic expressions for the bands, such as that of Fig. 5.45. The tight-binding method is based on expanding the crystal electronic eigenstates as linear combinations of eigenstates of the atomic one-electron effective Hamiltonian H^{at}:

$$|\mathcal{E}\rangle = \sum_{\mathbf{R}} \sum_{n} b_{\mathbf{R}n} T_{\mathbf{R}} |\phi_n\rangle. \tag{5.29}$$

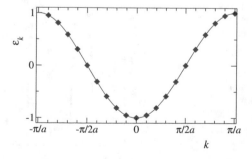

Fig. 5.45 As the N_n equally-spaced **k** points in the first BZ allowed by the entire-sample periodic boundary conditions become denser and denser in the $N_n \to \infty$ large-size limit, the electronic energies form the band continuum of a macroscopic crystal

In this expression, appropriate for a monoatomic crystal, $|\phi_n\rangle$ is the n-th single-electron eigenstate of the isolated atom, $T_\mathbf{R}$ is an operator which translates rigidly by a lattice translation $\mathbf{R}$, and $b_{\mathbf{R}n}$ are suitable coefficients. One usually selects a subset $n = 1, \ldots n_o$ of the atomic states, usually comprising a few valence orbitals only, and solves the one-electron Hamiltonian problem (C.36) in its variational matrix form, see Appendix C.5. The involved matrix elements of the 1-electron Hamiltonian are $H_{mn}(\mathbf{R}) = \langle \phi_m | H_1 T_\mathbf{R} | \phi_n \rangle$, and the overlap integrals are $B_{mn}(\mathbf{R}) = \langle \phi_m | T_\mathbf{R} | \phi_n \rangle$. The construction of the eigenstates and eigenvalues of Eq. (C.36) involves the (generalized) diagonalization of a square matrix whose size equals the number n_o of orbitals involved in each unit cell multiplied by the number of unit cells in the crystal. This prohibitively large size can be reduced to just n_o, by taking advantage of the discrete translational symmetry.

We decompose the one-electron Hamiltonian $H_1 = [T_{e1} + V_{\text{eff}}(\mathbf{r})]$ as

$$H_1 = H^{\text{at}} + \Delta U, \tag{5.30}$$

where H^{at} contains the kinetic energy T_{e1} plus the effective potential energy for an electron moving around one *isolated* ion placed at the origin $\mathbf{R} = \mathbf{0}$, and $\Delta U(\mathbf{r})$ is the total screened potential energy generated by all ions in the crystal, reduced by the single-ion contribution, see Fig. 5.46. In terms of the atomic levels E_m^{at} and matrix elements

$$C_{mn}(\mathbf{R}) = \langle \phi_m | \Delta U \, T_\mathbf{R} | \phi_n \rangle, \tag{5.31}$$

the tight-binding matrix secular problem (C.36) can be expressed as

$$\sum_{n\,\mathbf{R}} \left[E_m^{\text{at}} B_{mn}(\mathbf{R}) + C_{mn}(\mathbf{R}) \right] b_{\mathbf{R}n} = \mathcal{E} \sum_{n\,\mathbf{R}} B_{mn}(\mathbf{R}) \, b_{\mathbf{R}n}. \tag{5.32}$$

Fig. 5.46 The screened one-electron potential energy $V_{\text{eff}}(x)$ (*dashed*) experienced by an electron moving in the crystal, and the same potential to which the contribution of the ion at $\mathbf{R} = \mathbf{0}$ has been subtracted ($\Delta U(x)$, *solid*). The plot sketches a 1D cut along the x axis of a typical crystal

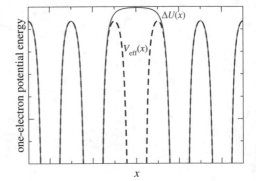

Inspired by Eqs. (5.24) and (5.28), we guess an appropriate $\mathbf{R}$-dependence of the coefficients $b_{\mathbf{R}n}$, to reconstruct the appropriate Bloch symmetry:

$$b_{\mathbf{R}n} = e^{i\mathbf{k}\cdot\mathbf{R}}\,\tilde{b}_n. \tag{5.33}$$

In practice, Eq. (5.33) imposes a specific amplitude and inter-cell phase relation to the wavefunction $\langle\mathbf{r}|\mathcal{E}\rangle$, as dictated by the selected $\mathbf{k}$. The wavefunction $\langle\mathbf{r}|\mathcal{E}\rangle$ satisfies Bloch's theorem, as is readily verified. Eigenstates are obtained by selecting appropriate combinations of the atomic orbitals, tuned by the coefficients $\tilde{b}_n$.

For given $\mathbf{k}$, we indicate by $|\mathcal{E}_{\mathbf{k}}\rangle$ the Bloch eigenstate that solves the Schrödinger equation

$$H_1\,|\mathcal{E}_{\mathbf{k}}\rangle = \mathcal{E}_{\mathbf{k}}\,|\mathcal{E}_{\mathbf{k}}\rangle, \tag{5.34}$$

where the eigenvalue $\mathcal{E}_{\mathbf{k}}$ is the band energy. We multiply at the left by $\langle\phi_m|$: Eq. (5.34) maps to a matrix secular equation for the expansion coefficients $\tilde{b}_n$:

$$\sum_n H_{\mathbf{k},mn}\,\tilde{b}_n = \mathcal{E}_{\mathbf{k}}\sum_n B_{\mathbf{k},mn}\,\tilde{b}_n. \tag{5.35}$$

For each fixed $\mathbf{k}$, Eq. (5.35) is the generalized eigenvalue problem for the $n_o \times n_o$ matrix $H_{\mathbf{k}}$, of the type of Eq. (C.36). In the matrix multiplication language, Eq. (5.35) reads simply

$$H_{\mathbf{k}}\cdot\tilde{b} = \mathcal{E}_{\mathbf{k}}\,B_{\mathbf{k}}\cdot\tilde{b}. \tag{5.36}$$

The n_o eigenvalues obtained for each $\mathbf{k}$ point form n_o bands, as $\mathbf{k}$ varies across the BZ. To map the smooth $\mathbf{k}$ dependence, the matrix diagonalization (5.36) needs to be carried out for several $\mathbf{k}$ points picked on a sufficiently fine discrete mesh covering the first BZ. $\mathcal{E}_{\mathbf{k}\,j}$ represents a given eigenvalue, corresponding to an eigenvector $\tilde{b}^j$ with components $\tilde{b}^j_n$. The solution eigen-coefficients $\tilde{b}^j_n$ depend on $\mathbf{k}$ and on the precise eigenvalue (i.e. band) considered. The tight-binding energy and overlap matrices of Eqs. (5.35, 5.36) are explicitly:

$$H_{\mathbf{k},mn} = \sum_{\mathbf{R}} e^{i\mathbf{k}\cdot\mathbf{R}}\,\langle\phi_m|H_1\,T_{\mathbf{R}}|\phi_n\rangle \tag{5.37}$$

$$B_{\mathbf{k},mn} = \sum_{\mathbf{R}} e^{i\mathbf{k}\cdot\mathbf{R}}\,\langle\phi_m|\,T_{\mathbf{R}}\,|\phi_n\rangle, \tag{5.38}$$

with the sums over $\mathbf{R}$ extending in principle over infinitely many lattice points. In practice, when $|\mathbf{R}|$ grows significantly beyond the typical atomic size (a few times a_0), the energy and overlap matrix elements decay rapidly to 0, due to the exponential decay of the atomic wavefunctions $\langle\mathbf{r}|\phi_n\rangle$ at large distance.

Using the decomposition (5.30), Eq. (5.35) can be written in the instructive form related to Eq. (5.32):

$$\sum_n \left[E_m^{\text{at}} B_{\mathbf{k},mn} + C_{\mathbf{k},mn} \right] \tilde{b}_n = \mathcal{E}_{\mathbf{k}} \sum_n B_{\mathbf{k},mn}\, \tilde{b}_n, \tag{5.39}$$

where the matrix $C_{\mathbf{k}}$ is the Fourier transform of the matrix element (5.31),

$$C_{\mathbf{k},mn} = \sum_{\mathbf{R}} e^{i\mathbf{k}\cdot\mathbf{R}}\, \langle \phi_m | \Delta U\, T_{\mathbf{R}} | \phi_n \rangle. \tag{5.40}$$

Assuming, for simplicity, that the overlap matrix is the identity ($B_{\mathbf{k},mn} \approx \delta_{mn}$), Eq. (5.39) takes the following explicit form:

$$\begin{pmatrix} E_1^{\text{at}} + C_{\mathbf{k},11} & C_{\mathbf{k},12} & C_{\mathbf{k},13} & \cdots \\ C_{\mathbf{k},21} & E_2^{\text{at}} + C_{\mathbf{k},22} & C_{\mathbf{k},23} & \cdots \\ C_{\mathbf{k},31} & C_{\mathbf{k},32} & E_3^{\text{at}} + C_{\mathbf{k},33} & \cdots \\ \vdots & \vdots & \vdots & \ddots \end{pmatrix} \cdot \begin{pmatrix} \tilde{b}_1 \\ \tilde{b}_2 \\ \tilde{b}_3 \\ \vdots \end{pmatrix} = \mathcal{E}_{\mathbf{k}} \begin{pmatrix} \tilde{b}_1 \\ \tilde{b}_2 \\ \tilde{b}_3 \\ \vdots \end{pmatrix}. \tag{5.41}$$

Figure 5.47 sketches the connection between the atomic and the solid-state spectra as the component atoms, sodium in this example, move together to form a crystal. Equation (5.41) provides a quantitative justification for the physics of Fig. 5.47: for very large interatomic separations, all $\mathbf{k}$-dependent part of the matrix elements $C_{\mathbf{k},mn}$

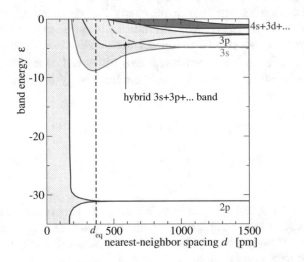

Fig. 5.47 The energy bands of bcc sodium, as a function of the nearest-neighbor interatomic spacing $d = a\sqrt{3}/2$. As the lattice parameter decreases, the atoms move closer, their separation becomes comparable to the size of the atomic wavefunctions, and the individual atomic levels overlap, originating energy bands. Further lattice compression, e.g. under the effect of pressure, makes all bandwidths increase further at the expense of interband gaps, with more bands overlapping

vanishes exponentially and the overlap matrix $B_{k,mn}$ does coincide with δ_{mn}, thus the eigenvalue problem (5.41) becomes trivially independent of k and the bands become flat and coinciding with the atomic eigenenergies E_m^{at}. As the atoms move closer together, $|C_{k,mn}|$ generally increase, thus leading to proper band dispersions with finite *bandwidth* (i.e. the difference $\mathcal{E}_{max} - \mathcal{E}_{min}$). Further shrinking of the lattice periodicity (as induced, e.g., by the application of hydrostatic pressure to the crystal) tends to broaden the band dispersion, with the forbidden energy gaps shrinking correspondingly. Similarly, for a given interatomic separation, since high up excited atomic states decay relatively slowly with distance (see, e.g., Fig. 2.6), these states produce generally larger $|C(k)_{mn}|$, and therefore broader band dispersions than the more localized deep (core) levels.

The tight-binding method, requiring a number of (small) matrix diagonalizations, is easily coded in a computer program, and it is employed routinely as a simple technique to obtain semi-quantitative band structures with a modest effort.

In the extreme limit where only $n_o = 1$ atomic s state $|\phi_n\rangle$ is retained in the atomic basis, this model is conveniently solved analytically, because matrices turn into numbers, and matrix products turn into ordinary products of numbers. The "eigenvector" can be taken $\tilde{b}_n = 1$, and Eq. (5.39) becomes

$$E_n^{at} B_{k,nn} + C_{k,nn} = \mathcal{E}_k B_{k,nn}, \qquad (5.42)$$

with solution

$$\mathcal{E}_{kn} = E_n^{at} + \frac{C_{k,nn}}{B_{k,nn}}. \qquad (5.43)$$

We retain the label n to $\mathcal{E}_k$ to remind us that it represents the band originated by the atomic state $|\phi_n\rangle$.

To further simplify Eq. (5.43), in the Fourier sums (5.38) and (5.40) of the B_k and C_k matrices, it is convenient to separate the $R = 0$ contribution from $R \neq 0$ terms. The $R = 0$ contribution to B_k equals unity, and that to C_k is a negative energy

$$\alpha_n = \langle \phi_n| \Delta U |\phi_n\rangle, \qquad (5.44)$$

representing the attraction that an electron in $|\phi_n\rangle$ at the origin experiences due to all surrounding atoms. For the $R \neq 0$ integrals contributing to B_k and C_k in (5.38) and (5.40) we use the notation:

$$\beta_n(R) = \langle \phi_n| T_R |\phi_n\rangle \qquad (5.45)$$
$$\gamma_n(R) = \langle \phi_n| \Delta U \, T_R |\phi_n\rangle. \qquad (5.46)$$

Observe that (i) by symmetry $\beta_n(-R) = \beta_n(R)$, $\gamma_n(-R) = \gamma_n(R)$, and (ii) for arbitrary Bravais lattice, when some R occurs in the R-sums, also $-R$ is present. One can then replace the complex exponentials by cosines, and rearrange the solution (5.43) for $\mathcal{E}_{kn}$ as:

$$\mathcal{E}_{\mathbf{k}n} = E_n^{\text{at}} + \frac{\alpha_n + \sum_{\mathbf{R}\neq 0} \cos(\mathbf{k}\cdot\mathbf{R})\,\gamma_n(\mathbf{R})}{1 + \sum_{\mathbf{R}\neq 0} \cos(\mathbf{k}\cdot\mathbf{R})\,\beta_n(\mathbf{R})}. \tag{5.47}$$

This relation expresses the tight-binding band energy as an explicit function of $\mathbf{k}$, in terms of a set of numerical parameters E_n^{at}, α_n, $\gamma_n(\mathbf{R})$, and $\beta_n(\mathbf{R})$. Equations (5.44), (5.45) and (5.46) provide the explicit recipe to compute these parameters in terms of overlap integrals of atomic wavefunctions. Note in particular that the "hopping energy" $\gamma_n(\mathbf{R}) < 0$, because the sign of the overlapping long-distance tail of the s atomic wavefunction is the same for both wavefunctions in Eq. (5.46), and the potential ΔU is attractive.

As observed above, due to the localization of these atomic wavefunctions, both $\gamma_n(\mathbf{R})$ and $\beta_n(\mathbf{R})$ become exponentially small for large $|\mathbf{R}|$. It therefore makes sense to ignore all the integrals for $|\mathbf{R}| > R_{\text{max}}$, which would bring in only negligible corrections to the band structure $\mathcal{E}_{\mathbf{k}n}$. In particular, consider the simplest approximation which provides a meaningful band structure depending on a minimal number of parameters: (i) neglect *all* of the $\beta_n(\mathbf{R})$'s (so that the denominator $B_{\mathbf{k},nn}$ becomes unity) and (ii) include only the $\gamma_n(\mathbf{R})$'s for the nearest neighbors (nn) of $\mathbf{0}$. In this extreme simplification, the expression (5.47) becomes

$$\mathcal{E}_{\mathbf{k}n} = E_n^{\text{at}} + \alpha_n + \gamma_n \sum_{\mathbf{R}\in(\text{nn})} \cos(\mathbf{k}\cdot\mathbf{R}), \tag{5.48}$$

where $\gamma_n = \gamma_n(\mathbf{R}_{\text{nn}})$.

Example 1 for a 1D chain, Eq. (5.48) yields a cosine-shaped band $\mathcal{E}_k = E^{\text{at}} + \alpha + 2\gamma\cos(a\,k)$. Since $\gamma < 0$, the minimum of the tight-binding band is reached at $\mathbf{k} = \mathbf{0}$, as sketched in Figs. 5.45 and 5.48.[16] The molecular orbitals of the benzene ring of 6 atoms (Fig. 3.8e) can also be described approximately by a 6-sites 1D-lattice tight binding, with 6 $\mathbf{k}$ points, see Eq. (5.27).

Example 2 in 2D, a square lattice has $\mathcal{E}_{\mathbf{k}} = E^{\text{at}} + \alpha + 2\gamma\left[\cos(a\,k_x) + \cos(a\,k_y)\right]$. Analogous expressions can be obtained for 2D lattices with different symmetries.

Example 3 for the fcc lattice Eq. (5.48) yields

$$\mathcal{E}_{\mathbf{k}} = E^{\text{at}} + \alpha + 2\gamma\left\{ \cos\left[\frac{a}{2}(k_x + k_y)\right] + \cos\left[\frac{a}{2}(k_y + k_z)\right] + \cos\left[\frac{a}{2}(k_x + k_z)\right] \right.$$
$$\left. + \cos\left[\frac{a}{2}(k_x - k_y)\right] + \cos\left[\frac{a}{2}(k_y - k_z)\right] + \cos\left[\frac{a}{2}(k_x - k_z)\right] \right\}$$
$$= E^{\text{at}} + \alpha + 4\gamma\times$$
$$\left[\cos\left(\frac{a}{2}k_x\right)\cos\left(\frac{a}{2}k_y\right) + \cos\left(\frac{a}{2}k_y\right)\cos\left(\frac{a}{2}k_z\right) + \cos\left(\frac{a}{2}k_z\right)\cos\left(\frac{a}{2}k_x\right)\right].$$

---

[16]The angular dependence of p or d atomic wavefunction can affect the phase of the relevant $\gamma_n(\mathbf{R})$, occasionally producing "reversed bands", with a maximum rather than a minimum at $\mathbf{k} = \mathbf{0}$.

Fig. 5.48 *Solid curve*: the characteristic cosine-like shape of the nearest-neighbor orthogonal ($\beta = 0$) 1D tight-binding band of s electrons, Eq. (5.48). *Dashed curve*: the slightly deformed band, Eq. (5.47), resulting from a rather large $\beta = 0.3$ accounting for the overlap integral of atomic wave functions centered at neighboring sites

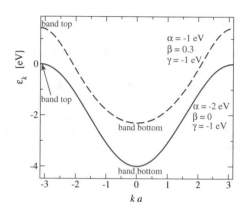

The above examples of approximate band dispersions confirm that the bandwidth increases proportionally to the off-diagonal hopping energy $|\gamma|$. Changes in the overlap integral (here β) at the denominator of Eq. (5.47) produce only minor band deformations, but do not overshadow the dominant bandwidth increase with $|\gamma|$. Observe also that the negative value of α reflects the fact that, on average, the energy of band states in a crystal is lower than the original level in the isolated atom, due to the extra attraction $\Delta U(\mathbf{r})$ of all other nuclei. If the considered band happens to be filled with electrons, this energy lowering contributes to the crystal cohesive energy. Even larger cohesive contributions $\propto |\gamma|$ arise from partly-filled bands.

The one-orbital approximation discussed here is appropriate for s bands, whenever the diagonal energy separation $\left| E_n^{at} - E_m^{at} \right|$ of $|\phi_n\rangle$ from all other atomic orbitals $|\phi_m\rangle$ is much larger than the dispersive part of $C_{\mathbf{k},mn}$. Under such conditions, the $|\phi_n\rangle$-derived s band remains isolated from all others. Indeed, for bands originating from inner s shells, the single-band tight-binding approximation is accurate and effective.

As soon as the inter-band gaps become comparable to the bandwidths and inter-band couplings, one needs to run a multi-orbital tight binding ($n_o > 1$). The bands obtained by a multi-band tight binding can be rather intricate, but they share several qualitative features with the single-band model. In particular, it remains true that the off-diagonal hopping matrix elements $\langle \phi_m | \Delta U \, T_{\mathbf{R}} | \phi_n \rangle$, and therefore the bandwidth and orbital hybridization, tend to increase when the lattice is squeezed. In the Na example of Fig. 5.47, at the equilibrium spacing $d_{eq} = 367$ pm, the strong overlap of the 3s, 3p,... bands indicates that a single-band tight-binding method is not especially well suited to describe the broad highly hybridized *conduction band*. The plane-waves method of Sect. 5.2.1.2 is usually considered to describe the conduction band of alkali metals better. Alternatively, by increasing the number n_o of the included orbitals one can improve the accuracy of the tight-binding model, at the expense of the extra computational cost required by the diagonalization of larger matrices. Also in polyatomic crystals, the size of the matrix problem (and correspondingly the number of resulting bands) equals the total number of orbitals contributed by all atoms in a unit cell. For further details related to crystal structures with several-atom basis, refer to Ref. [10].

5.2.1.2   The Plane-Waves Method

A practical alternative to the tight-binding method is the expansion of the Bloch
states on a plane-waves basis. Inside a crystal the effective one-electron potential
for the valence electrons is significantly screened, see Fig. 5.41: thus free-electron
eigenstates should approximate well the Bloch states for valence and conduction
bands everywhere except near the atomic nuclei. In the general discussion below, we
stick to a 3D formalism, but this problem can be visualized more easily in 1D, where
we restrict most of our illustrative examples.

With the variational mapping of a Schrödinger equation to a matrix problem,
Appendix C.5, we expand the electronic states in a basis:

$$|b\rangle = \sum_{\mathbf{k}'} b_{\mathbf{k}'} |\mathbf{k}'\rangle,$$

where $|\mathbf{k}'\rangle$ are momentum eigenstates, represented by plane waves, Eqs. (C.53) and
(C.56). This sum extends over all $\mathbf{k}'$ vectors, and represents therefore an integration.

The application of the Schrödinger equation to the candidate eigenket $|b\rangle$ and
multiplication at the left by $\langle \mathbf{k}|$ (implying a volume integration over $\mathbf{r}$) maps the
initial differential problem to an algebraic (matrix) equation for the wavefunction
Fourier components $b_{\mathbf{k}'}$:

$$H\,|b\rangle = \left[\frac{p^2}{2m_e} + V_{\text{eff}}\right]|b\rangle = \mathcal{E}\,|b\rangle$$

$$\langle \mathbf{k}|\left[\frac{p^2}{2m_e} + V_{\text{eff}}\right]|b\rangle = \mathcal{E}\,\langle \mathbf{k}|b\rangle$$

$$\sum_{\mathbf{k}'}\left[\frac{\hbar^2 k^2}{2m_e}\delta_{\mathbf{k},\mathbf{k}'} + \langle \mathbf{k}|V_{\text{eff}}|\mathbf{k}'\rangle\right] b_{\mathbf{k}'} = \mathcal{E}\sum_{\mathbf{k}'}\delta_{\mathbf{k},\mathbf{k}'}\,b_{\mathbf{k}'}, \tag{5.49}$$

where we use appropriate orthonormality $\langle \mathbf{k}|\mathbf{k}'\rangle = \delta_{\mathbf{k},\mathbf{k}'}$ of the plane waves and that
$\mathbf{p}\,|\mathbf{k}'\rangle = \hbar \mathbf{k}'|\mathbf{k}'\rangle$. The matrix elements

$$\langle \mathbf{k}|V_{\text{eff}}|\mathbf{k}'\rangle = \mathcal{N}\int e^{i(\mathbf{k}'-\mathbf{k})\mathbf{r}}\,V_{\text{eff}}(\mathbf{r})\,d^3\mathbf{r} = \tilde{V}_{\text{eff}}(\mathbf{k}-\mathbf{k}')$$

are the Fourier components of the potential, and $\mathcal{N}$ is the appropriate normalization,
see Eqs. (C.53) and (C.56).

Until this point, we made no mention of any lattice symmetry: indeed Eq. (5.49)
is the momentum representation of the Schrödinger equation. For a generic potential
V_{eff} this reciprocal-space formulation would provide no significant advantage over
real space, since the matrix indexes $\mathbf{k}$ of the eigenvalue problem (5.49) are smooth
quantities, taking continuously many values, exactly like $\mathbf{r}$. In contrast, as observed
in the general discussion of Eq. (5.11), the Fourier expansion of a *periodic potential* is

a *discrete Fourier series* over the reciprocal lattice, i.e. $\tilde{V}_{\text{eff}}(\mathbf{k} - \mathbf{k}')$ vanishes unless $\mathbf{k} - \mathbf{k}' = \mathbf{G}$, a vector of the reciprocal lattice. This means that in the continuous-indexed energy matrix of Eq. (5.49) most off-diagonal matrix elements vanish. In practice, given any $\mathbf{k}$ in the first BZ, the off-diagonal potential matrix elements connect the plane wave $|\mathbf{k}\rangle$ only to plane waves $|\mathbf{k}'\rangle$, displaced by a reciprocal-lattice vector: $\mathbf{k}' = \mathbf{k} - \mathbf{G}$. One can then address separately each subset of states originated from a given $\mathbf{k}$. The matrix form of Eq. (5.49) for one such sub-block is:

$$
\begin{pmatrix}
\ddots & \vdots & \vdots & \vdots & \\
\cdots & T(\mathbf{k} + \mathbf{G}_1) + \tilde{V}_{\text{eff}}(\mathbf{0}) & \tilde{V}_{\text{eff}}(\mathbf{G}_1 - \mathbf{G}_2) & \tilde{V}_{\text{eff}}(\mathbf{G}_1 - \mathbf{G}_3) & \cdots \\
\cdots & \tilde{V}_{\text{eff}}(\mathbf{G}_2 - \mathbf{G}_1) & T(\mathbf{k} + \mathbf{G}_2) + \tilde{V}_{\text{eff}}(\mathbf{0}) & \tilde{V}_{\text{eff}}(\mathbf{G}_2 - \mathbf{G}_3) & \cdots \\
\cdots & \tilde{V}_{\text{eff}}(\mathbf{G}_3 - \mathbf{G}_1) & \tilde{V}_{\text{eff}}(\mathbf{G}_3 - \mathbf{G}_2) & T(\mathbf{k} + \mathbf{G}_3) + \tilde{V}_{\text{eff}}(\mathbf{0}) & \cdots \\
& \vdots & \vdots & \vdots & \ddots
\end{pmatrix}
$$

$$
\cdot
\begin{pmatrix}
\vdots \\
b_{\mathbf{k}+\mathbf{G}_1} \\
b_{\mathbf{k}+\mathbf{G}_2} \\
b_{\mathbf{k}+\mathbf{G}_3} \\
\vdots
\end{pmatrix}
= \mathcal{E}_{\mathbf{k}}
\begin{pmatrix}
\vdots \\
b_{\mathbf{k}+\mathbf{G}_1} \\
b_{\mathbf{k}+\mathbf{G}_2} \\
b_{\mathbf{k}+\mathbf{G}_3} \\
\vdots
\end{pmatrix},
\tag{5.50}
$$

where $T(\mathbf{k}) = \hbar^2 k^2/(2m_e)$. The matrix in Eq. (5.50) must be diagonalized for each (fixed) $\mathbf{k}$ in the first BZ.

Diagonalization is of course trivial in the exceptional case when all off-diagonal matrix elements are identically null, i.e. for a constant potential $[\tilde{V}_{\text{eff}}(\mathbf{G} \neq \mathbf{0}) = 0]$: the eigenvalues in the solid are then simply the free-particle energies $\mathcal{E}_{\mathbf{k}}$ shifted by the constant potential $\tilde{V}_{\text{eff}}(\mathbf{0})$, and the eigenstates coincide with the original plane waves $|\mathbf{k}\rangle$. However, in any real-life crystal, many $\mathbf{G} \neq \mathbf{0}$ Fourier components of V_{eff} are nonzero: these components provide off-diagonal couplings between plane waves. The exact eigenstates of the problem are linear combinations of the plane waves differing by $\mathbf{G}$ vectors, obtained by the diagonalization of the full matrix in Eq. (5.50). As there are infinitely many $\mathbf{G}$ points, this is still an infinite matrix problem. In practice, one can cut the basis restricting it to a finite number n_{pw} of $\mathbf{G}$-spaced plane waves, and then diagonalize numerically a finite version of Eq. (5.50). This method is used routinely for standard band-structure calculations.

If the potential is "weak" (small $|\tilde{V}_{\text{eff}}(\mathbf{G} \neq \mathbf{0})|$), analytic information can be extracted out of Eq. (5.50). More precisely, whenever the off-diagonal elements $\tilde{V}_{\text{eff}}(\mathbf{G}_2 - \mathbf{G}_1)$ that couple a state $|\mathbf{k} + \mathbf{G}_1\rangle$ to *all* other free-particle states $|\mathbf{k} + \mathbf{G}_2\rangle$ are small compared to their diagonal separations $|T(\mathbf{k} + \mathbf{G}_1) - T(\mathbf{k} + \mathbf{G}_2)|$, the off-diagonal couplings act as a small perturbation to the diagonal energies. One can then apply the perturbative methods of Appendix C.9. At order zero, the band energies should coincide with the diagonal energies

$$\mathcal{E}_{\mathbf{k}} \approx \frac{\hbar^2 |\mathbf{k}|^2}{2m_e} + \tilde{V}_{\text{eff}}(\mathbf{0}). \tag{5.51}$$

The off-diagonal perturbation should act as a weak second-order effect, as long as the condition

$$\left| \tilde{V}_{\text{eff}}(\mathbf{G}_1 - \mathbf{G}_2) \right| \ll |T(\mathbf{k} + \mathbf{G}_1) - T(\mathbf{k} + \mathbf{G}_2)| \tag{5.52}$$

is verified. However, even in the favorable case that the Fourier components of V_{eff} are small, this condition (5.52) fails to apply at certain special $\mathbf{k}$ points, namely those where two or more $\mathbf{G}$-translated kinetic-energy parabolas get degenerate or nearly so: $T(\mathbf{k} + \mathbf{G}_1) \simeq T(\mathbf{k} + \mathbf{G}_2)$. At these points (highlighted in Fig. 5.49), the off-diagonal term $\tilde{V}_{\text{eff}}(\mathbf{G}_1 - \mathbf{G}_2)$ becomes dominating, and it displaces the actual band significantly away from the free-electron parabola Eq. (5.51). At $\mathbf{k}$ points where states $|\mathbf{k} + \mathbf{G}_1\rangle$ and $|\mathbf{k} + \mathbf{G}_2\rangle$ are degenerate or nearly so, approximate band energies can be calculated by diagonalizing the sub-matrix

$$\begin{pmatrix} T(\mathbf{k} + \mathbf{G}_1) + \tilde{V}_{\text{eff}}(\mathbf{0}) & \tilde{V}_{\text{eff}}(\mathbf{G}_1 - \mathbf{G}_2) \\ \tilde{V}_{\text{eff}}(\mathbf{G}_2 - \mathbf{G}_1) & T(\mathbf{k} + \mathbf{G}_2) + \tilde{V}_{\text{eff}}(\mathbf{0}) \end{pmatrix}.$$

The spectrum of a 2×2 matrix is discussed in Appendix C.5.2. The eigenenergies are provided in Eq. (C.39), with $E_L \to T(\mathbf{k} + \mathbf{G}_1) + \tilde{V}_{\text{eff}}(\mathbf{0})$, $E_R \to T(\mathbf{k} + \mathbf{G}_2) + \tilde{V}_{\text{eff}}(\mathbf{0})$, and $-\Delta \to \tilde{V}_{\text{eff}}(\mathbf{G}_1 - \mathbf{G}_2)$. Figure C.2 shows that due to the level "repulsion" produced by the off-diagonal elements, the band energies never come any closer than $\left| 2\tilde{V}_{\text{eff}}(\mathbf{G}_1 - \mathbf{G}_2) \right|$.

This model illustrates the tendency of the periodic components of the potential to open forbidden energy intervals in the parabolic free-electron dispersion. As shown in Fig. 5.50, in 1D, the periodic potential opens gaps at all $\mathbf{G}/2$ points, unless some Fourier component $|\tilde{V}_{\text{eff}}(\mathbf{G})|$ of the potential happens to vanish. In 3D, degeneracies of the kinetic term, and therefore $\tilde{V}_{\text{eff}}(\mathbf{G})$-related gaps, occur for all points $\mathbf{k}$ such that

Fig. 5.49 Graphical intersections of a few translated free-electron parabolas $T(\mathbf{k} + \mathbf{G}_1)$, $T(\mathbf{k} + \mathbf{G}_2)$,... In a neighborhood of the *highlighted intersection points*, even small off-diagonal $\tilde{V}_{\text{eff}}(\mathbf{G} \neq \mathbf{0})$ matrix elements can distort the parabolas quite significantly. In this figure, $a = 210$ pm, and $G_l = l\, 2\pi/a$

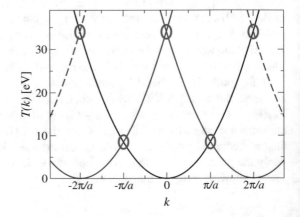

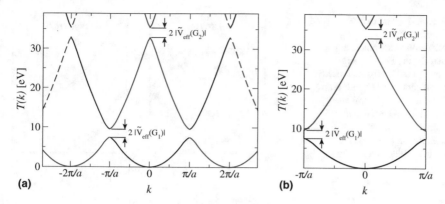

Fig. 5.50 **a** A $|2\tilde{V}_{\text{eff}}(G)|$-wide gap opens at each degeneracy point $k = G_l/2$: the parabolic bands distort in an entire neighborhood around that k point. This representation over a broad k range emphasizes that the bands are periodic functions of k, with the periodicity of the reciprocal lattice. **b** The same band structure, restricted to the first BZ

$|\mathbf{k} + \mathbf{G}| = |\mathbf{k}|$, for some reciprocal-lattice point G. This is the condition (5.19) for Bragg scattering. In 3D, the periodic potential does not always generate a proper band gap, i.e. a range of forbidden energy, since the forbidden energies for a $\mathbf{k}$-direction may well be allowed in some other $\mathbf{k}$-direction, see Fig. 5.54.

Both this method based on free electrons and the tight-binding method of Sect. 5.2.1.1 lead to single-electron spectra characterized by bands of allowed energy separated by gaps of forbidden energy, as expected for Bloch electrons. Both methods provide a physical meaning to the band index j of the electronic states in solids, see Eq. (5.26) and Fig. 5.42. In the tight-binding scheme, the band index contains indications about the (n, l) labeling of the main-component atomic state (which is an especially clear-cut concept for the narrow bands of inner shells, see Fig. 5.47). In the quasi-free-electron approach (especially fit for the wide bands related to the empty atomic levels), the index j of a given band simply counts how many free-electron-parabola-derived bands sit lower in energy.

5.2.2 Filling of the Bands: Metals and Insulators

The $T = 0$ (ground) state of a set of independent electrons in a periodic potential is obtained by filling the one-electron band states up to the Fermi energy μ, like in the free-fermion model described in Sect. 4.3.2.1. The Fermi energy separates filled (below) from empty (above) levels. Depending on the total number of electrons in the solid, the Fermi energy may end up either inside one (or several) energy band(s), or within a band gap, see Fig. 5.51. In the first case, the electrons in the partly filled band(s) close to the Fermi energy are ready to take up excitation energy, provided typically by an external electromagnetic field. In particular, an arbitrarily

Fig. 5.51 The two fundamental $T = 0$ band-filling schemes: **a** a *metal*—the Fermi energy level (dotted) crosses a partly filled conduction band; **b** an *insulator*—the Fermi energy sits in a gap between a completely empty conduction band and an entirely filled valence band

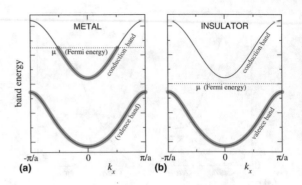

weak applied electric field can accelerate electrons, which can then conduct electric current. Such a solid is a *metal*. In contrast, electrons in completely filled bands remain "frozen" by Pauli's principle. Any dynamical response of these electrons requires an excitation across an energy gap. A solid where all bands are either entirely filled or empty, with the Fermi energy inside a gap, is an *insulator*. In solids, the highest completely filled band is usually called the *valence band*, while the lowest empty or partly-filled band is the *conduction band*.

This basic difference affects substantially all electronic properties of these two major classes of materials, even at finite temperature. The Fermi-Dirac distribution (4.97) applies to independent electrons at equilibrium in a solid pretty much like in a free gas, simply replacing the free-electron energies and plane-wave states with the band energies and Bloch states. Thus, a nonzero temperature in a metal generates a finite concentration of electrons above the chemical potential and holes below it, similarly to what happens around the Fermi sphere of a free-electron gas. Accordingly, the thermodynamics of electrons in metals is interesting and rich of physical consequences, including a characteristic T-linear contribution to the heat capacity of the solid, as discussed in Sect. 4.3.2.1. Instead, in an insulator with gap Δ between conduction and valence band, the average occupation of a valence-band state is $[n_v] \gtrsim 1 - \exp(-\Delta/[2k_B T])$, extremely close to 1 at low temperature. Correspondingly the average occupation of a conduction-band state is $[n_c] \lesssim \exp(-\Delta/[2k_B T])$, extremely small at low temperature. For a temperature much smaller than Δ/k_B, Pauli's principle "freezes" the electrons of an insulator in the filled bands. Their excitation requires a sizable energy, as can be provided e.g. by photons in spectroscopy. For a wide-gap insulator (e.g. Al_2O_3, $\Delta \simeq 5$ eV; C diamond, $\Delta \simeq 5.5$ eV; SiO_2, $\Delta \simeq 8.0$ eV; NaCl, $\Delta \simeq 8.97$ eV), any practical temperature is by far smaller than Δ/k_B. As a result, the band occupancies are indistinguishable from those at $T = 0$. For example, with a 4 eV gap at room temperature ($k_B T \approx 0.025$ eV), $[n_c] \approx e^{-80} \simeq 10^{-35}$. In contrast, at room temperature, small-gap insulators, usually called *semiconductors* (e.g. Ge, $\Delta \simeq 0.74$ eV; Si, $\Delta \simeq 1.17$ eV; GaAs, $\Delta \simeq 1.52$ eV), exhibit a measurable nonzero conduction associated to electrons thermally excited across the gap, as sketched in Fig. 5.52, and discussed in Sect. 5.2.2.2.

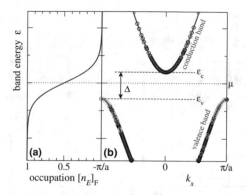

Fig. 5.52 **a** The finite-temperature Fermi-Dirac distribution with the chemical potential μ sitting in between the conduction-band bottom $\mathcal{E}_c$ and the valence-band top $\mathcal{E}_v$ **b** A sketch of the finite-T random occupations of the bands of a semiconductor near the gap. Occasional empty states (holes) in the valence band are highlighted as white circles

How can one predict the Fermi-energy position? In many-electron atoms and molecules, counting electrons in orbitals is sufficient. An infinite crystal boasts infinitely many band states and infinitely many electrons: to predict the Fermi energy level, we need some counting rule. Consider a crystal portion with volume $V = N_{n1}N_{n2}N_{n3}\,V_c$, extending for N_{n1}, N_{n2}, N_{n3} lattice repetitions in the $\mathbf{a}_1$, $\mathbf{a}_2$, $\mathbf{a}_3$ primitive directions. Discrete translational invariance is preserved by applying periodic boundary conditions $\psi_{\mathbf{k}j}(\mathbf{r}) = \psi_{\mathbf{k}j}(\mathbf{r} + N_{ni}\mathbf{a}_i)$ to this chunk of solid. To satisfy these artificial boundary conditions, the $\mathbf{k}$ label of the Bloch states is restricted to

$$\mathbf{k} = \frac{n_1}{N_{n1}}\,\mathbf{b}_1 + \frac{n_2}{N_{n2}}\,\mathbf{b}_2 + \frac{n_3}{N_{n3}}\,\mathbf{b}_3, \tag{5.53}$$

$$\text{with } n_i = -\frac{N_{ni}}{2} + 1, \ -\frac{N_{ni}}{2} + 2, \ \dots \ \frac{N_{ni}}{2} - 2, \ \frac{N_{ni}}{2} - 1, \ \frac{N_{ni}}{2}.$$

These $\mathbf{k}$ values are the lattice equivalent to those of Eq. (4.38), and the 3D generalization of Eq. (5.27). These $N_n = N_{n1}N_{n2}N_{n3}$ discrete $\mathbf{k}$ values become dense and fill the unit cell of the reciprocal lattice as the infinite real-space crystal is recovered for $N_{ni} \to \infty$, see Fig. 5.44. Each band (orbital) state has room for one electron with spin $|\uparrow\rangle$ and one with spin $|\downarrow\rangle$. Overall, each band has $2N_n$ spin-orbital states, therefore it has room for $2N_n$ electrons in total. The number of electrons present in this finite crystal portion equals N_n times the number n_{cell} of electrons per unit cell. If successive bands are separated by gaps, the $N_n\,n_{\text{cell}}$ electrons fill the $n_{\text{cell}}/2$ lowest bands, as illustrated in Fig. 5.53.

An even value of n_{cell} leads to $n_{\text{cell}}/2$ full bands, followed by empty bands above. In the example of Fig. 5.53a, each atom of solid neon (one atom per cell, in the real 3D fcc crystal too) provides $n_{\text{cell}} = 10$ electrons, i.e. a total of $10\,N_n$ electrons in

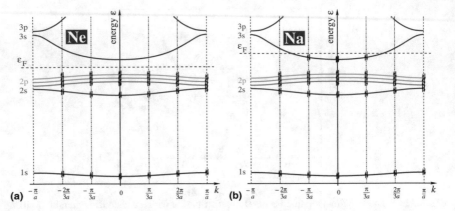

Fig. 5.53 The filling of the bands of hypothetical **a** neon and **b** sodium 1D crystals composed of $N_n = 6$ atoms. According to Eq. (5.53), the allowed k-points in the first BZ are $k = 0$, $\pm\frac{1}{6}\frac{2\pi}{a}$, $\pm\frac{2}{6}\frac{2\pi}{a}$, and $\frac{3}{6}\frac{2\pi}{a}$. The $-\frac{\pi}{a}$ point must be excluded, because it coincides with $\frac{\pi}{a}$. Individual bands are assumed not to overlap. Energies are purely qualitative and not to scale

the whole crystal. In a tight-binding language, $2\,N_n$ electrons fill the 1s band, $2\,N_n$ electrons fill the 2s band, and $6\,N_n$ electrons fill entirely the three 2p bands. The crystal has a total of 5 filled bands. The Fermi energy then lies in the gap between the filled 2p band and the empty 3s-3p bands above: the neon crystal is an insulator.

Odd n_{cell} leads to $(n_{\text{cell}} - 1)/2$ full bands plus one half-filled band. For example, Fig. 5.53b, $10\,N_n$ of the $n_{\text{cell}}\,N_n = 11\,N_n$ electrons that the sodium atoms contribute to the band states of their (in reality bcc) monoatomic crystal fill entirely the five 1s, 2s, and 2p bands. The remaining N_n electrons occupy the 50% low-energy spin-orbital states of the 3s band. This 3s band is then half filled. The Fermi energy cuts through this band: the sodium crystal is therefore a metal.

While it is evident that *any crystal with an odd* number of electrons per cell n_{cell} *is a metal*,[17] for even n_{cell} both insulators (like Ne) and metals are possible. A metallic solid occurs when the $(n_{\text{cell}}/2)$-th and the $(n_{\text{cell}}/2 + 1)$-th bands are not separated by a gap, as illustrated in Fig. 5.54. The alkaline-earth (group 2) and the end-of-the-transition (group 12: Zn, Cd, and Hg) elemental solids are examples of metals with even n_{cell}, precisely due to overlapping bands at the Fermi energy.

---

[17] When the independent-electrons approximation fails, as in *strongly-correlated materials*, insulating states, often accompanied by a magnetic order of the spins, can occur with odd n_{cell}.

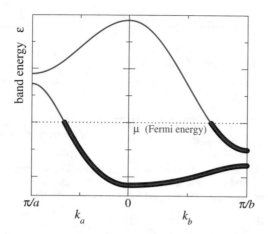

Fig. 5.54 Band overlap. The bands $\mathcal{E}_{\mathbf{k}\,j}$ of 2D and 3D crystals depend on the direction of $\mathbf{k}$. Even when in each direction two bands remain separated by an energy gap, an energy range forbidden in one direction k_a can be covered by either band along some other direction k_b. With such kind of band structure the Fermi energy can end up crossing several overlapping bands, thus yielding a metal, even if the number of electrons per cell is even

5.2.2.1 Metals

Metals exhibit a distinctive ability to conduct electric current. In practice, however, all solids, even insulators, conduct electricity to some degree. What does characterize uniquely the metallic state is a large electric conductivity which decreases as T is raised. In contrast, the small conductivity of insulators increases as T is raised, due to extra thermally-excited *charge carriers*. For metals at $T = 0$, a *Fermi surface* separates full from empty states in the $\mathbf{k}$ space, generalizing the Fermi sphere of the ideal Fermi gas.

To describe the motion of electrons in the periodic field of the ions, plus externally applied fields, we conveniently adopt a semiclassic approach based on the band theory. This kind of approach is entirely standard in QM: electrons are represented as wave packets, i.e. superpositions of Bloch states of a single band j, characterized by a peaked wave-number distribution. The wave-packet spatial extension is correspondingly large (much wider than the crystal lattice spacing, see Fig. 5.55). The following coupled equations govern the motion of the center of mass $\mathbf{r}$ of such wave packets [10]:

$$\frac{d}{dt}\mathbf{r} = \mathbf{v}_j(\mathbf{k}) = \frac{1}{\hbar}\nabla_{\mathbf{k}}\mathcal{E}_{\mathbf{k}\,j} \tag{5.54}$$

$$\hbar\frac{d}{dt}\mathbf{k} = -q_e\left[\mathbf{E}(\mathbf{r},t) + \mathbf{v}_j(\mathbf{k})\times\mathbf{B}_{\text{ext}}(\mathbf{r},t)\right]. \tag{5.55}$$

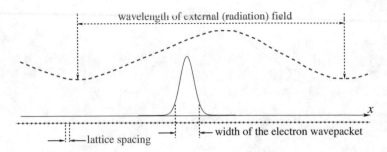

Fig. 5.55 A cartoon for the length scales of the semiclassic model of electron dynamics in crystals. The wavelength of externally applied fields (*dashed*) far exceeds the width in the electron wave packet (*solid*), and this wavepacket, in turn, extends across several lattice spacings

The external perturbing electric $\mathbf{E}(\mathbf{r}, t)$ and magnetic $\mathbf{B}_{\text{ext}}(\mathbf{r}, t)$ fields[18] are supposed to vary slowly on the scale of the wave-packet size (Fig. 5.55). A rigorous derivation of Eqs. (5.54) and (5.55) goes beyond the scope of the present text: here we suggest a few heuristic arguments to support their plausibility.

- The center-mass velocity $\mathbf{v}_j$ of the wave packet is the *group velocity* associated to the dispersion $\mathcal{E}_{\mathbf{k}\,j}$ of the Bloch waves. Equation (5.54) expresses the basic fact of wave mechanics that a wave packet with dispersion $\omega(\mathbf{k}) = \hbar^{-1}\mathcal{E}_{\mathbf{k}\,j}$ moves with the velocity given by its group velocity $\nabla_{\mathbf{k}}\,\omega(\mathbf{k})$. Note that in the special case of a free-electron dispersion $\mathcal{E}_{\mathbf{k}} = \hbar^2 k^2/(2m_e)$, Eq. (5.54) yields the usual relation $\mathbf{v}(\mathbf{k}) = \hbar\mathbf{k}/m_e$ between velocity and momentum.
- If the force associated to a static external electric potential $\phi_{\text{ext}}(\mathbf{r})$ acts on the band electron, then its total energy $\left[\mathcal{E}_{\mathbf{k}\,j} - q_e\,\phi_{\text{ext}}(\mathbf{r})\right]$ should remain constant along the semiclassic motion. To verify this conservation, we derive this total energy with respect to time, with the shorthand $\frac{df}{dt} \equiv \dot{f}$:

$$\frac{d}{dt}\left[\mathcal{E}_{\mathbf{k}\,j} - q_e\,\phi_{\text{ext}}(\mathbf{r})\right] = \nabla_{\mathbf{k}}\,\mathcal{E}_{\mathbf{k}\,j} \cdot \dot{\mathbf{k}} - q_e\,\nabla_{\mathbf{r}}\,\phi_{\text{ext}}(\mathbf{r}) \cdot \dot{\mathbf{r}} = \mathbf{v}_j(\mathbf{k}) \cdot \left[\hbar\dot{\mathbf{k}} + q_e\,\mathbf{E}(\mathbf{r})\right],$$

where we used Eq. (5.54). If Eq. (5.55) is satisfied as well, then this derivative indeed vanishes, as the vector $\hbar\dot{\mathbf{k}} + q_e\mathbf{E}(\mathbf{r}) = -q_e\mathbf{v}_j \times \mathbf{B}_{\text{ext}}(\mathbf{r})$ remains perpendicular to the velocity $\mathbf{v}_j(\mathbf{k})$. Energy is thus conserved.
- Equation (5.55) recalls the classical equation of motion of a particle with charge $-q_e$ moving under the action of the *external* electromagnetic fields $\mathbf{E}(\mathbf{r})$ and $\mathbf{B}_{\text{ext}}(\mathbf{r})$ only. Instead, the periodic forces produced by the lattice act through the band dispersion $\mathcal{E}_{\mathbf{k}\,j}$, generating the nontrivial $\mathbf{k}$-dependence of the velocity in Eq. (5.54). This means that, at variance with free electrons, the *"crystal momentum"* $\hbar\mathbf{k}$ *does not equal the real electron momentum.*

---

[18]The magnetic field $\mathbf{H}$ produced by currents *external* to the solid is related to the magnetic induction field $\mathbf{B}_{\text{ext}}$ of Eq. (5.55) by $\mathbf{H} = \epsilon_0 c^2\,\mathbf{B}_{\text{ext}}$.

- The semiclassic equations assume that the external fields are sufficiently weak to induce no inter-band transitions. Strong fields would make the semiclassic approximation fail, leading to electric or magnetic *breakdown*. Also, the single-band semiclassic equations hold until the frequency of any time-oscillating field does not approach inter-band gaps ($\omega \ll \Delta/\hbar$). Rapidly varying fields are applied in spectroscopy precisely with the purpose of inducing inter-band transitions.

According to the semiclassic equations, a completely filled band cannot contribute to either electric or heat current. The electric *current density* carried by a wave packet representing an electron moving in a volume V is $(-q_e)\mathbf{v}_j(\mathbf{k})/V$. The total electric current density carried by all electrons in a filled band j amounts to[19]

$$\mathbf{j} = \int_{BZ} (-q_e)\mathbf{v}_j(\mathbf{k})\, \frac{1}{4\pi^3}\, d^3\mathbf{k} = \mathbf{0}. \tag{5.56}$$

This result is due to the vanishing of the integral of the gradient of a periodic function over a full period. Indeed, due to Eq. (5.54), the integrand function ($\propto \nabla_\mathbf{k}\mathcal{E}_{\mathbf{k}\,j}$ is the gradient of a periodic function $\propto \mathcal{E}_{\mathbf{k}\,j}$, which is periodic with the reciprocal-lattice periodicity; and the first BZ is an entire cell of the reciprocal lattice. In physical terms, in a filled band for each electron carrying current in some direction another electron carries current the opposite way, totaling a vanishing current.

The same observations apply to the energy (heat) current

$$\mathbf{j}_\mathcal{E} = \int_{BZ} \mathcal{E}_{\mathbf{k}\,j}\mathbf{v}_j(\mathbf{k})\, \frac{1}{4\pi^3}\, d^3\mathbf{k} = \mathbf{0}, \tag{5.57}$$

by noting that the integrand is proportional to $\nabla_\mathbf{k}(\mathcal{E}_{\mathbf{k}\,j})^2$. Completely filled bands do not contribute to transport any more than completely empty bands. All *electric and thermal conductivity is to be attributed to partly filled bands*. This explains why no systematic increase of conductivity is observed in the crystalline elements for increasing Z (for example the conductivities of fcc Cu, Ag, and Au are quite similar), despite the largely different total number of electrons.

Coming to partly filled bands, note that Bloch states are *stationary* states of the Schrödinger equation in the perfect crystal: if a wave packet (made of Bloch states) representing an electron has a finite mean velocity (as occurs unless $\nabla_\mathbf{k}\mathcal{E}_{\mathbf{k}\,j} = \mathbf{0}$), then that velocity shall persist forever. Thus, this semiclassic theory based on Bloch states predicts that, once started, persistent currents should cross metals as if they were vacuum, even in the absence of any external electric field. In practice, no such persistent currents are observed.

---

[19]Similarly to Eq. (4.90), the n_i sum is turned into an integral by inserting the appropriate density of states. According to Eq. (4.38), the $\mathbf{k}$-density of states is $V/(2\pi)^3$. An extra factor $g_s = 2$ accounts for the spin degeneracy. The integration of $d^3\mathbf{k}\, V/(4\pi^3)$ over the entire BZ represents the summation over all states in band j.

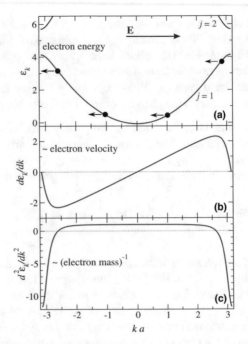

Fig. 5.56 The reciprocal-space motion of the electrons in a 1D crystal. Under the action of a constant uniform rightward electric field **E**, (leftward external force) the wave number k drifts at constant speed to the left, Eq. (5.55). At the first BZ boundary k is brought back into the first BZ by $k = -\pi/a \to \pi/a$, and then continues to move leftward and to traverse the BZ again and again. Correspondingly, **a** the electron band energy oscillates. The real-space electron **b** velocity and **c** acceleration also oscillate, because they are proportional to the first, Eq. (5.54), and second, Eq. (5.64), derivative of the band energy

Additionally, as illustrated in Fig. 5.56, according to Eq. (5.55) the semiclassic k-space motion cycles across the whole BZ under the action of a constant uniform electric field. Correspondingly, Eq. (5.54) yields an oscillating velocity generating positive and negative currents for the same amount of time. A DC electric field should then induce an AC current (*Bloch oscillations*) in a metal wire! This prediction is quite far from reality, too. Under ordinary conditions, real metals exhibit an entirely different response: *Ohm's law*[20] $\mathbf{j} = \sigma \mathbf{E}$. These inconsistencies of the model with observation is due to Bloch electrons, and wave packets thereof, being capable to travel forever through a perfect crystal, without any energy dissipation. Real crystals deviate from ideality because of

---

[20]The current I in a wire is directly proportional to the applied potential drop V: $I = R^{-1}V$. The proportionality coefficient R^{-1}, called conductance, depends on the length L and cross section A of the wire, but not on the current or potential drop. In terms of the current density $\mathbf{j}$ crossing perpendicularly the surface area A, $I = |\mathbf{j}|A$. In terms of the electric field **E**, the potential drop $V = L|\mathbf{E}|$. Thus Ohm's law can be expressed as $\mathbf{j} = L/(AR)\,\mathbf{E} = \sigma\,\mathbf{E}$, where the conductivity $\sigma = L/(AR)$ and resistivity $\rho = \sigma^{-1}$ are characteristic properties of the wire material.

- structural defects, as discussed in Sect. 5.1;
- the crystal atoms not being frozen at their ideal periodic equilibrium positions but actually vibrating around them, as we shall discuss in Sect. 5.3.

Both these discrepancies from the ideal-crystal picture are sources of *collisions* for conduction electrons.[21] To represent the effect of collisions as simply as possible, we shall assume that:

1. Each electron experiences an instantaneous collision at random, with probability τ^{-1} per unit time. The time τ, variously known as the *relaxation time*, the *collision time* or the *mean free time*, represents the average time that an electron travels freely between a collision and the next.
2. The electron emerges from a collision in a random $\mathbf{k}$ state, reflecting the (Fermi) distribution at the appropriate local temperature, and respecting Pauli's principle. All memory of the initial $\mathbf{k}$ prior to the collision is lost.
3. After a collision and until the next one, each electron moves according to the semiclassic equations of motion (5.54, 5.55).

As a result of frequent collisions, the external field induces only a weak perturbation to the thermal equilibrium distribution. When, following Eq. (5.55), a DC electric field pushes electrons in occupied states in the $-q_e\mathbf{E}$ direction, collisions rapidly transfer accelerated electrons from freshly occupied $\mathbf{k}$-states at one side of the *Fermi surface* back into states left empty at the opposite side. This tendency of collisions to re-establish thermal equilibrium, illustrated in Fig. 5.57a, quenches the free acceleration of the electrons, thus preventing the Bloch oscillations. After an initial transient, the net effect of the $\mathbf{E}$ field amounts to a small constant displacement of the filled states relative to the symmetric zero-field occupations, Fig. 5.57b. This displacement generates a steady current density $\mathbf{j}$ equal to the $\mathbf{k}$-space integration of $V^{-1}(-q_e)\mathbf{v}_j(\mathbf{k})$ extended to the region $\delta^3 k$ of unbalanced occupations. For a free-electron parabolic band, $\mathbf{j}$ can be estimated roughly as

$$\mathbf{j} = \int_{\delta^3 k} V^{-1}(-q_e)\mathbf{v}_j(\mathbf{k})\frac{V}{4\pi^3}d^3\mathbf{k} \simeq \int_{\delta^3 k}(-q_e)\hat{\mathbf{E}}v_F\, d^3\mathbf{k} \simeq -q_e\, v_F\,\hat{\mathbf{E}}\,(\delta^3 k), \quad (5.58)$$

where we drop numeric factors of order unity. The $\mathbf{k}$-space volume $(\delta^3 k)$ in between the equilibrium Fermi surface and its field-shifted replica, sketched in Fig. 5.58 for free electrons, can be estimated by observing that in the average time τ between collisions, each electron changes its wave vector by $\delta\mathbf{k} = \hbar^{-1}\tau\,(-q_e)\,\mathbf{E}$. Dropping again factors of order unity, the volume $(\delta^3 k)$ is approximately

$$(\delta^3 k) \simeq -|\delta\mathbf{k}|\,k_F^2 \simeq \frac{\tau}{\hbar}\,(-q_e)\,|\mathbf{E}|\,k_F^2,$$

[21] Electron-electron interactions are responsible for a small amount of electron scattering too, quite negligible compared to the main sources of collisions in ordinary metals at ordinary temperature.

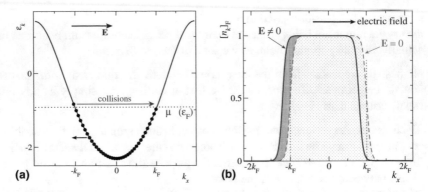

Fig. 5.57   a Band occupation under the competing action of (i) a rightward static electric field **E**, accelerating the electrons to the left— Eq. (5.55), and (ii) collisions with crystal defects and vibrations, which tend to reestablish equilibrium by scattering extra-energetic electrons from the left side of the Fermi surface prevalently into lower-energy states which have been left empty at the right side below k_F. **b** The steady occupation distribution in **k** space, shifted to the left by a rightward electric field. This shift δ**k**, here largely exaggerated, is proportional to both **E** and the average time τ between collisions, and accounts for ohmic electric-current transport

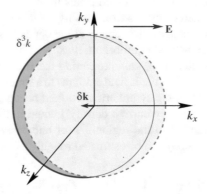

Fig. 5.58   The shift δ**k** of the Fermi surface (here represented by a sphere) induced by a rightward external electric field $\mathbf{E} = E\,\hat{\mathbf{x}}$. The highlighted **k**-space volume ($\delta^3 k$) is responsible for the velocity-distribution asymmetry supporting the net electric current carried by this partly-filled band

where the minus sign indicates that the shift is opposite to the **E** field. By substituting this expression and $v_F \simeq \hbar k_F/m_e$ (valid for free electrons) in Eq. (5.58), we obtain

$$\mathbf{j} \simeq (-q_e)\frac{\hbar k_F}{m_e}\,\hat{\mathbf{E}}\,\frac{\tau}{\hbar}\,(-q_e)\,|\mathbf{E}|\,k_F^2 \simeq \frac{q_e^2\,\tau}{m_e}\,k_F^3\,\mathbf{E} \simeq \frac{q_e^2\,\tau}{m_e}\,\frac{N}{V}\,\mathbf{E}, \qquad (5.59)$$

where we used the free-electron relation of k_F with the electron density, $k_F^3 = 3\pi^2 N/V$ (Sect. 4.3.2.1), dropping factors of order unity. Equation (5.59) agrees with Ohm's law, and predicts a conductivity

$$\sigma \simeq \frac{q_e^2 \tau}{m_e} \frac{N}{V}. \tag{5.60}$$

By measuring the resistivity $\rho = \sigma^{-1}$ and the conduction-electron number density N/V of metals and inverting Eq. (5.60), we infer the relaxation time τ. The values reported in Table 5.2, in the 10^{-13} s range, indicate that between two successive collisions an electron travels an average distance (called the electron's *mean free path*) $\ell \simeq \tau v_F \approx 100$ nm, namely hundreds of typical interatomic spacings.

For increasing temperature, the resistivity is observed to increase, indicating that the relaxation time decreases. Collisions become more frequent as thermal motion produces larger instantaneous displacements of the nuclei away from their equilibrium positions, as described in Sect. 5.3. The average time between collisions is indeed inversely proportional to the total number of phonons: $\tau^{-1} \propto \sum_\epsilon [n_\epsilon]_B$. At thermal energies $k_B T$ much larger than the characteristic phonon energies ($\epsilon \lesssim 100$ meV), the total number of phonons is proportional to T, as can be verified by expanding Eq. (4.96): $[n_\epsilon]_B = (e^{\beta\epsilon} - 1)^{-1} \simeq (\beta\epsilon)^{-1} = k_B T/\epsilon)$. Indeed, at high temperature metals do approximately exhibit a resistivity $\rho \propto T$. At low temperature, the phonon number decreases rapidly [see Eq. (4.116)], and phonon become irrelevant. Here crystal defects provide a residual T-independent scattering rate. Accordingly, at low temperature the T-linear regime turns into a T-independent (but sample-dependent) resistivity, as illustrated in Fig. 5.59 for Na. The low-temperature resistivity contribution of defects varies widely among different metals; is is especially large in disordered alloys.

Like we did for electric conductivity, we can estimate the *thermal conductivity* of electrons in a metal. When a thermal gradient $\nabla_r T = \hat{z} \, dT/dz$ is established across the sample, the distribution of the electrons reaching a given point inside

Table 5.2 The measured electrical resistivity $\rho = \sigma^{-1}$ of a few elemental metals [10]; the corresponding relaxation times are obtained from Eq. (5.60) through $\tau = m_e/(\rho q_e^2 N/V)$, inserting the relevant density N/V of conduction electrons (Data from Ref. [10])

Element	ρ [nΩm] at 77 K	ρ [nΩm] at 273 K	τ [10^{-14} s] at 77 K	τ [10^{-14} s] at 273 K
Na	8	42	17	3.2
K	14	61	18	4.1
Rb	22	110	14	2.8
Cu	2	16	21	2.7
Ag	3	15	20	4.0
Au	5	20	12	3.0
Mg	6	39	7.0	1.1
Al	3	25	6.5	0.8

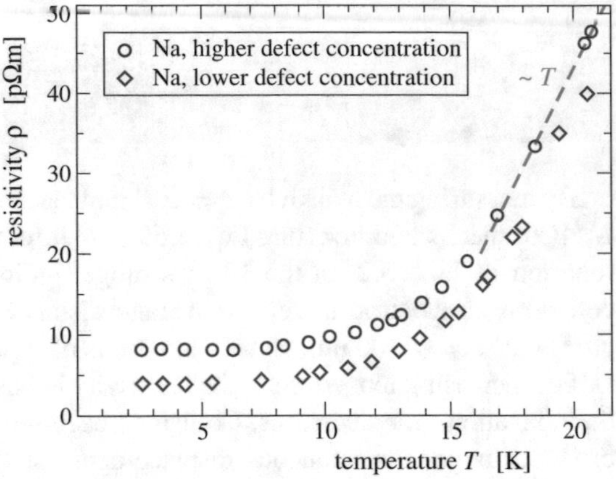

Fig. 5.59 The low-temperature resistivity of sodium, measured for two samples characterized by different concentrations of defects, leading to different low-T residual resistivity. At low temperature the tiny phonon contribution to resistivity grows as T^5, but it then turns rapidly into a T-linear increase (*dashed line*)

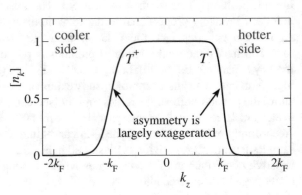

Fig. 5.60 The **k**-space distribution of electrons coming at a given spot in the metal from approximately a mean free path ℓ away exhibits a little asymmetry in the direction of $\nabla_{\mathbf{r}}T$. Heat transport is associated to electrons emerging from the hotter (right) region carrying (on average) higher energy than those coming from the cooler (left) region

the metal have a slightly different thermal broadening depending on whether they come from the high-temperature side or from the low-temperature side, see Fig. 5.60. The electrons coming to that point from a region at temperature $T + \Delta T$ carry an extra energy $-V^{-1}C_V\,\Delta T = -V^{-1}C_V\,\ell\,\hat{\mathbf{z}}\cdot\nabla_{\mathbf{r}}T$. Here C_V is the electron-gas heat capacity, Eq. (4.106), $\ell \simeq \tau v_F$ is the average distance traveled by the electrons since their previous collision, and the minus sign indicates that energy is transported in the direction opposite to the temperature gradient. The heat current density is obtained by multiplying this heat density by the speed v_F at which electrons travel:

$$\mathbf{j}_\varepsilon \simeq -V^{-1}Nk_B\,\frac{\pi^2}{2}\frac{k_BT}{\epsilon_F}\tau\,v_F\nabla_{\mathbf{r}}Tv_F \simeq -k_B^2\,T\tau\,\frac{v_F^2}{\epsilon_F}\frac{N}{V}\,\nabla_{\mathbf{r}}T \simeq -\frac{k_B^2\,T\,\tau}{m_e}\frac{N}{V}\,\nabla_{\mathbf{r}}T,$$
$$(5.61)$$

where we use the relation $\epsilon_F = m_e v_F^2/2$, and we drop factors of order unity. The expression (5.61) for the heat current is a direct-proportionality relation of the type $\mathbf{j}_\varepsilon = -\mathcal{K}\,\nabla_{\mathbf{r}}T$. The *thermal conductivity* is therefore

$$\mathcal{K} \simeq \frac{k_B^2\,T\,\tau}{m_e}\frac{N}{V}.$$
$$(5.62)$$

By comparing σ and $\mathcal{K}$ in Eqs. (5.62) and (5.60) we predict a ratio

$$\frac{\mathcal{K}}{\sigma} = \frac{\pi^2}{3} \frac{k_{\mathrm{B}}^2}{q_e^2} T \tag{5.63}$$

between thermal and electric conductivities according to the relaxation-time model considered. The factor $\pi^2/3$ in Eq. (5.62) is obtained by means of a careful analysis of the factors of order unity previously omitted. This relation predicts that good electrical conductors are also good heat conductors. The empirical observation of this fact is known as the *Wiedemann-Franz law*. Experimentally, for most metals and across a broad temperature range, the ratio $\mathcal{K}/\sigma$ follows Eq. (5.63) remarkably well, even better than one may expect, given the model idealizations. Indeed, measurements of the ratio $\mathcal{K}/(\sigma T)$ normalized to the universal expression $\pi^2 k_{\mathrm{B}}^2/(3q_e^2)$, yield: 0.868 (Na, 273 K), 0.950 (Au, 273 K), 0.966 (Au, 373 K), 1.08 (Pb, 273 K), 1.04 (Pb, 373 K), indicating that the deviations from Eq. (5.63) are fairly small.

This simple Fermi-gas+relaxation-time model can explain several other transport experiments, in both the DC and the AC regimes. The Hall effect deserves a special mention: an electric field arises perpendicularly to a current running through a conductor immersed in a magnetic field $\mathbf{B}_{\mathrm{ext}}$ also perpendicular to the current. As sketched in Fig. 5.61, this Hall field is generated by the charge accumulating on the conductor sides due the Lorentz force of Eq. (5.55). The Lorentz force acting on the carriers is independent of the sign of their charge. Therefore, the sign of the charge buildup and of the resulting transverse Hall field probes the sign of the charge carriers. In most metals, the resulting Hall field is, as in Fig. 5.61, consistent with negative charge carriers. However, a number of metals (e.g. beryllium and cadmium) exhibit a striking reversed Hall field, as if their carriers had a positive charge! This inversion is one of the most spectacular consequences of the deviations of the crystalline bands from a simple free-electron parabola.

The crystal periodic potential has the effect of replacing the free-electron dispersion (a parabola whose curvature $\hbar^2/m_e$ is fixed by the electron mass) with nontrivial bands. As illustrated in Fig. 5.56, electrons in a crystal band accelerate at different

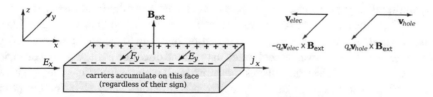

Fig. 5.61 A scheme of Hall's experiment: the Lorentz force $q\mathbf{v}_j(\mathbf{k}) \times \mathbf{B}_{\mathrm{ext}}$ in Eq. (5.55) deflects the charge carriers moving in a conductor immersed in a transverse magnetic field. This deflecting force F_y pushes carriers toward the sample front surface, regardless of the sign of the charge q of the carriers. In the steady state, the charge accumulated at the surface generates a transverse electric field E_y that balances the Lorentz force, and whose sign reflects the sign of the charge carriers. Here the charge signs are appropriate for negatively-charged electrons

rates compared to free electrons, due to the nontrivial **k**-dependent band curvature. According to Eqs. (5.54) and (5.55), the electron acceleration

$$\frac{d^2}{dt^2}\mathbf{r} = \frac{d}{dt}\mathbf{v}_j(\mathbf{k}) = \left[\nabla_{\mathbf{k}}\mathbf{v}_j(\mathbf{k})\right] \cdot \frac{d\mathbf{k}}{dt}$$
$$= \frac{1}{\hbar^2}\sum_{uw}\hat{\mathbf{e}}_u\left[\frac{\partial^2 \mathcal{E}_{\mathbf{k}\,j}}{\partial k_u \partial k_w}\right]\frac{d(\hbar k_w)}{dt} = \sum_{uw}\hat{\mathbf{e}}_u (m^*)_{uw}^{-1} F_w. \quad (5.64)$$

Here $\hat{\mathbf{e}}_1 = \hat{\mathbf{x}}$, $\hat{\mathbf{e}}_2 = \hat{\mathbf{y}}$ $\hat{\mathbf{e}}_3 = \hat{\mathbf{z}}$ indicate the Cartesian versors, and $\mathbf{F} = -q_e(\mathbf{E} + \mathbf{v}_j \times \mathbf{B}_{\text{ext}})$ represents the total external force acting on the electron. The final form of Eq. (5.64) expresses a sort of Newton equation, with an inverse mass *tensor* with components

$$(m^*)_{uv}^{-1} = \frac{1}{\hbar^2}\frac{\partial^2 \mathcal{E}_{\mathbf{k}\,j}}{\partial k_u \partial k_v}, \quad (5.65)$$

describing the band curvature. While this curvature is defined for any **k**, its relevant values are those obtained for **k** points at the Fermi surface, where the electrons active in transport sit. $|m^*|$ replaces the free-electron mass in Eqs. (5.60) and (5.62), thus accounting for the actual value of the acceleration of electrons close to the Fermi energy in a crystal-potential band. In 1D, m^* is proportional to the inverse of the band curvature, Fig. 5.56c. In 3D, the effective mass m^* is a suitable average over the tensor components $(m^*)_{uv}$, and may differ substantially from the free-electron mass m_e. In particular $|m^*|$ turns out significantly larger than m_e for narrow flat bands, characterized by a mild curvature, such as the bands of 3d electrons in transition metals or 4f electrons in lanthanide metals. According to Eqs. (5.60) and (5.62), a larger mass leads to smaller conductivity because, in between two collisions, external fields can accelerate heavier band electrons less than free electrons. Note however that the charge carrier mass cancels in the ratio $\mathcal{K}/\sigma$ of Eq. (5.63): the Wiedemann-Franz law should and does hold roughly independently of m^*.

Importantly, *negative* effective masses occur whenever the Fermi level sits in a region where the band curves downward (e.g. close to the BZ boundary of Fig. 5.57a). A negative-m^* electron accelerates in a direction opposite to the applied force: since its charge $(-q_e)$ is negative, it accelerates in the same direction as the external electric field. A negative-m^* electron behaves therefore in all ways as a particle with positive charge $(+q_e)$ and positive mass $|m^*|$, called a *hole*. Holes carry electric current in the same direction as the applied field (like genuine electrons), but produce reversed Hall effect, because they behave as positive charge carriers. Thus, a negative effective mass, i.e. hole conduction, accounts for the reversed Hall field of several metals.

5.2.2.2 Semiconductors

Semiconductors are insulators characterized by a ≈ 1 eV gap between the valence and conduction bands. For example, Fig. 5.62 sketches the bands of solid Si and Ge, semiconductors sharing the same crystal structure as C diamond but with larger lattice spacing a. In these elemental semiconductors, the Fermi energy sits inside the band gap between a full sp^3 bonding-type band and an empty sp^3 antibonding-type band. Note that the possibility of this band "splitting" is directly connected to the diamond crystal structure, and would not occur e.g. in a hypothetical monoatomic bcc or fcc Si. Precisely this band gap yields a large cohesive stability to the rarefied diamond structure of those crystals where the number of electrons matches the capacity of the bonding-type bands.[22] III–V (i.e. binary compounds of elements in groups 13 and 15, e.g. GaAs) and certain II-VI (compounds of elements in groups 2 or 12 and 16, e.g. BeSe and ZnS) compounds crystallize in a similar crystal structure, namely the zincblend structure, Fig. 5.25, with two different chemical species occupying the two geometrically inequivalent sites in the unit cell. These compounds are semiconductors, too. Other semiconductors exhibit different crystalline structures.

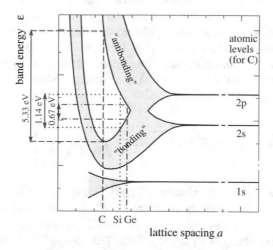

Fig. 5.62 Qualitative lattice-parameter dependence of the band energies for the elementary solids with the diamond structure. Starting from very large lattice spacing, as a is reduced, the narrow distinct s and p bands expand and combine into a hybrid sp^3 band. Eventually this hybrid band splits in a filled "bonding" band plus an empty "antibonding" band (similar to the bonding and antibonding states of CH_4). These bands move apart as the crystal shrinks ($a_{Ge} > a_{Si} > a_C$), leaving a bandgap $\Delta_{Ge} < \Delta_{Si} < \Delta_C$ in between. An uncommon bandgap increase under applied external pressure (and decrease with thermal expansion) is observed

---

[22]In contrast, elements with different numbers of valence electrons, e.g. Na, Mg, and Al, favor more compact crystal structures (bcc, hexagonal close-packed, and fcc respectively), which are energetically more stable. Their electrons would be insufficient to fill the bonding-type sp^3 band of the diamond structure, and that hypothetical structure would therefore be less stable.

Fig. 5.63 Direct versus indirect band gap in insulators and semiconductors. A direct gap can be probed spectroscopically by "direct" optical absorption/reflectivity. In contrast, absorption across an indirect gap requires some simultaneous phonon excitation/de-excitation, in order to fulfill the wave-number conservation in the process, see Sect. 5.2.3

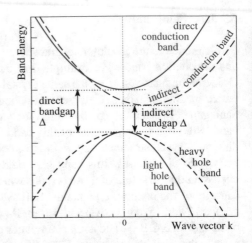

Due to structural and chemical differences, the electronic bands of different crystals exhibit qualitative and quantitative differences. In particular, the gap $\Delta = \mathcal{E}_c - \mathcal{E}_v$ between the top of the valence band and the bottom of the conduction band can be either direct (with the minimum $\mathcal{E}_c$ of $\mathcal{E}_{\mathbf{k}c}$ and the maximum $\mathcal{E}_v$ of $\mathcal{E}_{\mathbf{k}v}$ at the same $\mathbf{k}$ point, as in GaAs, GaN, and InP) or indirect (with these two extremes occurring at different $\mathbf{k}$ points, as in Si, Ge, GaP), see Fig. 5.63.

Transport in a pure (*intrinsic*) semiconductor is dominated by the thermal excitation of electrons from the valence into the conduction band. The average number density of such excited electrons is

$$N_c = \frac{[n_c]}{V} = \frac{1}{V} \int_{\mathcal{E}_c}^{\infty} g(\mathcal{E}) \, [n_{\mathcal{E}}]_{\mathrm{F}} \, d\mathcal{E} = \frac{1}{V} \int_{\mathcal{E}_c}^{\infty} g(\mathcal{E}) \, \frac{1}{e^{\beta(\mathcal{E}-\mu)} + 1} \, d\mathcal{E}, \qquad (5.66)$$

where $g(\mathcal{E})$ is the density of band states, of the type sketched in Fig. 5.64, and $\mathcal{E}_c$ is the conduction-band bottom. The chemical potential lies somewhere in the gap between conduction and valence band, e.g. close to the mid-gap energy. As a result $\mathcal{E}_c - \mu$ amounts to several times $k_{\mathrm{B}}T$, see Fig. 5.65. Therefore $e^{\beta(\mathcal{E}-\mu)} \gg 1$, thus we can approximate $[n_{\mathcal{E}}]_{\mathrm{F}} = (e^{\beta(\mathcal{E}-\mu)} + 1)^{-1} \simeq e^{-\beta(\mathcal{E}-\mu)} = e^{-\beta(\mathcal{E}_c-\mu)} \, e^{-\beta(\mathcal{E}-\mathcal{E}_c)}$. The first factor is the same for all states in the band: it reflects the exponential suppression of the electron occupation in the conduction band due to its distance from μ. The second factor is the standard Boltzmann occupation of a high-temperature ideal gas, as in Eq. (4.10): fermion statistics has little or no effect in such an extremely rarefied electron gas. The energies $\mathcal{E}$ can be estimated by expanding the conduction band around its minimum[23]

---

[23]For energies far above the band minimum $\mathcal{E}_{\mathbf{k}c}$, the quadratic expansion is inaccurate, but the statistical occupation factor suppresses the contribution of the higher-energy states anyway.

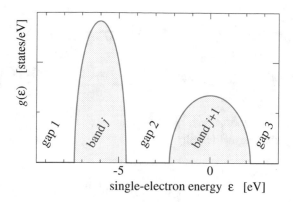

Fig. 5.64 A cartoon density of electronic states in a crystal. An actual solid usually exhibits narrow high-density bands at lower energy, and broader (often overlapping) lower-density bands at higher energy. The band boundaries show a characteristic $g(\mathcal{E}) \propto |\mathcal{E} - \mathcal{E}_{\text{boundary}}|^{1/2}$ energy dependence, a consequence of the quadratic $\mathbf{k}$ dependence of $\mathcal{E}_{\mathbf{k}\,j}$ near a band maximum or minimum, see Eq. (4.48). Inside the gaps, $g(\mathcal{E}) = 0$

Fig. 5.65 In an intrinsic semiconductor, at low temperature the chemical potential μ lies near the mid-gap energy $(\mathcal{E}_c + \mathcal{E}_v)/2$. The energy distances of μ to the edges of both conduction and valence band are much larger than $k_B T$. Accordingly, the electron occupation of individual conduction band states $[n_{\mathcal{E}_c}]_F$ is very small and that of valence states $[n_{\mathcal{E}_v}]_F$ is very close to 100%

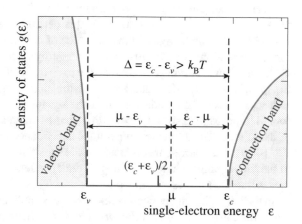

$$\mathcal{E} = \mathcal{E}_{\mathbf{k}\,c} = \mathcal{E}_c + \frac{1}{2} \sum_{uw} \left. \frac{\partial^2 \mathcal{E}_{\mathbf{k}\,c}}{\partial k_u \partial k_w} \right|_{\mathbf{k}^{\min}} (k_u - k_u^{\min})(k_w - k_w^{\min}) + \cdots$$

$$\simeq \mathcal{E}_c + \frac{\hbar^2}{2m_c^*} |\mathbf{k} - \mathbf{k}^{\min}|^2 + \cdots$$

Accordingly, the excitation energy $E = (\mathcal{E} - \mathcal{E}_c)$ in the second exponential factor takes the approximate form of the kinetic energy of a free particle with mass m^*, once the wave numbers are referred to $\mathbf{k}^{\min}$. As a result, the density of conduction states (excluding spin) goes as $g_{\text{tr}}(E) = m_c^{*3/2} V/(\sqrt{2}\pi^2\hbar^3) E^{1/2}$ [see Eq. (4.48)]. The calculation of the $\mathcal{E}$ integration in Eq. (5.66) is then identical to the calculation of the classical partition function $Z_{1\,\text{tr}} = V/\Lambda_c^3$ of Eq. (4.42). In detail, we obtain

$$N_c = \frac{e^{-\beta(\mathcal{E}_c-\mu)}}{V} \int_{\mathcal{E}_c}^{\infty} g(\mathcal{E}) e^{-\beta(\mathcal{E}-\mathcal{E}_c)} \, d\mathcal{E} \simeq \frac{e^{-\beta(\mathcal{E}_c-\mu)}}{V} \int_0^{\infty} g_s \, g_{\mathrm{tr}}(E) \, e^{-\beta E} \, dE$$

$$\simeq e^{-\beta(\mathcal{E}_c-\mu)} \frac{g_s \, Z_{1\,\mathrm{tr}}}{V} = e^{-\beta(\mathcal{E}_c-\mu)} \frac{2}{\Lambda_c^3} = e^{-\beta(\mathcal{E}_c-\mu)} \, 2 \left(\frac{m_c^* k_{\mathrm{B}} T}{2\pi \hbar^2} \right)^{3/2}, \quad (5.67)$$

where the factor $g_s = 2$ reflects the spin degeneracy.[24]

In Eq. (5.67), the thermal length

$$\Lambda_c = \sqrt{\frac{2\pi \hbar^2}{m_c^* k_{\mathrm{B}} T}} \simeq 6 \text{ nm},$$

at $T = 300$ K and taking $m_c^* \simeq 0.5 \, m_e$. For μ sitting at the middle of a 1.1 eV gap, i.e. $\mathcal{E}_c - \mu = 0.55$ eV, the exponential factor is in the order $e^{-\beta(\mathcal{E}_c-\mu)} \approx e^{-21} \simeq 6 \times 10^{-10}$. Thus Eq. (5.67) yields $N_c \approx 5 \times 10^{15}$ m^{-3}, a modest charge-carrier density compared to that of regular metals ($\approx 10^{28}$ m^{-3}). Note however that this carrier density varies exponentially with T^{-1} (e.g. for the same conditions, $N_c \approx 6 \times 10^{20}$ m^{-3} at $T = 600$ K). If this dependence is plugged into the expression (5.60) for conductivity in the presence of collisions, one expects a rather poor room-temperature conductivity, which is rapidly increasing with the exponential of T^{-1}, with slowly varying corrections due to (i) drifts of μ, (ii) the $\Lambda_c^{-3} \propto T^{3/2}$ term in Eq. (5.67), and (iii) a decreasing τ due to the increasing collision rate with phonons. Indeed, intrinsic semiconductors (e.g. the square data points in Fig. 5.66) exhibit a room-temperature resistivity exceeding that of metals (Table 5.2) by many orders of magnitude. The dramatic temperature dependence (approximately exponential in T^{-1}) of ρ makes pure semiconductors quite sensitive temperature sensors, especially in the few-K range where the almost-T-independent resistivity of metals, Fig. 5.59, makes the latter useless for this purpose.

In technology, semiconductors find countless applications mainly as "doped" crystals, called *extrinsic semiconductors*. Doping is realized typically by substitutional

---

[24] A very similar result is obtained for the number of *holes* in the valence band:

$$P_v = \frac{1}{V} \int_{-\infty}^{\mathcal{E}_v} g(\mathcal{E})(1 - [n_\mathcal{E}]_{\mathrm{F}}) \, d\mathcal{E} \simeq e^{-\beta(\mu-\mathcal{E}_v)} \frac{2}{\Lambda_v^3} = e^{-\beta(\mu-\mathcal{E}_v)} \, 2 \left(\frac{m_v^* k_{\mathrm{B}} T}{2\pi \hbar^2} \right)^{3/2}.$$

For a pure semiconductor, the requirement of charge neutrality, $N_c = P_v$, fixes the position of the chemical potential. Simplifying common factors, we have $e^{-\beta(\mathcal{E}_c-\mu)} m_c^{*3/2} = e^{-\beta(\mu-\mathcal{E}_v)} m_v^{*3/2}$, i.e. $e^{\beta(2\mu-\mathcal{E}_c-\mathcal{E}_v)} = (m_v^*/m_c^*)^{3/2}$. By taking the logarithm at both sides we obtain

$$\mu = \frac{1}{2}(\mathcal{E}_v + \mathcal{E}_c) + \frac{3}{4} k_{\mathrm{B}} T \ln \frac{m_v^*}{m_c^*}.$$

This relation confirms that indeed for $T \to 0$ the chemical potential sits at the middle of the gap. When $T > 0$, μ drifts gently toward the band characterized by the smaller effective mass, to compensate for the smaller density of states in that band, see Fig. 5.65.

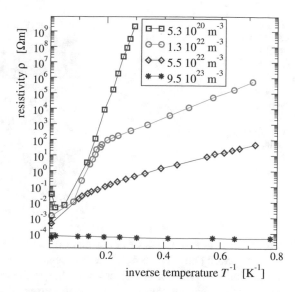

Fig. 5.66 The measured resistivity of antimony-doped germanium as a function of the inverse temperature, for a few donor-impurity concentrations: from 5.3×10^{20} m^{-3} (i.e. 0.012 parts per million atoms) to 9.5×10^{23} m^{-3} (i.e. 21 parts per million atoms). Comparing with Fig. 5.59 and Table 5.2, observe that the resistivity of this semiconductor is far larger than that of metals and that it changes far more dramatically with temperature (Data from Ref. [39])

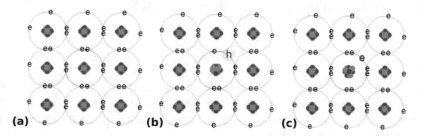

Fig. 5.67 Substitutional atoms of group 13 or group 15 replace a few Si or Ge atoms of the pure semiconductor (**a**), producing extrinsic semiconductors of p type (**b**) or n type (**c**) respectively. The square lattice is just a convenient pictorial for the actual diamond structure

impurities replacing a few of the perfect-crystal atoms, as sketched in Fig. 5.67. Pentavalent impurities, such as P or As replace Si/Ge atoms at the regular lattice sites, thus formally establishing four chemical bonds. As a result, the crystal structure and bands are not significantly modified by the rarefied impurities. On the other hand, each pentavalent atom carries one extra positive nuclear charge, which generates a potential well, which tends to bind an electron in addition to those in the filled valence band. This energy well produces a characteristic localized "impurity" state, associated to an energy level inside the band gap. At zero temperature, the extra electron provided by the pentavalent atom occupies this impurity state.

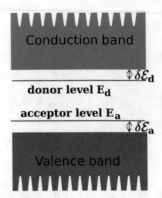

Fig. 5.68 Localized donor and acceptor levels are very shallow. They are usually located within few tens meV of the conduction-band minimum and valence-band maximum, respectively. Room temperature is hot enough to excite most of the charge carriers out of these localized states into the extended band states

The main feature of the impurity states of pentavalent dopants (*donors*) is their close vicinity to the conduction band (Fig. 5.68). Due to screening, the extra electron is bound to the impurity ion very weakly. The electron binding to the impurity can be roughly described as a particle with charge $-q_e$ and mass m_c^* attracted to the impurity nucleus by the screened Coulomb potential of the impurity ion $\simeq q_e/(4\pi\epsilon_0\varepsilon_r r)$, where ε_r is the relative permittivity of the pure semiconductor ($\varepsilon_r \simeq 12$ for Si, $\varepsilon_r \simeq 16$ for Ge). The bound-state energy levels of this 1-electron "atom" are given by Eq. (2.10), replacing the nuclear charge Z with ε_r^{-1} and the reduced mass μ with m_c^*. The ground-state energy equals a suitably-rescaled Rydberg energy:

$$\delta\mathcal{E} \simeq \frac{m_c^*}{m_e} \frac{1}{\varepsilon_r^2} \frac{E_{\text{Ha}}}{2}. \tag{5.68}$$

Typical values of m_c^* and ε_r lead to binding energies in the order $10^{-3}\, E_{\text{Ha}} \simeq 30\,\text{meV}$. Indeed, the separations $\delta\mathcal{E}_d = \mathcal{E}_c - E_d$ of the P and As impurity levels from the bottom of the conduction band of Si are observed 44 and 49 meV respectively (12 and 13 meV respectively in Ge).

A trivalent impurity (*acceptor*) carries a localized excess negative charge, as long as the valence (bonding) states are filled. The missing electron can become a bound *hole*, weakly attracted to the excess negative charge at the impurity. In the electron picture this bound hole manifests itself as an additional electronic level E_a slightly above the top of the valence band. The hole is bound when this localized level is half empty, making the impurity electrically neutral. When, at a small energy cost $\delta\mathcal{E}_a = E_a - E_v$, an electron is excited from the top of the valence band into the impurity level, the impurity hosts an excess charge $-q_e$. The missing electron in the valence band can be seen as a delocalized unbound hole with charge $+q_e$.

Impurity states are localized and do not contribute to transport: band electrons/holes are the only charge carriers in crystals. At $T = 0$, a uniformly doped semiconductor behaves as an insulator, with the Fermi level sitting near either E_a or E_d, according to whether the density N_a of acceptors or that N_d of donors is larger. As temperature is raised from 0, the bound charges get rapidly unbound into the

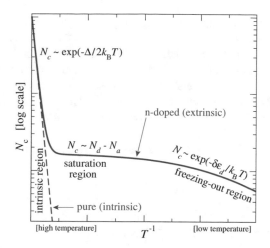

Fig. 5.69 A sketch of the temperature dependence of the majority carrier density in a n-doped semiconductor. The high-temperature regime is dominated by intrinsic carriers. In the intermediate "saturation" regime with almost constant $N_c \simeq N_d - N_a$, extrinsic carriers released from their impurities dominate N_c. The low-temperature decrease of N_c is due to the "freezing" of the electrons: at absolute zero they all get trapped at the impurity-bound states

band states: either in the valence band, if $N_a > N_d$ (*p doping*), or in the conduction band, if $N_d > N_a$ (*n doping*). The binding energy of the impurity levels is far smaller than the band gap: as a result, it is far easier to thermally excite an electron into the conduction band from a donor level (or a hole into the valence band from an acceptor level) than to excite an electron across the entire gap Δ, from valence to conduction band. At room temperature, the probability that electrons unbind from the impurity levels is high. As T is raised, the chemical potential moves slowly away from $E_{a/d}$ toward the middle of the gap. This leads to a much larger concentration of majority carriers (electrons for n doping, holes for p doping) than in the intrinsic semiconductor at the same temperature. In this "saturation" regime the density of majority carriers changes slowly and remains close to the net concentration of impurities,[25] as in Fig. 5.69. This carrier concentration, plugged into Eq. (5.60), is compatible with the doping and temperature dependence of resistivity shown in Fig. 5.66. In the saturation regime, doped semiconductors conduct like poor metals. In particular, p-doped semiconductors exhibit reversed Hall field, indicating hole conduction. The minority carrier density (electrons in the conduction band for p doping; holes in the valence band for n doping) is extremely small, and grows rapidly with T.

Doped semiconductors deliver a vast range of applications mostly as inhomogeneous devices, i.e. crystals where the impurity concentrations vary in space. Advanced techniques allow industries and labs to tailor the doping degree with sub-μm accuracy, across crystals with lateral size exceeding several cm. This technology is at the basis of the industry of electronics, where semiconductor devices act as the

---

[25]For n doping an electron density $N_c \simeq N_d - N_a$. For p doping a hole density $P_v \simeq N_a - N_d$.

active components which manipulate (e.g. amplify or shape) electric signals. Count-less other applications of inhomogeneous semiconductor include sensors, light pro-duction, data storage and manipulation... Since the 1950s, semiconductor devices have been a leading area of research and development, a common playground of fundamental QM, solid-state physics, materials science, industrial engineering, and electronics. Specific courses delve in this vast field. Here we only sketch the princi-ple of functioning of the simplest inhomogeneous extrinsic semiconductor: the *p-n junction*.

Consider a piece of semiconductor with ideal step-like p-n dopant densities

$$N_a(x) = \begin{cases} N_a, & x < 0 \\ 0, & x > 0 \end{cases}$$

$$N_d(x) = \begin{cases} 0, & x < 0 \\ N_d, & x > 0 \end{cases}, \qquad (5.69)$$

as a function of some displacement x across the sample (Fig. 5.70a). This is what one could conceptually (not practically) construct by assembling a p-doped and a n-doped piece of semiconductor. As required by thermodynamic equilibrium, the chemical potential in the two separate sections of the semiconductor, initially significantly different, must become the same after this contact is realized. Starting with each portion in a homogeneous neutral situation, the equilibration of the chemical potential is realized by the diffusion of electrons from the n side (where their concentration is higher) into the p side (where their concentration is smaller), and of holes in the opposite direction, like when a wall separating oxygen and nitrogen is removed from the middle of a vessel and the two gases mix. As diffusion proceeds, the resulting charge transfer builds up an electric field opposing further diffusive currents until a steady configuration is reached, with the electric forces on the charge carriers balancing the effect of diffusion.

The electric field $E_x = -d\phi/dx$ with the local densities of electrons and holes satisfy a set of coupled equations including Maxwell's $dE_x/dx = \rho/\epsilon_0$, Eqs. (5.67), (5.54), and (5.55). Here we summarize their qualitative outcome. Because the carriers are highly mobile, in the equilibrium configuration the carrier densities are very low wherever E_x has an appreciable value, see Fig. 5.70b. The resulting *depletion layer* is typically $10-10^3$ nm thick, depending mainly on the dopant concentrations and the relative permittivity of the semiconductor. The impurity-ion charges remain uncompensated in the depletion layer, thus producing the electric charge-density profile illustrated in Fig. 5.70c. This double layer of charge generates the electric field sketched in Fig. 5.70d. This field is associated to a finite electric potential drop (Fig. 5.70e), which compensates the difference in bulk chemical potential

$$q_e \Delta\phi_0 \equiv q_e[\phi(+\infty) - \phi(-\infty)] = \mu_n - \mu_p. \qquad (5.70)$$

Here μ_p and μ_n indicate the chemical potential in homogeneous bulk p or n semicon-ductors, prior to the construction of the junction. This potential barrier shifts rigidly

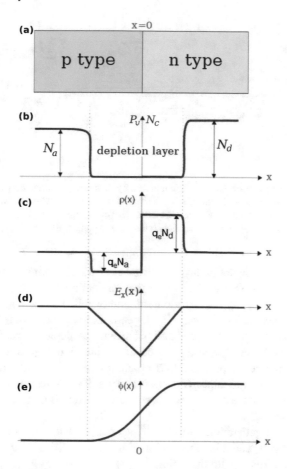

Fig. 5.70 **a** A sharp p-n junction at thermodynamic equilibrium (no voltage bias). **b** The qualitative profile of the hole density $P_v(x)$ (*left*) and electron density $N_c(x)$ (*right*). These quantities approach rapidly the bulk values $P_v(x \ll 0) \simeq N_a$, $N_c(x \gg 0) \simeq N_d$ that guarantee charge neutrality far away from the junction. In the region near the junction, both carrier concentrations are strongly suppressed: hence the name *depletion region*. **c** The total electric charge density $\rho(x)$: the uncompensated densities N_a and N_d of the ionized impurities dominate the double layer, with a net negative charge at the p side of the junction and a net positive charge at its n side. **d** This double layer of charge produces a leftward electric field $E_x(x)$, which sweeps the carriers away, maintaining the depletion. **e** The electric potential $\phi(x)$ consistent with $E_x(x) = -d\phi/dx$

the bands (and impurity levels) away from the junction, as illustrated in Fig. 5.71a. When metal electrodes are deposited on both sides of the junction and joined by a wire, at thermodynamic equilibrium the chemical potential aligns everywhere inside the circuit to a common value: no net current flows in the circuit, since the individual potential drops at the interfaces cancel algebraically.

In detail, at the p-n junction the net current vanishes due to the cancellation of four contributions:

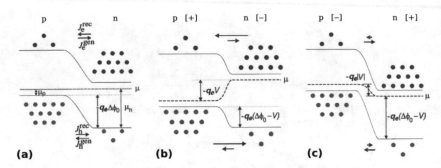

Fig. 5.71 The energy barrier $-q_e\phi(x)$ encountered by negatively-charged carriers, as they move across a p-n junction. **a** At equilibrium, no external bias. **b** In the presence of a forward bias ($V > 0$). **c** In the presence of a reverse bias ($V < 0$). *Horizontal arrows* represent generation currents (nearly independent of V) and recombination currents (changing exponentially in $q_e V / k_B T$)

- At the p side of the depletion layer, thermal excitations promote a few "minority" electrons out of the valence band into the conduction band: these electrons are immediately "swept" by the electric field, and expelled to the n side (Fig. 5.71a). This electron *generation current* J_e^{gen} depends exponentially on temperature, but only weakly on the potential drop and the size of the depletion region.
- A few (majority) electrons at the n side of the junction acquire enough thermal energy to overcome the potential barrier and arrive in the p side. In the p region, these electrons turn into minority carriers and are likely to recombine with (majority) holes in the valence band: the corresponding current is therefore called *recombination current* J_e^{rec}.
- Similarly, a hole generation current J_h^{gen} accounts for thermal holes swept from the n to the p side, and a hole recombination current J_h^{rec} accounts for thermally excited holes diffusing into the n side.[26]

At equilibrium these currents cancel in pairs ($J_e^{gen} = J_e^{rec}$, and $J_h^{gen} = J_h^{rec}$): no net current flows through the junction.

Imagine now to cut the external shorting wire, and to insert a voltage generator that adds a tunable electric potential V to the potential drop across the semiconductor, with the sign convention of Fig. 5.72a. Compared to the depletion layer, the bulk p and n regions are low-resistance conductors, thanks to their relevant carrier density. Accordingly, practically all the potential drop V associated to the external applied field occurs across the depletion region, as illustrated in Fig. 5.71b, c. The nonzero external field associated to $V \neq 0$ takes the junction away from thermodynamic equilibrium: a net current density j_x (thus a current I) is established through the junction. To understand semi-quantitatively the $I - V$ characteristic of the p-n junction, observe that the recombination currents depend very strongly on the electric potential drop across the depletion layer: J_e^{rec} is proportional to the number of carriers

---

[26]The four quantities $J_e^{rec/gen}$, $J_h^{rec/gen}$ are defined as the (positive) norm of the corresponding number current-density vectors, and are measured in units of $s^{-1}m^{-2}$.

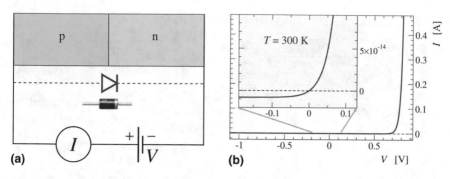

Fig. 5.72 **a** Circuit scheme for measuring the $I - V$ characteristics of a p-n junction. The sign convention has forward bias $V > 0$ increasing the electric potential at the p side and producing a fast-rising current $I > 0$; $V < 0$ (reverse bias) generates a tiny essentially V-independent reverse current $I < 0$. The diode electronic symbol plus a packaged silicon diode are also depicted under the p-n junction. **b** The ideal-diode $I - V$ characteristic according to Eq. (5.73)

acquiring sufficient energy to overcome the potential barrier. V modifies this barrier, so that $J_e^{rec} \propto \exp\left[-q_e(\Delta\phi_0 - V)/(k_B T)\right]$. The equilibrium requirement at $V = 0$ ($J_e^{rec} = J_e^{gen}$) and the fact that the generation current is almost independent of V fixes the proportionality constant:

$$J_e^{rec} = J_e^{gen} \exp\left(\frac{q_e V}{k_B T}\right). \tag{5.71}$$

The total current density carried by conduction electrons is therefore

$$J_e = J_e^{rec} - J_e^{gen} = J_e^{gen}\left[\exp\left(\frac{q_e V}{k_B T}\right) - 1\right]. \tag{5.72}$$

A similar analysis, with a similar result, can be carried out for holes. Compared to electrons, the hole currents J_h^{gen} and J_h^{rec} move in opposite directions, but carry positive charge, thus their contribution adds up to that of electrons, to give a total electric current density

$$j_x = q_e \left(J_e^{gen} + J_h^{gen}\right)\left[\exp\left(\frac{q_e V}{k_B T}\right) - 1\right]. \tag{5.73}$$

This ideal $I - V$ characteristic, reported in Fig. 5.72b, is highly asymmetric and nonlinear: basically the p-n junction operates as a rectifier, allowing electric current to circulate in one direction, like in the old diodes based on the thermionic emission of electrons in vacuum. For this reason, a two-terminal semiconductor device consisting of a single p-n junction is named a *diode*. Real diodes exhibit $I - V$ curves (Fig. 5.73) similar to Fig. 5.72b. Deviations include a reverse breakdown regime at large negative voltage, and resistive current limitation at large positive voltage.

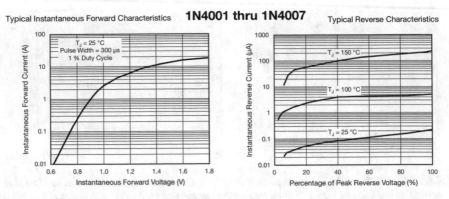

Fig. 5.73 The $I - V$ characteristics of a family of commercial silicon diodes. Observe the deviations from the ideal behavior of Eq. (5.73): (i) The exponential current increase as a function of forward bias V ends at $V \approx 0.8$ V, above which the current is limited by the series resistance of the homogeneous regions of the doped semiconductor. (ii) The reverse current $-I$ is significantly larger than predicted by Eq. (5.73) and increases gently with the reverse voltage $-V$, rather than remaining constant for $-V > 4k_B T/q_e \simeq 0.1$ V. (iii) A breakdown knee is observed as the reverse voltage exceeds the rated peak reverse voltage (50 to 1000 V for this family of diodes) (From the web site https://www.diotec-usa.com/; copyright 2010 by Diotec)

Carefully tailored commercial p-n junctions are built for numerous applications, of which we mention a few [40]. In the light-emitting diode (LED) a forward electrical bias causes both species of charge carrier—holes and electrons—to be "injected" into the depletion region, where their spontaneous radiative recombination generates IR or visible light, depending on the semiconductor gap amplitude, see Appendix B.3. A laser diode is similar to a LED, with the addition of an optical cavity surrounding the p-n junction to enhance stimulated emission. A photovoltaic cell is the reverse of a LED: light induces extra electron-hole pairs in the depletion layer; these carriers drift following the local electric field; the resulting charge separation reduces the junction electric-potential drop from its equilibrium value $\Delta\phi_0$, and sustains a current in an external circuit. The result is the conversion of radiation into electric power.

Beside 2-terminal devices, semiconductor single crystals are doped to form multiple p-n junctions. The simplest 3-terminal device is the bipolar *transistor*, a two-junctions sandwich, where the voltage (and a tiny current) across one junction controls the (usually larger) current through the second junction, thus providing amplification. Amplification is more often implemented by means of the constructively different field-emission transistor (FET). Diodes, transistors, and other semiconductor electronic components provide the ability to manipulate electronic signals. These components (by the thousands, or usually by the millions) are routinely *integrated* in $\sim$cm-sized semiconductor (usually silicon) single crystals, to provide the complex functionality at the core of modern microelectronics.

5.2.3 Spectra of Electrons in Solids

The experimental investigation of the energies of electrons in solids relies on several techniques. Here we discuss briefly three basic spectroscopies: photoemission, X-ray absorption/emission, and optical absorption/reflectivity.

Figure 5.74 illustrates the conceptual setup of a *photoemission* experiment. The kinetic-energy E_{kin} distribution $I(E_{kin})$ of the photoemitted electrons measures the density of *occupied states* of the crystal. This intensity is nonzero inside the bands, but vanishes in the forbidden energy gaps and above the Fermi level, see Fig. 5.64. The spectrum is shifted rigidly by the photon energy $\hbar\omega$. The observed continuous spectra probe the filled bands of the materials, essentially up to the chemical potential. Angle-integrated photoemission measures the density of occupied band states $g(\mathcal{E})[n_{\mathcal{E}}]_F$. Moreover, in metals, photoemission in the region $\mathcal{E} \simeq \mu$ probes the Fermi edge, i.e. the step-like drop of $[n_{\mathcal{E}}]_F$ close to μ, see Fig. 5.75. Angle-resolved photoemission provides further information about the **k**-dependence of the band energy $\mathcal{E}_{\mathbf{k}j}$. In insulating samples, a possible source of systematic error in photoemission spectra is the charge buildup caused by electron removal, that produces a macroscopic electric field outside the sample. Further factors limiting the quality of the spectra are: secondary collisions of the emitted electron while still inside the solid, and the short

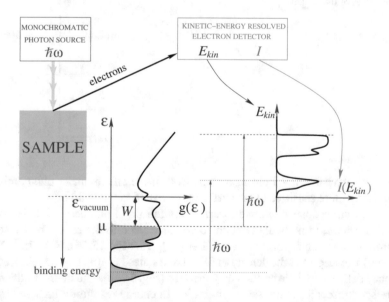

Fig. 5.74 The scheme and interpretation of a photoemission experiment. An electron initially in a band state of the solid absorbs a photon and acquires its energy $\hbar\omega$. If the energy of the resulting final state is sufficient, the electron can move out of the solid into the surrounding vacuum. A detector measures the kinetic-energy ($E_{kin} = \mathcal{E} + \hbar\omega$) distribution $I(E_{kin})$ of the photoemitted electrons, thus mapping the density of *occupied* states of the solid, up to the Fermi energy

Fig. 5.75 The observed photoemission spectrum of a gold film. "Binding energy" should indicate $\mathcal{E} = (\hbar\omega - E_{kin})$, as in the left scale of Fig. 4.13. However, as is common practice, here the scale zero is shifted from the vacuum level to the Fermi edge. The weak intensity between 0 and 1.5 eV represents the broad metallic (mainly 6s) band; the intense structures in the 2 − 7 eV range are mainly associated to the narrow 5d bands

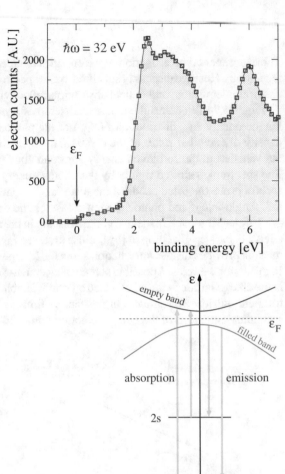

Fig. 5.76 An elementary interpretation of X-ray absorption and emission spectra in terms of transitions between the core levels and the band states of the crystal

penetration of visible/UV light in metals. Both these effects make photoemission spectroscopy more sensitive to the surface layers than to the solid bulk.

X-ray spectroscopies circumvent certain limitations of photoemission. Resonant *X-ray absorption*, in particular, excites core electrons into the empty bands states, thus probing their density of states $g(\mathcal{E})(1 - [n_\mathcal{E}]_F)$, see Figs. 5.76 and 5.77a.[27] *X-ray emission*, instead, probes the density of filled band states $g(\mathcal{E})[n_\mathcal{E}]_F$, whence electrons can decay to fill a core hole previously generated (Fig. 5.77b). The main disadvantage of core spectroscopies is the severe limitation in energy resolution imposed by the broad short-lived core-hole states (recall Sects. 1.2 and 2.2.8).

Optical excitations are usually probed by means of optical *reflectivity*, rather than absorption, since solids are often too opaque to transmit significant visible or UV

---

[27] The interpretation of these spectra is complicated by the presence of the core hole, which acts as a localized extra charge and distorts the bands as if an impurity was located at that site.

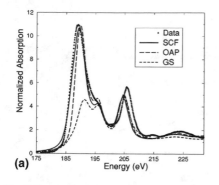
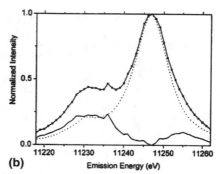

Fig. 5.77 **a** Points: the X-ray absorption spectrum of cubic boron nitride (BN) near the B 1s edge, probing the empty band levels of this material. Continuous curves: model calculations. Reprinted figure with permission from A. L. Ankudinov, B. Ravel, J. J. Rehr, and S. D. Conradson, Phys. Rev. B **58**, 7565 (1998). Copyright (1998) by the American Physical Society. **b** The 4d→2p$_{3/2}$ and 3p→2s emission spectra of Pt excited at respectively 11610 eV (dotted) and 14010 eV (connected points), probing the density of occupied states [41]. Reprinted figure with permission from F. M. F. de Groot, M. H. Krisch, and J. Vogel, Phys. Rev. B **66**, 195112 (2002). Copyright (2002) by the American Physical Society

radiation across any practical sample thickness. Both reflectivity and absorption are related to the material's complex dielectric permittivity function $\varepsilon_r(\omega)$, that accounts for both the slowing down and the attenuation of the radiation field in the material.

Due to intra-band electron excitations, in the low-frequency limit, the reflectivity of metals approaches 100%, see Fig. 5.78.[28] At larger energy, in the $\sim 2 - 10$ eV region, inter-band transitions dominate the reflectivity. Different colors of different metals (Cu, Ag, Au, Fe) indicate different spectral reflectivities, associated to their individual band structures.

Inter-band transitions conserve energy and momentum (or wave vector):

$$\hbar\omega = \mathcal{E}_{\mathbf{k}'\,j'} - \mathcal{E}_{\mathbf{k}\,j}, \qquad \mathbf{q} = \mathbf{k}' - \mathbf{k}, \tag{5.74}$$

where $\hbar\omega$ and $\mathbf{q}$ are the photon energy and wave vector. As usual in a crystal, the initial and final $\mathbf{k}$ and $\mathbf{k}'$ are defined up to a reciprocal lattice vector $\mathbf{G}$, and can therefore be reported inside the first BZ. Importantly, at optical wavelengths $\lambda \approx 500$ nm,

---

[28]In the long-wavelength small-photon-energy limit, the DC regime is approached, where a good conductor "short circuits" the electric-field component of radiation, thus efficiently suppressing it inside the material, and reflecting it outside. This $R \to 1$ limit describes the reflectivity of an infinitely-thick metal layer. Note however that the absorption coefficient α decreases slowly to 0 as the photon frequency decreases down into the radio-wave region. As a combined result of these two observations, in practice through a sufficiently thin metal layer, of thickness not exceeding a few times α^{-1}, radio-waves penetrate with detectable intensity and are less than 100% reflected. This is the reason why we can operate cell phones even inside e.g. a metallic elevator cab.

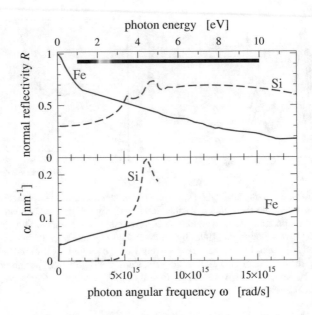

Fig. 5.78 The normal-incidence reflectivity R and the absorption coefficient α of a metal (iron, data from Ref. [42]) and of a semiconductor (silicon, data from Ref. [43]) at room temperature, for electromagnetic radiation in the IR-visible-UV region. In the static ($\omega \to 0$) limit, the reflectivity of iron approaches 100%, as typical of metallic intra-band transitions. In the same limit, the reflectivity of semiconductors and insulators approaches a constant value related to their static relative permittivity by $R = (\varepsilon_r^{1/2} - 1)^2/(\varepsilon_r^{1/2} + 1)^2 \simeq 30\%$ for Si, which has $\varepsilon_r \simeq 12$. The wiggles in the visible-UV region originate from inter-band transitions. In the far-UV and X-ray region both R and α decrease smoothly toward 0, except for specific peaks at the core-shell energies, see Fig. 2.22

the photon wave vector $|\mathbf{q}| \approx 10^7$ m^{-1}, while typical wave vectors of electrons in a solid are in the order of the size of the first BZ, i.e. approximately the reciprocal lattice spacing, in the $10^9 - 10^{10}$ m^{-1} region. Therefore in the conservation (5.74) the photon wave vector $\mathbf{q}$ can be neglected, implying $\mathbf{k}' \simeq \mathbf{k}$. In words: optical transitions are "vertical" in $\mathbf{k}$ space, as sketched in Fig. 5.79.[29]

Unlike metals, pure semiconductors and insulators have no intra-band excitations, and are therefore basically transparent at low frequency, with very weak absorption up to photon energies at least as large as the gap Δ. In direct-gap semiconductors, the absorption threshold measures the gap amplitude directly. However, when the gap is indirect (see Fig. 5.63), the onset of vertical transitions occurs at an energy larger than the actual gap. In silicon this difference is remarkable: $\Delta \simeq 1.17$ eV but very weak interband absorption occurs before $\sim$3 eV, see Fig. 5.78. With the assistance of

[29]In insulators and intrinsic semiconductors, the conduction electrons and valence holes generated by photon absorption can carry current as if they were originated by high temperature or by doping: this phenomenon, named "*photoconductivity*", is exploited in light sensors.

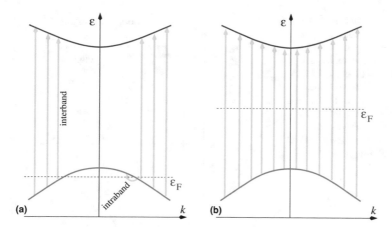

Fig. 5.79   A scheme of the excitations probed in optical absorption/reflectivity experiments of crystalline solids. **a** A metal; **b** a direct-gap insulator. Transitions are vertical due to wave-vector conservation

phonons providing the missing wave vector, weak "indirect" nonvertical transition right across the band gap can still occur.[30]

This simplified picture relating optical properties to differences in electronic band energies relies on the assumption that the photoexcited electron and hole behave as independent (quasi-)particles moving in the conduction and valence band, respectively. While this approximation may work in semiconductors, in wide-gap insulators the attraction between the electron and the hole cannot be neglected because it is screened poorly by the other electrons. This electron-hole attraction can modify strongly the optical absorption near and below the gap [12, 44], usually forming *excitons*, i.e. bound electron-hole pairs. An excitonic peak located well below the gap is prominent e.g. in the UV spectrum of LiF, see Fig. 5.80.

Impurities in otherwise perfect insulators provide localized states inside the band gap: transitions involving these impurity states often appear with strong intensity at photon energies far below Δ. For example, pure synthetic Al_2O_3 (corundum) is transparent to visible light ($\Delta \simeq 5$ eV), while impure Cr- or Ti-doped Al_2O_3 scatters characteristic red or blue light: in these doped forms Al_2O_3 gemstones are commonly called "ruby" or "sapphire", respectively, see Fig. 5.10c–d.

---

[30]The phonon energy is negligible compared to typical electronic energies. The need of an assisting phonon makes indirect transitions far less likely: for this reason, in applications involving light detection/production/control (optoelectronics) direct-gap semiconductors are usually preferred.

Fig. 5.80 Same as
Fig. 5.78, but for the
wide-bandgap insulator LiF.
An excitonic state, i.e. the
bound state of an electron in
the conduction band and a
hole in the valence band, is
responsible for the intense
peak below the bandgap
energy Δ (*dashed line*) (Data
from Ref. [45])

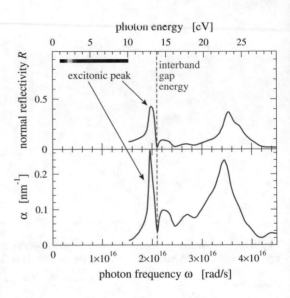

5.3   The Vibrations of Crystals

In Sect. 5.2 the atomic nuclei were assumed to sit quietly at their more-or-less ideal
crystalline equilibrium positions. This static-ion model is fine for describing many
properties of solids. However, several other observed properties including elasticity,
heat capacity, and heat transport depend on the dynamics and thermodynamics associated to atomic vibrations. The reader will observe a similarity with the investigation
of the internal motions of molecular gases—Sects. 3.3 and 4.3.1.2.

5.3.1   The Normal Modes of Vibration

Let us go back to the Eq. (3.9) for the motion of the nuclei. We assume that the
adiabatic potential $V_{ad}(R)$ of Eq. (3.10) has a single well-defined minimum.[31] Like
we did for molecules in Eq. (3.20), we expand $V_{ad}(R)$ around its minimum R_M. As
for molecules, the first-order term proportional to $\nabla V_{ad}(R)$ vanishes at R_M because
the minimum is a stationary point:

---

[31] For many materials, a unique absolute minimum R_M of $V_{ad}(R)$ represents the perfect crystal.
However, the actual configuration of a solid often deviates substantially, as it remains trapped in a
defective configuration, represented by one of many deep local minima occurring in the configurations space somewhere around R_M. The thermal and quantum tunneling rate through the barrier
leading from the defective configuration to the perfect crystal is often negligible, so that the defected
configuration is metastable. The vibrations around a moderately defective local minimum might
not differ too much from those around the crystalline absolute minimum.

$$V_{ad}(R) = V_{ad}(R_M) + \frac{1}{2} \sum_{\alpha\beta} \frac{\partial^2 V_{ad}}{\partial R_\alpha \partial R_\beta}\bigg|_{R_M} u_\alpha u_\beta + O(u^3). \tag{5.75}$$

In Eq. (5.75) we introduce the vector u with components $u_\alpha = R_\alpha - R_{M\alpha}$, collecting the atomic displacements away from the minimum. We label the atomic nuclei $j = 1, 2, \ldots N_n$, their Cartesian components $\xi = x, y, z$, and we introduce the collective index $\alpha = j, \xi$, which spans $N_n \times 3$ values. Explicitly:

$$\begin{pmatrix} u_{1x} \\ u_{1y} \\ u_{1z} \\ u_{2x} \\ u_{2y} \\ u_{2z} \\ u_{3x} \\ \vdots \\ u_{N_n x} \\ u_{N_n y} \\ u_{N_n z} \end{pmatrix} = \begin{pmatrix} R_{1x} \\ R_{1y} \\ R_{1z} \\ R_{2x} \\ R_{2y} \\ R_{2z} \\ R_{3x} \\ \vdots \\ R_{N_n x} \\ R_{N_n y} \\ R_{N_n z} \end{pmatrix} - \begin{pmatrix} R_{M1x} \\ R_{M1y} \\ R_{M1z} \\ R_{M2x} \\ R_{M2y} \\ R_{M2z} \\ R_{M3x} \\ \vdots \\ R_{MN_n x} \\ R_{MN_n y} \\ R_{MN_n z} \end{pmatrix}. \tag{5.76}$$

Call Φ the $3N_n \times 3N_n$ (Hessian) matrix of the second derivatives of $V_{ad}(R)$: the element $\Phi_{\alpha\alpha'} = \Phi_{j\xi\, j'\xi'}$ in the second-order term of the expansion (5.75) amounts to minus the change in the ξ-force acting on atom j due to a unit displacement of atom j' in direction ξ'. By neglecting all terms of third and higher order in u in Eq. (5.75) we obtain the *harmonic approximation*. This approximation is meaningful to describe small oscillations around equilibrium, thus it is usually appropriate for solids at low temperature, but it is expected to fail e.g. when the solid is highly deformed, or near melting.

The Schrödinger equation (3.9) could be solved exactly for a multi-dimensional quadratic potential. Unfortunately, this fully quantum approach involves significant mathematical intricacies. It is preferable to first solve the *classical* Newton-Hamilton equations for the motion of the ions, and introduce QM at a later stage. The classical equation of motion for the vector u of the $3N_n$ displacements is:

$$M_n \frac{d^2 u}{dt^2} = -\nabla_u V_{ad}(R_M + u) \tag{5.77}$$

$$\simeq -\nabla_u \left[V_{ad}(R_M) + \frac{1}{2} \sum_{\alpha'\beta'} \Phi_{\alpha'\beta'} u_{\alpha'} u_{\beta'} \right] = -\frac{1}{2} \nabla_u \sum_{\alpha'\beta'} \Phi_{\alpha'\beta'} u_{\alpha'} u_{\beta'}.$$

For simplicity, here we assume that all atoms have the same mass M_n. Equation (5.77) can be written for a generic α-component of u, by noting that in the double sum u_α appears either as $u_{\alpha'}$, when $\alpha' = \alpha$ or as $u_{\beta'}$, when $\beta' = \alpha$. The two derivative terms involve a identical sum (on an index, say, β) and cancel the factor $1/2$:

$$M_n \frac{d^2 u_\alpha}{dt^2} = - \sum_\beta \Phi_{\alpha\beta} \, u_\beta. \tag{5.78}$$

We have obtained a set of $3N_n$ equations for $3N_n$ coupled harmonic oscillators. Couplings are introduced by the off-diagonal matrix elements of Φ, i.e. by the forces that the displacement of an atom produces on other atoms.

A *normal mode* is a periodic motion with all atoms oscillating at a single frequency. To investigate normal modes, we substitute $u(t) = \bar{u} \, e^{i\omega t}$, where $\bar{u}$ is a time-independent still undetermined vector. The actual displacements will be identified with the real part of the complex solutions for $u(t)$. By evaluating the time derivative, we transform the differential equation (5.78) for $u(t)$ into

$$M_n \, \omega^2 \, \bar{u}_\alpha = \sum_\beta \Phi_{\alpha\beta} \, \bar{u}_\beta. \tag{5.79}$$

This is an algebraic equation for $\bar{u}$, and it can be rewritten as

$$\sum_\beta \left(\omega^2 \, \delta_{\alpha\beta} - \frac{1}{M_n} \, \Phi_{\alpha\beta} \right) \bar{u}_\beta = 0. \tag{5.80}$$

In this form, we recognize the equation for the diagonalization of the matrix $D = M_n^{-1} \, \Phi$, the so-called dynamical matrix. The eigenvalues of D are the values ω^2 which make the determinant $|\omega^2 \, I - D|$ vanish. The eigenvector corresponding to each ω^2 is the nonzero vector $\bar{u}$ satisfying Eq. (5.80). With its $3N_n$ components, $\bar{u}$ describes the relative amplitudes and phases of the displacements of all N_n atoms in the 3D space. Different masses can be accounted for by a simple generalization of the dynamical matrix.[32]

The present formalism is entirely general: it applies equally well to the determination of the normal modes of a polyatomic molecule (see, e.g., Fig. 3.12), or of a crystal, or an amorphous solid. In each case, the calculation of the normal frequencies and displacements requires the diagonalization of a $3N_n \times 3N_n$ matrix.

For a crystal, we take advantage of the lattice symmetry to greatly simplify the diagonalization of D. This simplification is akin to the one provided by Bloch's theorem to the description of the electronic motions in a perfect crystal, Sect. 5.2. Consider first the case of a single atom per unit cell. We construct the following wave-type ansatz: $\bar{u}_\alpha \equiv \bar{u}_{j\xi} = e^{i\mathbf{k} \cdot \mathbf{R}_j} \, \epsilon_\xi$ is the ξ component of the displacement of atom j sitting in the cell identified by lattice translation $\mathbf{R}_j$. Here ϵ_ξ is the $\xi = x, y,$

---

[32] In practice, replacing M_n^{-1} with $M_{n\alpha}^{-1/2} \, M_{n\beta}^{-1/2}$ in the expression for the dynamical matrix $D_{\alpha\beta}$.

or z component of a still undetermined 3D vector. From Eq. (5.80), we obtain the following equation for ϵ:

$$\sum_{\xi'} \left[\omega^2 \delta_{\xi\,\xi'} - \tilde{D}_{\xi\,\xi'}(\mathbf{k}) \right] \epsilon_{\xi'} = 0. \tag{5.81}$$

In Eq. (5.81) we introduce the Fourier-transformed dynamical matrix D:

$$\tilde{D}_{\xi\,\xi'}(\mathbf{k}) = \sum_{j} e^{-i(\mathbf{R}_j - \mathbf{R}_{j'})\cdot\mathbf{k}}\, D_{j\xi\,j'\xi'}. \tag{5.82}$$

This 3×3 matrix is independent of the choice of $\mathbf{R}_{j'}$ due to the lattice-translational invariance of the dynamical matrix:

$$D_{j\xi\,j'\bar{\xi}} = D_{j''\xi\,j'''\bar{\xi}} \qquad \text{if} \quad \mathbf{R}_j - \mathbf{R}_{j'} = \mathbf{R}_{j''} - \mathbf{R}_{j'''}. \tag{5.83}$$

Equation (5.81) shows that the normal modes of a monoatomic crystal are the result of the diagonalization of a 3×3 $\mathbf{k}$-dependent matrix. As a consequence, for each $\mathbf{k}$ point in the first BZ, 3 normal modes and associated frequencies $\omega_s(\mathbf{k})$ (labeled by an index $s = 1, 2, 3$) are obtained. The number of $\mathbf{k}$ points in the BZ is precisely N_n, see Eq. (5.53): overall we retrieve the expected $3N_n$ normal modes.

The example of the vibrations of a string, drawn in Fig. 5.81, suggests that for a given $\mathbf{k}$ (here, a given wavelength), three mutually orthogonal oscillations are observed: one *longitudinal* "compression-dilation" mode associated to displacements parallel to the string, plus two *transverse* modes associated to perpendicular displacements. Likewise, in 3D solids for each $\mathbf{k}$, the three vibrational frequencies correspond to one (mainly) longitudinal and two (mainly) transverse modes. The string example also suggests that, for small $|\mathbf{k}|$ (long wavelength), the frequencies of these modes should be proportional to $|\mathbf{k}|$. Indeed, in this limit, each frequency exhibits an "acoustic" dispersion $\omega(\mathbf{k}) \simeq v_s|\mathbf{k}|$, like sound waves in air. In crystals, three generally different and $\hat{\mathbf{k}}$-dependent sound velocities v_s, one for each *acoustic branch* are observed. Figure 5.82 sketches these dispersion branches. The zero-frequency acoustic modes at $\mathbf{k} = 0$ are the "infinite-wavelength vibrations" representing the free $\hat{\mathbf{x}}$, $\hat{\mathbf{y}}$, and $\hat{\mathbf{z}}$ rigid translations of the solid.

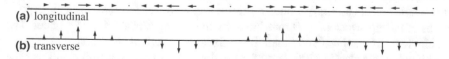

(a) longitudinal

(b) transverse

Fig. 5.81 A string exhibits three normal modes of vibration, for each fixed $\mathbf{k}$, i.e. for each given wavelength $2\pi/|\mathbf{k}|$, with $\mathbf{k}$ pointing along the string. **a** A longitudinal mode, with displacements parallel to $\mathbf{k}$. **b** Two transverse modes with displacements perpendicular to $\mathbf{k}$: the drawn one, plus a similar one with displacements orthogonal to the plane of the figure

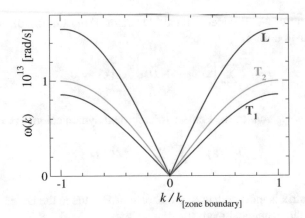

Fig. 5.82 Typical phonon dispersion curves along a generic direction in **k**-space for a crystal with 1-atom per cell. The two lower curves track the frequencies of the mainly transverse modes, the upper curve, the mainly longitudinal mode. The slopes of the three branches for $|\mathbf{k}| \simeq 0$ are the three sound velocities v_s. In some high-symmetry **k**-space direction the two transverse phonon branches could be degenerate

If n_d atoms occupy each crystal unit cell,[33] the problem complicates slightly because of the need to label the displacements of individual atoms within the cell: now ξ spans a total $3n_d$ different components, rather than just 3 as for the monoatomic crystal. The eigenvector components ϵ_ξ to be determined are $3n_d$. These eigenvectors are the result of the diagonalization of a $3n_d \times 3n_d$ dynamical matrix $\tilde{D}$, still far smaller than the original $3N_n \times 3N_n$ problem. As a result, for each **k**, we have $3n_d$ modes. Three of these modes are analogous to the acoustic modes of a monoatomic crystal, and represent deformation waves with all atoms in each unit cell oscillating essentially in phase. The remaining $(3n_d - 3)$ modes represent atomic vibrations that leave the center of mass of each cell immobile. These *optical modes* can be visualized by analyzing the simple model problem of a linear chain with springs of alternating strength connecting nearest-neighboring point masses. At all **k**, these $(3n_d - 3)$ optical dispersion branches remain characteristically far from $\omega = 0$, as illustrated in Fig. 5.83 for a crystal with $n_d = 2$ atoms per cell.

In summary, the harmonic normal modes of a general crystal have frequencies $\omega_s(\mathbf{k})$, and are labeled by **k** plus an index $s = 1, 2, \ldots 3n_d$ identifying the dispersion branch. The normal displacements form a basis for the classical dynamics of the whole crystal: arbitrary atomic motions, e.g. wave packets, are linear superpositions of the eigenvectors of $\tilde{D}$ multiplied by the appropriate $e^{i[\mathbf{k}\cdot\mathbf{R}-\omega_s(\mathbf{k})]t}$ factor. The coefficients of the linear combination are fixed by the initial condition.

Now that the classical problem of normal modes is under control, recall that the atomic positions are quantum degrees of freedom rather than classical ones. They should therefore follow Schrödinger's equation (3.9) rather than Newton's equation (5.77). Accordingly, we must upgrade the normal-mode classical displacements to

---

[33] So that the total number of atoms $N_n = n_d \times N_{cell}$. N_{cell} is the number of cells in the crystal.

Fig. 5.83 Typical phonon dispersion curves along a general direction in **k**-space for a crystal with $n_d = 2$ atoms per cell. For small $|\mathbf{k}|$, the three acoustic dispersion branches are approximately linear in $|\mathbf{k}|$. The three upper curves (optical branches) are characterized by $\omega_s(\mathbf{0}) > 0$. They can become quite flat if interactions between cells are much weaker than the intra-cell ones

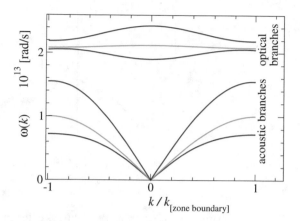

quantum operators, replacing the $3N_n$ classical oscillations with $3N_n$ quantum harmonic oscillators. A basis state for the Hilbert space of the whole vibrating crystal is then characterized by the number $v_{\mathbf{k}s}$ of excitations (*phonons*) of each oscillator, with frequency $\omega_s(\mathbf{k})$. In state $|\{v_{\mathbf{k}s}\}\rangle$, the vibrational energy of the whole crystal, viewed as a huge polyatomic molecule, Eq. (4.110), is

$$E_{\text{vib}}(\{v_{\mathbf{k}s}\}) = \sum_{\mathbf{k}s} \hbar\omega_s(\mathbf{k})\left(v_{\mathbf{k}s} + \frac{1}{2}\right). \tag{5.84}$$

Except for less trivial dispersions $\omega_s(\mathbf{k})$ in place of $\omega(\mathbf{k}) = c|\mathbf{k}|$, these vibrations resemble closely the oscillations of the electromagnetic fields in a cavity, whose thermodynamics we studied in Sect. 4.3.2.2. This close similarity suggests the name *phonons* for the quanta of these harmonic oscillators.[34] The creation of a phonon with given crystal wave vector **k** in branch s simply means raising the vibrational quantum number $v_{\mathbf{k}s}$ of the oscillator $(\mathbf{k}, s)$ by one. This excitation results in the total vibrational energy increasing by $\hbar\omega_s(\mathbf{k})$.

Phonon dispersions are investigated by means of several techniques, including *neutron inelastic scattering*, ultrasound echo, Raman spectroscopy, and IR absorption. All these techniques involve excitations or deexcitation of one or several phonon states. The outcome of these experiments generally agrees with sophisticated models based on the evaluation and diagonalization of the dynamical matrix D. Figures 5.84 and 5.85 show examples of phonon dispersions of real materials measured with inelastic scattering.

---

[34]Their formal similarity however should not make us forget that phonons are the quanta of the vibrations of atoms in crystals, while photons represent the oscillations of electromagnetic fields.

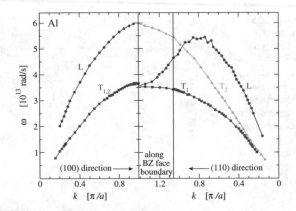

Fig. 5.84 Inelastic neutron scattering measurements of the phonon dispersion of crystalline (fcc) Al, along two k-space lines: (100) and (110). Note that the two transverse branches T_1 and T_2 remain degenerate along the (100) line, but exhibit distinct frequencies along (110)

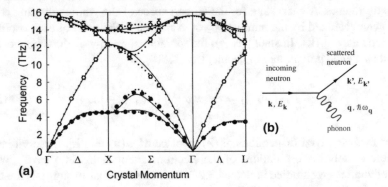

Fig. 5.85 **a** The observed phonon dispersion of silicon, measured by inelastic neutron (*circles*) and X-ray (*solid line*) scattering. Note the presence of both acoustic and optical branches, as appropriate for a (diamond) structure with a basis of $n_d = 2$ atoms per cell. The letters mark special points in **k**-space, e.g. Γ is **k** = **0**. Along high-symmetry paths (Γ-X and Γ-L) pairs of transverse branches are degenerate, while this degeneracy is resolved along other directions. Reprinted figure with permission from M. Holt, Z. Wu, H. Hong, P. Zschack, P. Jemian, J. Tischler, H. Chen, and T.-C. Chiang, Phys. Rev. Lett. **83**, 3317 (1999). Copyright (1999) by the American Physical Society. **b** The scheme of the inelastic scattering event

Optical phonons often appear as absorption peaks in the IR spectrum of otherwise transparent insulators. In ionic crystals, e.g. NaCl and LiF (Fig. 5.86), this absorption is especially intense, because optical displacements of oppositely-charged ions are strongly electric-dipole coupled to the electromagnetic radiation field.

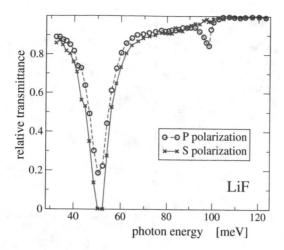

Fig. 5.86 Observed transmittance of IR radiation by a 200 nm-thick LiF film at room temperature. Radiation impinges on the film at $\sim 30°$ from vertical. The electric field is parallel/perpendicular to the plane of incidence for of P/S-polarized radiation, respectively. The peaks measure the frequencies of the transverse and longitudinal optical phonons, at energy $\hbar\omega \simeq 50$ meV and $\hbar\omega \simeq 100$ meV respectively. Momentum conservation [Eq. (5.74) and Fig. 5.85b] implies that IR radiation absorption probes the $\mathbf{k} \simeq \mathbf{0}$ region of the optical phonon branches (Data from Ref. [46])

5.3.2 *Thermal Properties of Phonons*

Phonons are essential to understand the thermal properties of solids. Indeed, vibrations dominate the heat capacity of all solids.[35] By further developing the analogy of photons and phonons, i.e. by applying the statistics of the harmonic oscillator as in Eq. (4.62), the total internal energy of the harmonic crystal is

$$U_{\text{vib}} = U_0 + \sum_{\mathbf{k}s} \epsilon_{\mathbf{k}s} [n_{\epsilon_{\mathbf{k}s}}]_{\text{B}} = U_0 + \sum_{\mathbf{k}s} \frac{\epsilon_{\mathbf{k}s}}{\exp\left(\frac{\epsilon_{\mathbf{k}s}}{k_{\text{B}}T}\right) - 1}, \qquad \text{with } \epsilon_{\mathbf{k}s} = \hbar\omega_s(\mathbf{k}).$$

$$(5.85)$$

U_0 includes the zero-point energy $\sum_{\mathbf{k}s} \hbar\omega_s(\mathbf{k})/2$, which is a material-dependent constant, irrelevant for thermodynamics. For each oscillator we use the result of Eq. (4.63) to obtain $\frac{\partial}{\partial T} U_{\text{vib}}$, namely the total heat capacity:

---

[35]With these noteworthy exceptions: (i) in metallic solids, at very low temperature (typically few K) the tiny electronic contribution (Fig. 4.15) dominates over the phonon heat capacity; (ii) occasionally, localized magnetic moments or impurity states may add a contribution which peaks at low temperature and may even exceed the contribution of the phonons.

Fig. 5.87 The measured molar heat capacity at constant volume for a few solids. At high temperature (but limited by melting), the $C_{V\text{vib}}$ of most solids comes close to the universal $3N_A k_B \simeq 24.9 \, \text{J} \, \text{mol}^{-1} \text{K}^{-1}$ Dulong-Petit value

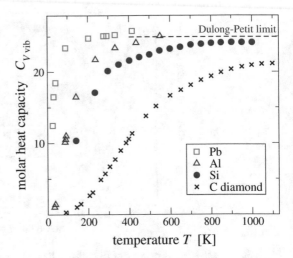

$$C_{V\text{vib}} = k_B \sum_{\mathbf{k}s} \frac{x_{\mathbf{k}s}^2 \exp(x_{\mathbf{k}s})}{(\exp(x_{\mathbf{k}s}) - 1)^2} = k_B \sum_{\mathbf{k}s} \left[\frac{x_{\mathbf{k}s}/2}{\sinh(x_{\mathbf{k}s}/2)} \right]^2, \text{ with } x_{\mathbf{k}s} = \frac{\hbar\omega_s(\mathbf{k})}{k_B T}.$$
(5.86)

Equations (5.85) and (5.86) depend on the details of the phonon dispersions. In practice however, experimentally crystals exhibit a fairly universal T-dependence of the molar specific heat (once the tiny T-linear electronic contribution of metals is subtracted) illustrated in Fig. 5.87: a $\propto T^3$ raise at low temperature (see also Fig. 4.15), and stabilization to a constant $\simeq 3N_A k_B$ (Dulong-Petit) limit at high temperature. To capture this T dependence, a popular simplified model is often adopted: the *Debye model*. This model replaces the details of the phonon dispersions by a fictitious linear dispersion (Fig. 5.88) reproducing the following minimal quantitative properties of the phonons of the material under consideration:

1. The average speed of sound v_s^* is adjusted to the average slope of the acoustic branches close to $\mathbf{k} = \mathbf{0}$. In this way, v_s^* provides the correct factor Ξ in the low-energy density of oscillator energies $g(\epsilon) = \Xi\epsilon^2$.[36] This makes the model fit the low-T limit, where only low-energy oscillators are excited significantly.
2. The total number of vibrational modes equals $3N_n$, which guarantees the correct high-temperature limit, in accordance with classical equipartition.

The latter point is implemented by replacing the sum over $\mathbf{k}$ and s in Eq. (5.85) with a sum over $\mathbf{k}$ extended over a spherical region of radius k_D (the Debye wave vector) and a factor $g_s = 3$ representative of the three acoustic branches. The eventuality of optical branches is effectively accounted for by the value of the cutoff k_D, which is fixed precisely by imposing that the total number of modes is $3N_n$. The allowed $\mathbf{k}$ points of a finite sample of N_n atoms in a $L \times L \times L$ cube are the discrete values

---

[36] In view of the speed dependence in Eq. (4.114), v_s^* is obtained as the angular average of the inverse cube of the three speeds of sound: $3v_s^{*-3} = \int (v_{s1}^{-3} + v_{s2}^{-3} + v_{s3}^{-3}) \, d\Omega_{\hat{\mathbf{k}}}/(4\pi)$.

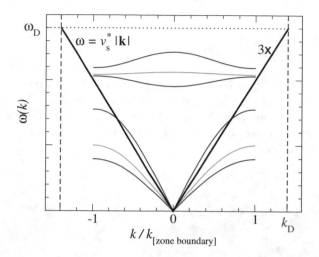

Fig. 5.88 The Debye model approximates the phonon branches of a real solid with an effective linear dispersion $\omega \simeq v_s^* |\mathbf{k}|$. The sound velocity v_s^* is taken as a suitable average of the $\mathbf{k} \to 0$ slopes of the three acoustic branches over all directions. The sum over the real phonon modes is replaced by an integration over a sphere of radius k_D and multiplication by a factor 3 representing the three acoustic branches, which are assumed degenerate. The wave-number cutoff k_D is of the same order as the size of the BZ. k_D is made sufficiently large so that integration over the Debye sphere counts correctly even the vibration modes associated to optical branches (if any)

$\mathbf{k} = \frac{2\pi}{L} \mathbf{n}$ [see Eqs. (4.38), (5.53)]. This observation allows us to determine the radius n_D and thus k_D of the sphere of allowed oscillators, by equating:

$$3N_n = g_s \sum_{|\mathbf{n}|<n_D} 1 \simeq g_s \, 4\pi \int_0^{n_D} n^2 \, dn = g_s \, \frac{4\pi}{3} n_D^3. \tag{5.87}$$

As a result $n_D = \left(\frac{3}{4\pi} N_n\right)^{1/3}$, and therefore $k_D = \frac{2\pi}{L} n_D = \left(6\pi^2 N_n / V\right)^{1/3}$. This wave vector corresponds to a Debye cutoff energy

$$\epsilon_D = \hbar\omega_D = \hbar v_s^* k_D = \hbar v_s^* \left(6\pi^2 \frac{N_n}{V}\right)^{1/3}. \tag{5.88}$$

The quadratic energy-density of photon oscillators, Eq. (4.114), takes a very similar form for phonons with a linear dispersion $\epsilon = \hbar\omega = \hbar v_s^* |\mathbf{k}|$:

$$g_s \, g_{\text{ph}}(\epsilon) = \begin{cases} 9 N_n \, \epsilon^2 / \epsilon_D^3 & \text{for } 0 \leq \epsilon \leq \epsilon_D \\ 0 & \text{elsewhere} \end{cases}, \tag{5.89}$$

where $9N_n/\epsilon_D^3 = 3/2 V/(\pi^2 \hbar^3 v_s^{*3})$. With this approximate (Debye) density of oscillators, the internal energy Eq. (5.85) becomes

$$U_{\text{vib}} = U_0 + \int_0^{\epsilon_D} g_s\, g_{\text{ph}}(\epsilon)\, \frac{\epsilon}{e^{\frac{\epsilon}{k_B T}} - 1}\, d\epsilon = U_0 + \frac{9\, N_n}{\epsilon_D^3} \int_0^{\epsilon_D} \frac{\epsilon^3}{e^{\frac{\epsilon}{k_B T}} - 1}\, d\epsilon. \tag{5.90}$$

Equation (5.90) reproduces formally the expression (4.115) for photons, except for (i) a factor $g_s = 3$ rather than 2, (ii) the speed of sound v_s^* in place of the speed of light c, and (iii) the cutoff ϵ_D in the energy integration. By introducing a dimensionless integration variable $x = \epsilon/k_B T$, and a Debye temperature $\Theta_D = \epsilon_D/k_B$, Eq. (5.90) is rewritten conveniently as

$$U_{\text{vib}} = U_0 + \frac{9 N_n k_B\, T^4}{\Theta_D^3} \int_0^{\Theta_D/T} \frac{x^3}{e^x - 1}\, dx. \tag{5.91}$$

At low temperature $T \ll \Theta_D$, the integration endpoint Θ_D/T can be approximated with infinity, and the resulting dimensionless integral $\int_0^\infty x^3/(e^x - 1)\, dx = \pi^4/15$ is also involved in the total blackbody energy, Eq. (4.115). In this limit, the internal energy is therefore

$$U_{\text{vib}}(T \ll \Theta_D) \simeq U_0 + 9 N_n k_B \frac{T^4}{\Theta_D^3} \frac{\pi^4}{15} = U_0 + \frac{3\pi^4}{5} N_n k_B \frac{T^4}{\Theta_D^3}. \tag{5.92}$$

Derivation with respect to T yields a heat capacity

$$C_{V\text{vib}}(T \ll \Theta_D) \simeq \frac{12\pi^4}{5} N_n k_B \left(\frac{T}{\Theta_D} \right)^3, \tag{5.93}$$

reproducing the experimental behavior at low temperature.

For large temperature, the ratio Θ_D/T is small, and so is the variable x in Eq. (5.91). The integrand function is then expanded as $x^3/(e^x - 1) \simeq x^2$. In this limit, the integral in Eq. (5.91) yields approximately $\frac{1}{3}(\Theta_D/T)^3$, thus the internal energy is

$$U_{\text{vib}}(T \gg \Theta_D) \simeq U_0 + 9 N_n k_B \frac{T^4}{\Theta_D^3} \frac{1}{3} \left(\frac{\Theta_D}{T} \right)^3 = U_0 + 3 N_n k_B T. \tag{5.94}$$

This form matches the classical equipartition limit, and yields a heat capacity

$$C_{V\text{vib}}(T \gg \Theta_D) \simeq 3 N_n k_B, \tag{5.95}$$

which reproduces the Dulong-Petit limit.

The general expression for the heat capacity, valid at arbitrary temperature, derives directly from Eqs. (5.86) and (5.89):

$$C_{V\text{vib}}(T) = 9N_n k_B \left(\frac{T}{\Theta_D}\right)^3 \int_0^{\Theta_D/T} \left[\frac{x^2/2}{\sinh(x/2)}\right]^2 dx. \tag{5.96}$$

The Debye curve (5.96) interpolates between the low- and high-temperature limits, see Fig. 5.89. The T dependence of $C_{V\text{vib}}(T)$ agrees closely with most experimental data of crystalline solids. This agreement on *thermodynamics* is striking, for a model which gathers all the complications of the phonon dispersions in a single parameter Θ_D (or, equivalently, ϵ_D, or ω_D) that can be determined purely through the measurement of *mechanical* properties of the solid, see Eq. (5.88).

The success of the Debye model highlights its parameter Θ_D as the main temperature scale characterizing the phonons of a crystal. Values of Θ_D, e.g. those listed in Table 5.3 for a few elementary crystals, range in the same region as the typical vibrational temperature Θ_{vib} of diatomic molecules (Sect. 4.3.1.2).

The Debye heat capacity, Fig. 5.89, exhibits a "quantum-unfreezing" increase with T, similar to what we observed for a single quantum oscillator, Fig. 4.8. The main difference, the low-temperature $C_{V\text{vib}} \propto T^3$ dependence in the solid [rather than $C_{V\text{vib}} \propto \exp(-\Theta_{\text{vib}}/T)$], is due to the presence of a continuum of oscillators

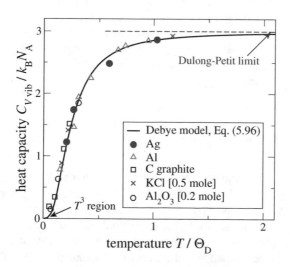

Fig. 5.89 The measured heat capacity (in units of $R = k_B N_A$) of five crystalline solids. Temperature is rescaled by the Debye temperature Θ_D of each material, so that all data follow the universal (Debye) curve of Eq. (5.96) (*solid line*). For a better comparison, here $C_{V\text{vib}}$ data refer to $N_n = N_A$ atoms in total: this represents 1 mole of Ag, Al, and C, but 0.5 mole of KCl (2 atoms per chemical formula), and 0.2 mole of Al_2O_3 (5 atoms per formula)

Table 5.3   The Debye temperature of a few elemental solids

Element	Θ_D [K]	Element	Θ_D [K]	Element	Θ_D [K]
Ag	225	C (diamond)	1,860	Na	156
Al	400	Cu	339	Ni	450
Au	165	Ge	374	Pt	240

Table 5.4   Compared analytic expressions for the low- and high-temperature heat capacity of several thermodynamic systems relevant for the physics of matter

Thermodynamic system	Low temperature	High temperature
Ideal spin-s system	$N k_B \left(\frac{g\mu_B B}{k_B T} \right)^2 \exp\left(-\frac{g\mu_B B}{k_B T} \right)$	$\frac{1}{3} s(s+1) N k_B \left(\frac{g\mu_B B}{k_B T} \right)^2$
Rotations of molecules	$3 N k_B \left(\frac{\hbar^2}{I k_B T} \right)^2 \exp\left(-\frac{\hbar^2}{I k_B T} \right)$	$N k_B$
Vibrations of molecules	$N k_B \left(\frac{\hbar\omega}{k_B T} \right)^2 \exp\left(-\frac{\hbar\omega}{k_B T} \right)$	$N k_B$
(Debye) vibrations of crystals	$\frac{12\pi^4}{5} N k_B \left(\frac{T}{\Theta_D} \right)^3$	$3N k_B$
Electromagnetic fields	$\frac{4\pi^2}{15} V k_B \left(\frac{k_B T}{\hbar c} \right)^3$	(holds at any temperature)
Gas of ideal Fermions	$\frac{\pi^2}{2} N k_B \frac{T}{T_F}$	$\frac{3}{2} N k_B$

with arbitrarily low frequency: even at $T \ll \Theta_D$, each oscillator with $\hbar\omega < k_B T$ remains quantum-unfrozen and contributes $\sim k_B$ to $C_{V\text{vib}}$.

It is instructive to compare the temperature dependence of the heat capacity of the Debye model to that of other thermodynamic systems studied earlier on. In particular, the reasons for all low- and high-T temperature dependences collected in Table 5.4 deserve a careful understanding.

5.3.3   Other Phonon Effects

In the absence of electronic mechanisms, phonons are the responsible of heat transport in insulating solids. With arguments of the type employed for electronic heat transport in Sect. 5.2.2.1, it is possible to evaluate the heat conducted by phonons, and to account quantitatively for the observational data [10, 11].

The macroscopic elastic properties of solids are directly related to the small-$\mathbf{k}$ region of the acoustic phonon branches, i.e. to the sound velocities. Rigid materials tend to have larger sound velocities than soft easily-deformable solids.

Thermal expansion is associated to crystal vibrations, too. However harmonic phonons do not account for any expansion, because $\langle v_{\mathbf{k}s} | u_{\mathbf{k}s} | v_{\mathbf{k}s} \rangle = 0$ for any phonon number $v_{\mathbf{k}s}$, thus at any temperature $[u_{\mathbf{k}s}] = 0$: a temperature raise only increases the average amplitude of oscillation around the same fixed equilibrium separation.

We conclude that thermal expansion arises entirely as a consequence of the *anharmonicity* of the adiabatic potential, i.e. of terms $O(u^{n \geq 3})$ in Eq. (5.75), which are present in real solids but are neglected in the harmonic theory of phonons.

Finally, the instantaneous displacements of the atoms away from their ideal crystal positions associated to thermal excitations affect structural investigations carried out by means of X-ray- and neutron-diffraction experiments. Similarly to U_{vib}, the mean amplitude $[u_\alpha^2]$ of the random thermal atomic displacements increases with temperature. At low temperature this increase is slow; and then for $T \gtrsim \theta_D$, $[u_\alpha^2] \propto T$. These random movements leave the diffracted-peak positions unchanged, but remove part of the intensity away from the peaks, and transfer it to a diffuse background.

Problems

A ⋆ marks advanced problems.

5.1 In the approximation of free non-interacting electrons at temperature 0 K compute the mean speed $\langle |v| \rangle$ of a conduction electron in solid aluminum, knowing that X rays with wavelength $\lambda = 100$ pm are diffracted by its fcc structure at a minimum angle $2\theta = 24.7°$, and recalling that every Al atom contributes 3 electrons to the conduction band.

5.2 Solid titanium has a density 4507 kg/m^3. Sound waves propagate through titanium at an average speed $v_s = 4.14$ km/s. In the Debye approximation, determine the low-temperature molar heat capacity of the phonons of titanium, specifically evaluating the parameters α and A which express its temperature dependence in the relation $C_V = A T^\alpha$. Evaluate also the molar heat capacity at a such a high temperature that the classical limit provides a good approximation for C_V. Finally, determine at what temperature the heat capacity reaches 1% of its maximum value.
[Recall the relation connecting the Debye cutoff frequency to the average speed of sound v_s and the number density N_{at}/V of the atoms in the solid: $\omega_D = v_s \left(6\pi^2 N_{at}/V\right)^{1/3}$.]

5.3 For the energy of the conduction band of a hypothetical one-dimensional cesium crystal, take the expression $\varepsilon_k = -A - B \cos(ka)$, where $A = 2.9$ eV, $B = 4.1$ eV, k is the wave vector, and $a = 125$ pm. The expression for ε_k assumes an energy scale whose zero refers to an electron at rest far from the solid. Evaluate the maximum wavelength of electromagnetic radiation capable to photoemit electrons from such a solid.

5.4 ⋆ Consider a one-dimensional crystal with lattice spacing a. Estimate the lowest-energy band gap at the BZ boundary in the approximation of quasi-free electrons (plane-waves basis), and assuming that the effective periodic potential acting on the electron takes the form

$$V(x) = E\left(\cos\frac{2\pi x}{a} + \frac{1}{3}\cos\frac{6\pi x}{a}\right),$$

where $E = 6.8$ eV. If each atom contributes two electrons, is the crystal a metal or an insulator? Is it opaque or transparent for visible light?

5.5* The structures of three different crystalline solids A, B, and C involve 2 different atoms (total mass $M_1 + M_2 = 120$ a.m.u.) occupying each primitive cell of the respective periodic lattices. A is simple-cubic (sc); B is face-centered cubic (fcc); C is body-centered cubic (bcc). Neutrons with 10 meV kinetic energy diffract over powder samples of A, B, and C, forming diffraction rings at several angles. For all three samples the smallest diffracted angle is $2\theta = 32.4°$. Evaluate the side a of the conventional cubic cells for each one of the 3 solids, determine which solid has the largest density, and the value of this density.

5.6 The vibrational contribution to the molar heat capacity C_V of aluminum is 0.0239 J/(mol K) and 0.0808 J/(mol K) at temperatures 10 K and 15 K, respectively. Knowing that X rays of wavelength $\lambda = 73$ pm are diffracted by its fcc structure at a minimum angle $2\theta = 18°$, evaluate the mean speed of sound v_s in this solid.
[Recall the relation connecting the Debye cutoff frequency, the speed of sound v_s and the number density N_{at}/V of the atoms in the solid: $\omega_D = v_s \left(6\pi^2 N_{at}/V\right)^{1/3}$.]

5.7* Focus on the phonons characterized by rigid displacements of each atomic plane perpendicular to one of the main diagonals of the conventional cubic cell of bcc iron. Restrict further to the displacements u_j perpendicular to the planes themselves, i.e. longitudinal phonons in the (111) direction. Assume an elastic interaction between neighboring planes, i.e. a restoring force acting on each atom in the j-th plane given by $F_j = C(u_{j+1} - u_j) + C(u_{j-1} - u_j)$, and indicate with a the equilibrium distance between near-neighbor planes. Write the dispersion relation of such phonons, and evaluate their maximum frequency, in the hypothesis that $C = 65.6$ N/m.

5.8 The phenomenon of Bloch oscillations expected in an ideal crystal could be observed even in a real crystal, provided that the mean collision time τ of the conduction electrons is long enough. Assuming that a uniform constant electric field $|\mathbf{E}| = 1$ V/m is applied to a sodium metal sample, estimate the minimum value of τ for an electron to stand a chance to traverse the entire BZ without colliding. Assume for simplicity a one-dimensional monoatomic crystal of lattice spacing 200 pm.

5.9 Compute the smallest nonzero Bragg diffraction angle θ observed in neutron-elastic-scattering experiments. The neutron kinetic energy is 22 meV. The sample is microcrystalline aluminum, with face-centered cubic (fcc) structure, density 2700 kg m^{-3}, and atomic mass number $A = 27$.

5.10 The following molar heat capacity of gallium arsenide (^{70}Ga ^{75}As, density $\rho = 5320$ kg/m^3) were measured:

T (K)	C_V [J mol^{-1} K^{-1}]
1.0	8.33×10^{-5}
2.0	6.67×10^{-4}
3.0	2.25×10^{-3}

Based on these data, evaluate the average sound velocity according to the Debye model and extrapolate the molar heat capacity at $T = 1000$ K.
[Recall the Debye-model expression for the vibrational internal energy *per atom*:

$$U_1 = 9k_B \frac{T^4}{\Theta_D^3} \int_0^{\Theta_D/T} \frac{x^3 dx}{e^x - 1} \quad \text{and the result} \quad \int_0^\infty \frac{x^3 dx}{e^x - 1} = \frac{\pi^4}{15} .]$$

5.11 Assume that the conduction band of a cesium crystal follows this dispersion:

$$\epsilon(\mathbf{k}) = -B - C \cos\left(\left[\frac{\pi}{50}\right]^{1/3} a \,|\mathbf{k}|\right).$$

Here $B = 2.1$ eV, $C = 2$ eV, and the vacuum energy outside the crystal sits at the reference level $\epsilon = 0$. The cesium crystal structure is body-centered cubic (bcc), with conventional cubic cell side $a = 614.1$ pm. Evaluate: (a) the work function of this solid, and (b) the effective mass of the electrons at the Fermi level, also expressing it as a fraction of the mass m_e of free electrons in vacuum.

5.12 Let the band energy of the 5s electrons of a hypothetical one-dimensional strontium crystal be

$$\varepsilon_k = A + B \cos(ka) + B \cos(2ka)/(2\sqrt{3}),$$

in an energy scale whose zero level represents an electron at rest far away from the solid. Here $A = -5.5$ eV, $B = -3.9$ eV, k is the wave vector and $a = 167$ pm is the lattice spacing. Determine the maximum wavelength of the electromagnetic radiation that can photoemit electrons from such a solid.

5.13 The diffraction pattern of a copper fcc crystal sample exhibits a spot at the Bragg angle $\theta = 32.76°$ at temperature 300 K. With the same X-ray wavelength, this spot moves to $\theta = 32.00°$ when the sample is heated to 1280 K. Based on these data, evaluate the linear expansion coefficient α of copper. [The coefficient α is defined by $\alpha = \frac{1}{\Delta T} \frac{\Delta l}{l}$, with $l = \frac{l_1 + l_2}{2}$.]

5.14 Consider a cube-shaped NaCl sample, with side $L = 1$ mm. NaCl is a fcc crystal with conventional-cell side $a = 564$ pm, and 2 atoms per *unit cell*. The phonon spectrum involves acoustic and optical branches. Describe the acoustic branches following the Debye model (sound velocity $v_s = 2$ km/s). Account for the optical branches modeling them approximately as a set of "Einstein" oscillators with a unique angular frequency $\omega_E = 20$ meV$/\hbar$. The number of such oscillators is three times the number of unit cells in the sample. Evaluate its total heat capacity at temperatures $T = 1$ K and $T = 1000$ K, motivating the adopted approximations. [Recall that $\int_0^\infty dx\, x^3/(e^x - 1) = \pi^4/15$.]

Appendix A
Conclusions and Outlook

The adiabatic separation sketched in Sect. 3.1 is a central tool for understanding the combined dynamics of electrons and nuclei. It works well in such contexts as close-shell molecules and wide-gap insulators, where an excitation gap "protects" the electronic ground state against transitions induced by atomic motions.

Metals violate this adiabatic paradigm since both electrons and vibrations involve arbitrarily small excitation energies. For this reason, even at low temperature, each electron near the Fermi surface has a significant chance of colliding with a phonon (Sect. 5.2.2.1): as a result, in contrast to the a diabatic assumption, the overall electronic state of a metal changes all the time. Precisely the absence of a gap in the electronic spectrum makes the neglect of the nonadiabatic terms of Eq. (3.6) not particularly well justified: the adiabatic approximation seems therefore rather ill-grounded in metals. In fact, despite these difficulties, it still makes sense to apply the adiabatic concepts to metals, since phonon scattering involves only a small fraction of the electrons: those within meVs of the chemical potential. All other "deeper" electrons remain "frozen" in their adiabatic state due to the forbidden (by Pauli's principle) "sea" of already occupied states sitting at near energies. This protection is equivalent to an effective gap, in the order of the energy distance of these deeper electrons from the chemical potential. This vast majority of (deeper) electrons is the main responsible of the crystal cohesion, and thus of the existence of a fairly well-defined adiabatic potential acting on the ions. It is precisely this adiabatic potential that provides a stable crystal structure and phonons even for a metal. The frequent collisions of Fermi-surface electrons with the vibrating crystal are then described as *electron-phonon coupling*, associated to the nonadiabatic terms, and result in a relatively weak perturbation to the adiabatic motion. Eventually however, the ground state of over one third of the elemental metals, and of many metallic compounds cannot be simply interpreted as an adiabatically separated state, but rather a *superconducting state*. While these materials remains solid (due to the cohesion produced by their deeper electrons), in superconductors a correlated motion of the phonons

N. Manini, *Introduction to the Physics of Matter*, Undergraduate Lecture Notes in Physics, https://doi.org/10.1007/978-3-030-57243-3

and the Fermi-surface electrons yields peculiar collective thermodynamic and transport properties (including vanishing resistivity and magnetic-field expulsion), quite distinct from those of the ordinary resistive metallic state.

Furthermore, in many materials, electron-electron correlations introduce new physics beyond the independent-electron approximation. Magnetism, basically a consequence of the same Coulomb-induced correlations as those resulting in atomic Hund's rules (Sect. 2.2.9.3), is a macroscopic deviation from the band picture, where, strictly speaking, it finds no place (see Fig. 5.53).

The present introductory course leaves out a number of important topics, including the coexistence of different phases, order parameters, phase transitions, chemical reactions, amorphous and nanostructured solids, the liquid, superfluid, and plasma states of matter, and all sorts of *collective behavior* which makes the study of the thermodynamics of matter fascinating. The account of experimental techniques is also radically limited to a few providing the most fundamental and transparent data sets. In the lab, readers are likely to come across several other popular techniques. A solid understanding of the material in the present textbook should provide the reader with the conceptual tools for understanding the investigated physical phenomena, and estimating the resolution and the limits of several specific classes of experimental techniques.

Appendix B
Applications: Light Sources and Lighting

With the exception of γ-ray generation in nuclear and subnuclear events, regular matter, namely the motion of electrons and nuclei, is responsible for the production, scattering, modification, and absorption of electromagnetic radiation. Electromagnetic spectroscopies—Sect. 1.2—investigate specific quantitative aspects of this matter-radiation interaction in the lab. The production of electromagnetic radiation, especially in the visible range (see Fig. 1.2) is an important, often taken for granted, ingredient of everyday life, not just of lab applications.

Colored and, even better, monochromatic light finds applications in spectroscopy, but also; e.g., in advertising panels, in traffic lights, in theater and cinema. Lasers—Sect. 4.4.1—are the best sources if highly monochromatic and coherent light is desired. On the other hand, coherent light is associated to speckles, which produce visually unpleasant sensations. Incoherent colored light is usually produced by means of fluorescent lamps or semiconductor diodes.

White light is preferable for general illumination purposes. Currently three main light sources are used for white-light generation: incandescent, fluorescent, and light-emitting diodes.

B.1 Incandescent Light Sources

Incandescent sources are hot surfaces emitting a continuum spectrum similar to blackbody radiation (see Sect. 4.3.2.2 and in particular Fig. 4.17).

For millennia, humans relied on incandescent light sources to illuminate their activities. The daylight solar spectrum is generated by the hot partly ionized gas at the Sun surface. This light fits the blackbody spectrum of a hot incandescent surface kept at $\sim$5,800 K, interspersed with atomic absorption lines from the tenuous solar atmosphere above the photosphere. Human eyes are trained to recognize this spectrum as "natural" white light.

© The Editor(s) (if applicable) and The Author(s), under exclusive license
to Springer Nature Switzerland AG 2020
N. Manini, *Introduction to the Physics of Matter*, Undergraduate Lecture Notes
in Physics, https://doi.org/10.1007/978-3-030-57243-3

Until the late 19th century, flames provided night lighting, e.g., by candles. In flames most light is emitted by soot particles kept hot by a highly exothermic combustion reaction. Given the intrinsic fire hazard represented by flames, mankind was eager to abandon them and switch to electric lighting as soon as it became available.

Incandescent electric light bulbs have been developed and commercialized from the 19th century. Early light bulbs were based on carbon filaments, kept glowing hot by running an electric current through them. Due to poor vacuum and rapid sublimation of the filament, these early bulbs lasted tens to hundreds of hours only. The filament temperature had to be kept quite low, well under 2,000 K. As a result these early bulbs emitted a rather reddish-hue blackbody spectrum peaked in the infrared region.

The major breakthrough for this technology was the introduction of tungsten filaments in the early 20th century. Tungsten filaments join the advantages of a higher operation temperature and a far longer useful life. As a result, they produce a yellowish-hue blackbody spectrum closer to the blackbody spectrum produced by the Sun.

A problem common to carbon- and tungsten-filament lamps is the progressive darkening of the inner surface of the bulb due to an accumulating film of sublimated filament material. Further improvements of this technology involved (i) the introduction of a low-pressure inert gas in the bulb, that suppresses the sublimation of tungsten; (ii) the double coiling of the filament, which promotes rapid redeposition of a significant fraction of the sublimated atoms; (iii) the doping of tungsten with traces of impurities which improve its stability; (iv) the admixture of halogens (usually iodine and bromine) with the inert gas. The halogen reacts with the sublimated tungsten before it reaches the bulb surface. The resulting halide molecules move around in the inert gas filling, with little chance of being adsorbed to the bulb surface. At some point, they reach the hot region near the filament, where collisions dissociate them, releasing tungsten back onto the filament and freeing the halogen. With these improvements, longer filament lifetimes of 1,000–2,000 h and higher filament temperatures, up to 3,300 K, can be achieved.

The useful visible photons are just a small fraction of the entire blackbody spectrum. Incandescent light sources waste a large fraction of the input power as heat in the near-infrared spectrum. Indeed, a 3,300 K blackbody spectrum peaks at 0.80 eV, namely ~50% of the energy of the lowest-energy red photons at the IR-visible boundary.[1] Overall, the efficiency of incandescent bulbs ranges from 2 to 3.5% only.

B.2 Fluorescent Lamps

Another type of light source was developed in parallel to incandescent lamps, starting from the 1850's. Photons are generated by electronic decay transitions of atoms in the gas phase, like in emission spectroscopy (see Fig. 1.4). Similarly to what occurs

---

[1] A hypothetical blackbody source whose emission peak sits at 2.4 eV, the middle of the visible spectrum, should be kept at the astronomical temperature $\simeq$9,870 K.

in atmospheric lightning, when electrons accelerated by an electric field hit atoms at a sufficiently high energy, inelastic collisions can occur. Part of the electron kinetic energy is converted into excitation energy of the atom. As a result, the atom is then either promoted to an excited state, or ionized. Shortly afterward, these excited states decay, emitting photons. As discussed in Chap. 2, these atomic transitions are sharply monochromatic. They are therefore useful for generating colored light, but they are far from optimal when standard white light is needed, as for general lighting. Inert atoms, such as noble gases, have the most intense of these emission peaks in the UV region. Reactive atoms such as alkali elements come with transitions in the visible or near IR, but their chemical reactivity makes them unsuitable for the construction of durable bulbs. In practice, most fluorescent lamps adopt mercury vapor as the primary photon source. The strongest emitted lines are the $6s6p\,^3P_1 \rightarrow 6s^2\,^1S_0$ transition at $\lambda = 253.7$ nm ($\hbar\omega = 4.89$ eV), and the $6s6p\,^1P_1 \rightarrow 6s^2\,^1S_0$ at $\lambda = 185$ nm, ($\hbar\omega = 6.70$ eV), both in the near UV.

The soft-UV photons generated by the Hg emission lines require a conversion to visible light. In most applications, these primary photons excite solid-state phosphors that coat the inside of the glass tube containing the Hg vapor. These phosphors, typically rare-earth oxides, convert the acquired energy, transforming part of it into phonon excitations, and release fluorescence photons mainly in the visible range, with the mechanism illustrated in Fig. 3.14a. The spectral quality of the fluorescent light is determined by the admixture of phosphors in the coating.

The development of efficient fluorescent lamps lagged behind incandescence lamps due to several extra technical complications, that needed to be overcome, e.g. the triggering of the electron discharge in the low-pressure gas, its stabilization against runaway ionization, the control of the erosion of the metal electrodes injecting the electric current in the gas. Commercial fluorescent lamps became available from the 1930s. Despite their higher complication and multiple energy-conversion steps, power efficiency marks a net improvement against incandescence. The downconversion (done by phosphors) of the primary UV photons to lower-energy visible photons is the main source of energy loss (approximately 60%) in modern fluorescent lamps. The combined photon downconversion plus other losses brings the overall efficiency of fluorescent lamps to 10−12%, several times better than the best incandescent sources. The useful life of fluorescent lamps approaches 10,000 h.

B.3 Light-Emitting Diodes (LEDs)

The most recent major class of light sources takes advantage of the emission of photons in electronic interband transitions in semiconductors. LEDs are regular p-n junctions engineered to exhibit an especially wide depletion region. When a direct bias maintains a current through them, the recombination of electrons and holes in the depletion region can occur radiatively, thus generating photons, a process called *electroluminescence*. The energy of these photons is not monochromatic, but the spectrum peaks near the bandgap energy of the semiconductor involved: in the IR, visible, or UV.

Indirect-gap semiconductors, see Fig. 5.63, such as Si and Ge are ill suited for LEDs. The reason is that most holes sit near the maximum of the valence band, and most conduction electrons sit near the minimum of the conduction band, at different **k** points. The conservation of momentum, Eq. (5.74), strongly suppresses electroluminescence in these crystals, and other nonradiative electron-hole recombination processes compete with the photon generation, thus highly reducing efficiency. In contrast, in direct-gap semiconductors radiative transitions dominate, making them ideal for LEDs.

After early attempts in the first half of the 20th century, improved solid-state techniques allowed the first GaAs infrared light-emitting diodes to be commercialized in the 1960's. Subsequent materials-science research led to LEDs based on wider-gap semiconductors such as $GaAs_{1-x}P_x$), leading to the commercialization of red, orange, and yellow LEDs. Various technical improvements on both the solid-state electronics and the optics sides led to improved efficiency in light emission and color availability. LEDs as colored sources became widely adopted in consumer electronics in the 1970s when mass production made them affordable. Among a wide range of technical improvements realized over the following decades, a major breakthrough was the development of crystal-growing techniques for GaN, a semiconductor with a 3.4 eV-wide band gap, suitable to emit blue light. Given the possibility to emulate white light combining red, green and blue light, the development of efficient blue-light LEDs has opened a route to general-purpose lighting based on LEDs [47].

White light-emitting diodes are produced following two basic strategies. One is to combine different LEDs that emit three primary colors—red, green, and blue—and mix these colors to form white light. The other is to convert colored light from a blue or UV LED to broad-spectrum white light by means of a phosphor material, like in a fluorescent lamp. The multi-LED scheme boasts a better quantum efficiency and a tunable color output; on the other hand, it suffers stability, cost, and light-whiteness problems. Therefore its adoption is currently limited to specific applications. The less efficient but simpler, cheaper, and more stable phosphor method is currently widely employed in the industrial production of white LEDs.

From the 2010s, LED lamps have reached a technological maturity, power efficiency, and price level that has pushed them toward a rapid replacement of fluorescent lamps. Advantages of LEDs include: power efficiency in the $15-25\%$ region, small size and high shock resistance; quick light up time $< 1\,\mu s$; long useful life in the 35,000–50,000 h range; low operating temperature. Ultimately these advantages are to be traced to solid-state operation close to room-temperature thermodynamic equilibrium, in avoidance of both high temperatures and electron-matter collisions. Out-of-equilibrium states are restricted to the electron occupations in the p-n junction, and they are reached "adiabatically", with the smooth injection of electrons in the conduction band and holes in the valence band, guided by carefully tailored impurity-concentration gradients.

Appendix C
Elements of Quantum Mechanics

A student in physics will probably appreciate the present textbook better after taking a full semester in QM, and/or studying substantial parts of some formal QM textbook [5, 15–17]. For the reader's convenience, we summarize here the essential concepts and ingredients of QM needed for digesting the physics of matter in the present volume. The present level of treatment is quite basic and compact, aiming at providing the minimal conceptual tools. Most subtle epistemological implications[2] and mathematical derivations are omitted.

C.1 Bras, Kets and Probability

In QM, a physical state is represented by a unit-norm vector—a *ket* $|j\rangle$ in Dirac notation [17]—in an abstract complex Hilbert space of vectors. The symbol j identifying uniquely the quantum state stands for a full set of *quantum numbers*. A *bra* $\langle q'|$ is the linear operator which, given any state $|j\rangle$, yields its projection $\langle q'|j\rangle$ along the "direction" representing state $|q'\rangle$. This projection (or *overlap*) is a complex number. The bra-ket application defines an inner product between vectors in the Hilbert space. The special case $\langle j|j\rangle$ of the overlap of a vector with itself produces a real nonnegative number, which represents the squared norm of the vector itself. When a vector represents a physical state, its norm equals 1. As long as $|j\rangle$ is not the null vector, it can be normalized by dividing it by its norm: $|j\rangle/(\langle j|j\rangle)^{1/2}$. The basic property of kets in a Hilbert space is linearity: $a|j\rangle + a'|j'\rangle$ is also a ket, for arbitrary complex numbers a and a'. The transformation of a ket into a bra is antilinear, in the sense that

---

[2]QM can be far from intuitive. Two great gurus of QM used to say: *For those who are not shocked when they first come across quantum theory cannot possibly have understood it.* [Niels Bohr, quoted in Heisenberg, Werner (1971). Physics and Beyond. New York: Harper and Row. p. 206.]; *I think I can safely say that nobody understands quantum mechanics.* [Richard Feynman, in The Character of Physical Law (1965)].

© The Editor(s) (if applicable) and The Author(s), under exclusive license
to Springer Nature Switzerland AG 2020
N. Manini, *Introduction to the Physics of Matter*, Undergraduate Lecture Notes
in Physics, https://doi.org/10.1007/978-3-030-57243-3

the bra associated to $a|j\rangle + a'|j'\rangle$ is $a^*\langle j| + a'^*\langle j'|$, where a^* indicates the complex conjugate to a.

Any observable, i.e. any quantity which can be measured in a physical experiment, is associated with a self-adjoint linear operator Q from the Hilbert space into itself. The fact that Q brings the Hilbert space into itself means that for any ket $|j\rangle$, $Q|j\rangle$ is a (generally unnormalized) ket too. The *linearity* of Q means simply that $Q(a|j\rangle + a'|j'\rangle) = aQ|j\rangle + a'Q|j'\rangle$ for any complex numbers and kets involved. The adjoint $X^\dagger$ of a linear operator X is a generalization of the complex conjugation which comes about when converting $X|j\rangle$ to the corresponding bra: $X|j\rangle \rightarrow \langle j|X^\dagger$. A *self-adjoint* operator Q is such that $Q^\dagger = Q$. In practice the self-adjointness means that $\langle j|Q|j'\rangle = \langle j'|Q|j\rangle^*$ for any $|j\rangle$ and $|j'\rangle$.

The values which can turn up as the result of a single measurement of the observable associated to Q are the *eigenvalues* of the operator Q, namely those special numbers q such that $Q|q\rangle = q|q\rangle$. Due to the self-adjointness property of Q, its eigenvalues are *real* quantities. Moreover, the set $\{|q\rangle\}$ of all possible *eigenstates* of Q, is a complete set of orthogonal states. This means that any state in the Hilbert space can be written as a linear combination $|j\rangle = \sum_q a_q|q\rangle$. Assuming that all $|q\rangle$ are properly normalized ($\langle q|q\rangle = 1$), the coefficients in the linear combination are $a_q = \langle q|j\rangle$.

These coefficients bear an important physical significance. Prepare the system in state $|j\rangle$. The probability that a measurement of Q yields the eigenvalue q is the square module of the overlap of the initial state $|j\rangle$ with the relevant eigenket of Q: $P(q) = |\langle q|j\rangle|^2$. In case q is degenerate, i.e. the same eigenvalue is associated to several orthonormal eigenkets $|q_i\rangle$, then one needs to sum over these eigenkets:

$$P(q) = \sum_{\substack{i \\ q_i \equiv q}} |\langle q_i|j\rangle|^2 .   \tag{C.1}$$

The probability distribution $P(q)$ needs to be properly normalized. Indeed, the sum of $P(q)$ over all possible outcomes q of a measurement must be

$$1 = \sum_q P_j(q) = \sum_{q'} |\langle q'|j\rangle|^2 = \sum_{q'} \langle j|q'\rangle\langle q'|j\rangle = \langle j| \left[\sum_{q'} |q'\rangle\langle q'| \right] |j\rangle .   \tag{C.2}$$

The quantity in square brackets is an operator from the Hilbert space into itself, and the completeness of the basis $\{|q'\rangle\}$ guarantees that it coincides with the identity I. This leaves $\langle j|j\rangle$ which we already assumed to be properly normalized to unity.

C.1.1   *Wavefunction and Averages*

We obtain an important example of this sort of overlaps when for Q we select the position operator R, with eigenvalues r and eigenkets $|r\rangle$. The overlap $\psi_j(r) = \langle r|j\rangle$ is called the *wavefunction* of state $|j\rangle$ at point r. $\psi_j(r)$ is a complex number. $|\psi_j(r)|^2$ is the probability density $P_j(r)$ that the particle in an initial state $|j\rangle$ is found at r when its position is measured. Since the eigenvalues of the position operator are continuous quantities, the sum in Eq. (C.2) must be replaced by an integral. Thus the normalization of the probability distribution in r'-space is

$$1 = \int P_j(r)\,dr = \int |\langle r|j\rangle|^2\,dr = \int |\psi_j(r)|^2\,dr\,. \qquad (C.3)$$

Going back to a general observable Q, typically one can prepare the quantum system every time in the same initial state $|j\rangle$, repeatedly measure the observable Q, and average over these measurements. Given the individual probabilities for all possible outcomes q' of the measurement of Q, quantum theory provides a straightforward formula to predict this average value $\langle Q\rangle$:

$$\langle Q\rangle = \sum_q q\,P_j(q) = \sum_q q|\langle q|j\rangle|^2 = \sum_q \langle j|q\rangle q\langle q|j\rangle = \sum_q \langle j|q\rangle\langle q|q|q\rangle\langle q|j\rangle\,. $$
$$(C.4)$$

Now, we substitute the eigenvalue equation $q|q\rangle = Q|q\rangle$,

$$\langle Q\rangle = \sum_q \langle j|q\rangle\langle q|Q|q\rangle\langle q|j\rangle = \sum_q \sum_{q'} \langle j|q\rangle\langle q|Q|q'\rangle\langle q'|j\rangle\,. \qquad (C.5)$$

In the last passage we have added a lot of null terms, namely all those with $q \neq q'$. Now, we recognize the identity I at both sides of Q:

$$\langle Q\rangle = \langle j|\left[\sum_q |q\rangle\langle q|\right] Q \left[\sum_{q'} |q'\rangle\langle q'|\right] |j\rangle = \langle j|Q|j\rangle\,. \qquad (C.6)$$

We conclude that the average of an observable Q is provided by the diagonal matrix element of its operator on the probed state. Be aware that $\langle Q\rangle$ need not coincide with any of the observations! For example, if the outcome of an observable is $+1/2$, with 50% probability, and $-1/2$, also with 50% probability, then its average equals zero. However no measurement would ever yield 0.

C.2   Position, Momentum, and Translations

QM is about the motion of particles. The simplest way to move a particle is to translate it. Consider an eigenstate of position $R|\mathbf{r}'\rangle = \mathbf{r}'|\mathbf{r}'\rangle$, where $\mathbf{r}' = (x', y', z')$. A translation by δx in the $\hat{\mathbf{x}}$ direction modifies $|x', y', z'\rangle$ into $|x' + \delta x, y', z'\rangle$. Indicate with $\mathcal{T}(\mathbf{x})$ the operator for an infinitesimal translation by $\mathbf{x} = \hat{\mathbf{x}}\,\delta x$:

$$|\mathbf{r}'\rangle \;\rightarrow\; \mathcal{T}(\mathbf{x})|\mathbf{r}'\rangle = |\mathbf{r}' + \mathbf{x}\rangle = |x' + \delta x, y', z'\rangle\,. \tag{C.7}$$

This result can be generalized to an arbitrary state $|j\rangle$, by expanding it on the basis of the position eigenstates:

$$|j\rangle \;\rightarrow\; \mathcal{T}(\mathbf{x})|j\rangle = \mathcal{T}(\mathbf{x}) \int d^3r' |\mathbf{r}'\rangle\langle \mathbf{r}'|j\rangle = \int d^3r'\,[\mathcal{T}(\mathbf{x})|\mathbf{r}'\rangle]\psi_j(\mathbf{r}')\,. \tag{C.8}$$

We apply Eq. (C.7) to the square bracket, and rewrite the integral in terms of a shifted variable $\mathbf{r}$:

$$\mathcal{T}(\mathbf{x})|j\rangle = \int d^3r' |\mathbf{r}' + \mathbf{x}\rangle\psi_j(\mathbf{r}') = \int d^3r\,|\mathbf{r}\rangle\psi_j(\mathbf{r} - \mathbf{x})\,. \tag{C.9}$$

We conclude that a translation by $\mathbf{x}$ acts on the wavefunction of a state by substituting its argument $\mathbf{r}$ with $\mathbf{r} - \mathbf{x}$.

Translation must conserve the normalization of arbitrary states:

$$\langle j|\mathcal{T}(\mathbf{x})^\dagger \mathcal{T}(\mathbf{x})|j\rangle = \langle j|j\rangle\,. \tag{C.10}$$

This condition is guaranteed by the requirement that the product

$$\mathcal{T}(\mathbf{x})^\dagger \mathcal{T}(\mathbf{x}) = \mathcal{I}\,, \tag{C.11}$$

the identity in the Hilbert space. According to Eq. (C.11), $\mathcal{T}(\mathbf{x})$ is a unitary operator, i.e. $\mathcal{T}(\mathbf{x})^\dagger = \mathcal{T}(\mathbf{x})^{-1}$. The following additional properties apply to infinitesimal translations:

$$\mathcal{T}(\mathbf{x})^{-1} = \mathcal{T}(-\mathbf{x})\,, \qquad \mathcal{T}(0) = \mathcal{I}\,, \qquad \mathcal{T}(\mathbf{x})\mathcal{T}(\mathbf{x}') = \mathcal{T}(\mathbf{x} + \mathbf{x}')\,. \tag{C.12}$$

The properties of Eqs. (C.11) and (C.12) are satisfied provided that

$$\mathcal{T}(\mathbf{x}) = \mathcal{I} - i\mathbf{K}\cdot\mathbf{x}\,, \tag{C.13}$$

for a suitable "generating" vector $\mathbf{K}$ of self-adjoint operator components $K_{x/y/z}$. We neglect terms of order $O(|\mathbf{x}|^2)$.

Translations, and their "generators" $\mathbf{K}$ are related to the position operator as follows:

$$\mathbf{R}\,\mathcal{T}(\mathbf{x})|\mathbf{r}'\rangle = \mathbf{R}|\mathbf{r}' + \mathbf{x}\rangle = (\mathbf{r}' + \mathbf{x})|\mathbf{r}' + \mathbf{x}\rangle\,, \tag{C.14}$$

$$\mathcal{T}(\mathbf{x})\mathbf{R}|\mathbf{r}'\rangle = \mathbf{r}'\mathcal{T}(\mathbf{x})|\mathbf{r}'\rangle = \mathbf{r}'|\mathbf{r}' + \mathbf{x}\rangle\,. \tag{C.15}$$

By subtracting these two relations we obtain

$$(\mathbf{R}\,\mathcal{T}(\mathbf{x}) - \mathcal{T}(\mathbf{x})\,\mathbf{R})\,|\mathbf{r}'\rangle = \mathbf{x}|\mathbf{r}' + \mathbf{x}\rangle \simeq \mathbf{x}|\mathbf{r}'\rangle\,, \tag{C.16}$$

where in the last passage we neglect a quantity of order $O(|\mathbf{x}|^2)$. We introduce the notation $[A, B] = AB - BA$, called the *commutator* of two operators. As Eq. (C.16) holds for any $|\mathbf{r}'\rangle$, which form a basis of the Hilbert space, it implies the relation

$$\mathbf{R}\,\mathcal{T}(\mathbf{x}) - \mathcal{T}(\mathbf{x})\,\mathbf{R} \equiv [\mathbf{R}, \mathcal{T}(\mathbf{x})] = \mathbf{x}\,I \tag{C.17}$$

for the operators. We proceed to substitute Eq. (C.13) into Eq. (C.17), obtaining

$$- i\mathbf{R}\,\mathbf{K}\cdot(\mathbf{x}) + i\mathbf{K}\cdot(\mathbf{x})\,\mathbf{R} = \mathbf{x}\,I\,. \tag{C.18}$$

This relation holds for the 3 components R_u of vector $\mathbf{R}$. We generate 9 independent equations by selecting successively the vector $\mathbf{x}$ along the three versors $\hat{\mathbf{x}}_v$:

$$R_u\,K_v - K_v\,R_u \equiv [R_u, K_v] = i\,\delta_{uv}\,I\,. \tag{C.19}$$

These fundamental commutation relations acquire a central role in QM, once one realizes that the *wave-vector operator* $\mathbf{K}$ is related to the *momentum* operator $\mathbf{P}$ by

$$\mathbf{P} = \hbar\mathbf{K}\,, \tag{C.20}$$

where $\hbar = 1.0546 \times 10^{-34}$ m^2 kg s^{-1} is the Planck constant. To highlight the role of momentum as the generator of translation, we rewrite Eq. (C.13) as

$$\mathcal{T}(\mathbf{x}) = I - i\hbar^{-1}\mathbf{P}\cdot\mathbf{x}\,. \tag{C.21}$$

The commutation relations (C.19) are rewritten in terms of momentum as

$$[R_u, P_v] = i\,\hbar\,\delta_{uv}\,I\,. \tag{C.22}$$

C.2.1 Compatible and Incompatible Observables

Position and momentum components are examples of non-commuting observables. To understand what this non-commutation implies, we need to clarify first the properties of *commuting observables*. Suppose that A and B are operators with $A|a'\rangle = a'|a'\rangle$, $B|b'\rangle = b'|b'\rangle$, and $[A, B] = 0$. Then we prove that if $a' \neq a''$ then

$\langle a'|B|a''\rangle = 0$. This is an immediate consequence of the definition of commutator:

$$0 = \langle a'|[A, B]|a''\rangle = \langle a'|AB - BA|a''\rangle = (a' - a'')\langle a'|B|a''\rangle. \qquad (C.23)$$

If all eigenvalues of A are nondegenerate, this means that the basis of eigenstates of A is diagonal for observable B, too. Even if A has some degenerate eigenvalue, nothing prevents us from diagonalizing B in the degenerate Hilbert subspace, with the result of constructing a basis of *simultaneous eigenstates* of both operators:

$$A|a', b'\rangle = a'|a', b'\rangle; \qquad B|a', b'\rangle = b'|a', b'\rangle. \qquad (C.24)$$

This result can be generalized to any number of commuting observables. The measurement of any of them does not affect the measurement of the others. After the measurement of A, the QM system ends up in a common eigenstate $|a', b'\rangle$. Any sequence of successive measurements of B and/or A will keep the system in the same eigenstate $|a', b'\rangle$. For this reason, commuting observables are said to be *compatible*.

The situation is quite different for *non commuting observables*, such as position and momentum components. First of all, no common basis of eigenkets can exist if $[A, B] \neq 0$. If there was one, then

$$AB|a', b'\rangle = Ab'|a', b'\rangle = a'b'|a', b'\rangle; \qquad BA|a', b'\rangle = Ba'|a', b'\rangle = a'b'|a', b'\rangle. \qquad (C.25)$$

By subtracting these relations, one proves that $[A, B]$ is diagonal and vanishes on a complete basis, which would mean that $[A, B]$ is the null operator, contrary to our assumption. Note that this statement does not prevent A and B from sharing one or several common eigenkets: it just means that such common eigenkets (if any) cannot form a complete basis of the Hilbert space.

Nonzero commutators affect fluctuations. Given any physical state, we define the standard deviation of an arbitrary operator A as

$$\sigma_A^2 = \langle (A - \langle A\rangle)^2\rangle = \langle A^2\rangle - \langle A\rangle^2, \qquad (C.26)$$

where the averages are taken over the given state. If the physical state is an eigenstate of A, clearly σ_A^2 vanishes. Otherwise σ_A^2 measures the uncertainty or fluctuation to be expected in the outcome of repeated measurements of A realized on that same physical state. Equipped with this definition, it is possible to prove that for any physical state and for any two observables A and B,

$$\sigma_A^2 \sigma_B^2 \geq \frac{1}{4}|\langle [A, B]\rangle|^2, \quad \text{or } \sigma_A \sigma_B \geq \frac{1}{2}|\langle [A, B]\rangle|. \qquad (C.27)$$

This *uncertainty relation* expresses the fact that *incompatible observables* cannot both be measured with infinite precision.

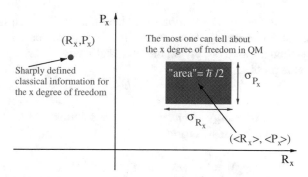

Fig. C.1 A sketch of Heisenberg's uncertainty principle, in the phase-space slice of the $\hat{x}$ component of the position and momentum of a particle. For classical mechanics (*left*) an individual point in this space represents the full information about (R_x, P_x). In contrast, QM (*right*) can only resolve phase-space regions with a minimum "area" $\hbar/2$

The example of position and momentum is especially straightforward, since the commutator of corresponding components is a constant $i\hbar$ times the identity operator, see Eq. (C.22). As a consequence of Eq. (C.27), on any physical state *Heisenberg's uncertainty relation*

$$\sigma_{R_u}\,\sigma_{P_v} \geq \frac{1}{2}\,\hbar\,\delta_{uv} \tag{C.28}$$

holds. Accordingly, for different components $u \neq v$, e.g. R_x and P_y, which are compatible observables, one can generate states with arbitrarily small uncertainty for both observables. In contrast, for a matching component $u = v$, e.g. R_x and P_x, namely incompatible observables, a state characterized by a small position uncertainty σ_{R_x} has necessarily a large $\sigma_{P_x} \geq \hbar/(2\sigma_{R_x})$, i.e. a largely fluctuating momentum, and vice versa. This result of QM contrasts with classical mechanics which is based on a phase-space continuous tracking of each particle's both position and momentum, with an in principle arbitrary precision. QM corrects this picture, leading to a fuzzy nature of the phase space, with no possibility of identifying sharp points in each pair of conjugate coordinates (R_u, P_u), introducing a minimum uncertainty "area" in the order of $\hbar/2$, see Fig. C.1.

C.3   Schrödinger's Equation

If one leaves a quantum mechanical system undisturbed, it evolves in time in a predictable manner dictated by the Schrödinger equation associated to its total-energy (or *Hamiltonian*) operator H:

$$i\hbar\frac{d}{dt}|\xi(t)\rangle = H|\xi(t)\rangle\,. \tag{C.29}$$

Here we focus on the common situation of a Hamiltonian H which is constant in time. We have a standard strategy to solve the time-dependent Schrödinger equation (C.29). This method is based on solving first the "stationary" Schrödinger equation, i.e. the eigenvalue problem

$$H |E_j\rangle = E_j |E_j\rangle . \tag{C.30}$$

In the resulting basis of *energy eigenkets* $|E_j\rangle$, the solution of Eq. (C.29) can be written in the following expansion form:

$$|\xi(t)\rangle = \sum_j |E_j\rangle \langle E_j|\xi(0)\rangle e^{-i E_j t/\hbar}. \tag{C.31}$$

Here the time-independent complex coefficients $\langle E_j|\xi(0)\rangle$ are precisely the expansion coefficients of the initial state $|\xi(0)\rangle$ on this basis.

Consider the noteworthy situation where the system is prepared in a pure energy eigenstate $|\xi(0)\rangle = |E_{\bar{j}}\rangle$. In such a special case, the time evolution involves a single rotating phase factor $\exp(-i E_{\bar{j}} t/\hbar)$. Such an overall phase factor cannot affect the quantum average of any physical quantity: $\langle \xi(t)|Q|\xi(t)\rangle = \langle \xi(0)|Q|\xi(0)\rangle = \langle E_{\bar{j}}|Q|E_{\bar{j}}\rangle$. This property justifies the qualification of the Hamiltonian eigenkets $|E_j\rangle$ as *stationary* states. According to Eq. (C.31), an isolated quantum system prepared in one of its stationary states remains unchanged (up to an irrelevant rotating phase) forever. As discussed in Sect. 1.2, the excited states of real-life time-independent quantum system would eventually decay under the action of weak interactions with the surrounding environment (e.g. the electromagnetic fields).

A more instructive condition is obtained when the initial state is a linear combination of two energy eigenstates: $|\xi(0)\rangle = a_1 |E_{j_1}\rangle + a_2 |E_{j_2}\rangle$. According to Eq. (C.31), the time evolution of this ket is

$$\begin{aligned}
|\xi(t)\rangle &= a_1 e^{-i E_{j_1} t/\hbar} |E_{j_1}\rangle + a_2 e^{-i E_{j_2} t/\hbar} |E_{j_2}\rangle \\
&= e^{-i E_{j_1} t/\hbar} \left[a_1 |E_{j_1}\rangle + a_2 e^{-i (E_{j_2}-E_{j_1}) t/\hbar} |E_{j_2}\rangle \right].
\end{aligned}$$

In the final expression, we have extracted an irrelevant overall phase factor $e^{-i E_{j_1} t/\hbar}$, which has no effect on any observable quantity. The relative phase $(E_{j_2} - E_{j_1}) t/\hbar$ of the two contributors of the linear combination is the crucial actor. This phase rotates at an angular speed $(E_{j_2} - E_{j_1})/\hbar$ proportional to the energy difference of the two states. At all times that are integer multiple of the rotation period $2\pi\hbar/(E_{j_2} - E_{j_1})$, the evolving ket reproduces its initial condition $|\xi(0)\rangle$ (up to an overall phase).

In contrast, when three or more energy eigenkets are involved in the initial condition, no similar return to the initial ket usually occurs. It could only occur in the unlikely event that *all* relative frequencies $(E_j - E_{j'})/\hbar$ of the eigenkets involved in the initial state are mutually commensurate.

C.4 Symmetry

Symmetries can usually be represented by unitary operators in the Hilbert space: they transform kets into symmetry-modified kets. For example, a mirror reflection plane σ_h takes a ket initially localized at the right side of the plane and moves it to the left side. A second application of this symmetry operator leads back to the original state, thus $\sigma_h \, \sigma_h = I$. Other symmetry operators behave differently. For example, a $120°$ rotation C_3 around a given axis must be applied 3 times before it leads the kets back to their initial location: $C_3 \, C_3 \, C_3 = I$. In other cases no number of successive applications of a symmetry operation leads back to I, as occurs e.g. with a 1 radian rotation.

The symmetry operators of a given QM system usually form a group. A group is a set closed for composition (for any two symmetry operations A and B, also $B \, A$ is a symmetry operation), that contains a neutral element (any system is trivially symmetric under the "no change" operator I), and such that for each A there exists an inverse operator A^{-1} such that the composition $A^{-1} \, A = A \, A^{-1} = I$.

Symmetry groups can contain a finite number of elements. For example, the symmetry operations for a water molecule in its equilibrium geometry (see Fig. 3.12b) are 4, namely: the identity I; a $180°$ rotation C_2 around an axis bisecting the $\widehat{HOH}$ angle; a reflection through the molecular plane; and a perpendicular reflection plane through the C_2 axis.

Other symmetry groups contain an infinite number of elements. A relevant example is the group of the discrete translations of a crystalline solid (the Bravais-lattice translations) described in Sect. 5.1.1. Other important infinite groups are (i) the group of symmetry operations of a linear (e.g. diatomic) molecule, which contains the infinitely many rotations around the molecular axis, (ii) the group of symmetry operations of an atom, which includes all possible rotations around its center of mass, and (iii) the group of all continuously many translations of a free particle.

For a given symmetry group, the fact that a QM system has that symmetry is expressed by the fact that any given state $|j\rangle$ has the same energy whether or not it is transformed by any of the group symmetry operators A:

$$\langle j | A^{\dagger} H A | j \rangle = \langle j | H | j \rangle . \tag{C.32}$$

As this equation holds for any $|j\rangle$ in the Hilbert space, the corresponding equality of operators must hold:

$$A^{\dagger} H A = H , \tag{C.33}$$

or, using the property $A^{\dagger} = A^{-1}$ of unitary operators,

$$H A = A H , \text{ or } [H, A] = 0 . \tag{C.34}$$

In words, all A operators in the group are compatible with H.

If A is also an observable (a self-adjoint operator $A^\dagger = A$, which as A is unitary implies $A^{-1} = A^\dagger = A$), as discussed in Sect. C.2.1, H and A can be diagonalized simultaneously. The energy eigenstates are then also symmetry eigenstates. For example, this occurs with the $L \leftrightarrow R$ symmetry operation σ_h of homonuclear dimers, Sect. 3.2.1: in that system, all one-electron energy eigenstates are either symmetric (bonding) or antisymmetric (antibonding) combinations of $|L\rangle$ and $|R\rangle$.

For more complicated symmetry groups, whose elements are not (all) self-adjoint, symmetry provides a labeling of the energy eigenstates, too. In these cases, this labeling is given by the irreducible representation of the symmetry group. For the mathematical definition of irreducible representations and how they label the energy eigenstates, the reader is referred to any textbook on group theory, e.g. Ref. [48]. We only need to retain here that many groups have multi-dimensional irreducible representations, thus leading to degenerate energy eigenstates. For example, the degeneracies of the p, d, f,...electronic one-electron states are related to the 3, 5, 7,...-dimensional irreducible representations of the full rotational symmetry group of the one-electron Hamiltonian of the atom.

C.5   Variational Methods

Start from an exact energy eigenstate $|E_j\rangle$ of an arbitrary time-independent Hamiltonian H, and modify it by adding a small "deviation" $|\delta\rangle$, such that $\langle\delta|\delta\rangle \ll 1$. The average energy of the unnormalized modified state $|\phi\rangle = |E_j\rangle + |\delta\rangle$ is:

$$
\begin{aligned}
\bar{E}(|E_j\rangle + |\delta\rangle) = \bar{E}(|\phi\rangle) &= \frac{\langle\phi|H|\phi\rangle}{\langle\phi|\phi\rangle} = \frac{((\langle E_j| + \langle\delta|)H(|E_j\rangle + |\delta\rangle))}{((\langle E_j| + \langle\delta|)(|E_j\rangle + |\delta\rangle))} \\
&= \frac{\langle E_j|H|E_j\rangle + \langle E_j|H|\delta\rangle + \langle\delta|H|E_j\rangle + \langle\delta|H|\delta\rangle}{\langle E_j|E_j\rangle + \langle E_j|\delta\rangle + \langle\delta|E_j\rangle + \langle\delta|\delta\rangle} \\
&= \frac{E_j + 2\operatorname{Re}\langle\delta|H|E_j\rangle + \langle\delta|H|\delta\rangle}{1 + 2\operatorname{Re}\langle\delta|E_j\rangle + \langle\delta|\delta\rangle} \\
&= \left(E_j + 2E_j \operatorname{Re}\langle\delta|E_j\rangle + \langle\delta|H|\delta\rangle\right) \times \\
&\quad \left(1 - 2\operatorname{Re}\langle\delta|E_j\rangle - \langle\delta|\delta\rangle + 4(\operatorname{Re}\langle\delta|E_j\rangle)^2 + O(|\delta\rangle)^3\right) \\
&= E_j + \langle\delta|H|\delta\rangle - E_j\langle\delta|\delta\rangle + O(|\delta\rangle)^3 = E_j + \langle\delta|H - E_j|\delta\rangle + O(|\delta\rangle)^3 \\
&= E_j + O(|\delta\rangle)^2 .
\end{aligned}
$$

Observe that the terms linear in the distortion $|\delta\rangle$ cancel out. Accordingly, changes to the mean energy of the modified state are at least quadratic in the norm of $|\delta\rangle$. This argument applied to the ground state shows that it is a quadratic *minimum* of the energy as a function of the ket; excited states are *stationary saddle points*.

Given a set of linearly independent states $\{|\phi_j\rangle\}$ (a priori unrelated to H) in the Hilbert space, one can build linear combinations $|\phi\rangle = \sum_l c_l|\phi_l\rangle$ of such states. The best approximation to the energy eigenstates $|E_j\rangle$ and eigenenergies E_j of H within the space spanned by the $\{|\phi_j\rangle\}$ is realized by tuning the coefficients $\{c_l\}$ in such a way that they implement the stationary property discussed above: the linear-order variation of the average energy $\bar{E}(|\phi\rangle) = \bar{E}(\{c_l\}) = \langle\phi|H|\phi\rangle/\langle\phi|\phi\rangle$ must vanish. $\bar{E}$ should vary at least quadratically as the coefficients $\{c_l\}$ are displaced away from their optimal values. To determine the values of the coefficients making $\bar{E}$ *stationary*, we nullify the gradient of $\bar{E}$ for variations of c_j^*:

$$0 = \frac{\partial \bar{E}(|\phi\rangle)}{\partial c_j^*} = \frac{\langle\phi_j|H|\phi\rangle}{\langle\phi|\phi\rangle} - \frac{\langle\phi|H|\phi\rangle\,\langle\phi_j|\phi\rangle}{(\langle\phi|\phi\rangle)^2} = \sum_l \frac{\langle\phi_j|H|\phi_l\rangle}{\langle\phi|\phi\rangle} c_l - \bar{E} \sum_l \frac{\langle\phi_j|\phi_l\rangle}{\langle\phi|\phi\rangle} c_l .$$
(C.35)

We introduce the shorthand $H_{jl} = \langle\phi_j|H|\phi_l\rangle$ for the matrix element, and $B_{jl} = \langle\phi_j|\phi_l\rangle$ for the overlap ($B_{jl} = \delta_{jl}$ if orthonormal states $\{|\phi_j\rangle\}$ are selected). In this notation Eq. (C.35) becomes:

$$\sum_l H_{jl}\, c_l = \bar{E} \sum_l B_{jl}\, c_l .$$
(C.36)

This variational method transforms the original abstract Schrödinger problem (C.30) into an algebraic problem, namely the generalized secular problem of the calculation of the eigenvalues $\bar{E}$ and eigenvectors of a numerical matrix $\{H_{jl}\}$; the overlap matrix $\{B_{jl}\}$ provides the relevant "metric tensor". Note that, given a Hilbert subspace generated by N_ϕ states $\{|\phi_j\rangle\}$, this method provides N_ϕ approximate eigenstates and eigenenergies. In particular, the lowest eigenvalue of the matrix problem (C.36) is an upper bound of the exact ground-state energy: $\bar{E}_0 \geq E_0$.

The formulation (C.36) is widely used in most computer simulations of QM problems. An orthonormal basis ($B_{jl} = \delta_{jl}$) is often adopted due to conceptual simplicity and numerical convenience.

C.5.1 One State

An especially simple application of the variational method is obtained for $N_\phi = 1$. In this case, the variational inequality $H_{00} \equiv \bar{E}_0 \geq E_0$ expresses the trivial fact that the average energy of an arbitrary state is greater than or equal to the ground-state energy. The single trial state $|\phi\rangle$ is often made depend on one or several adjustable parameters $a_1, a_2, \ldots$ When such parameters are available, they can be "optimized" with the target of minimizing the average energy $\bar{E}_0 = H_{00}$:

$$E_0 \leq \min_{a_1,a_2,\ldots} \bar{E}_0 \equiv \min_{a_1,a_2,\ldots} H_{00} \equiv \min_{a_1,a_2,\ldots} \langle\phi_{a_1,a_2,\ldots}| H |\phi_{a_1,a_2,\ldots}\rangle ,$$
(C.37)

thus realizing the best approximation (within this class of parameterized states) to the true ground state.

The HF method of Sect. 2.2.5 is an example of this single state approach. The parameters being optimized in the HF method are the single-electron wavefunctions themselves, which parameterize the permutation-antisymmetric many-body state.

C.5.2 Two States

The solution of Eq. (C.36) for two states is quite instructive. With the problem of Sect. 3.2.2 at hand, we name the two states $|\phi_1\rangle = |R\rangle$ and $|\phi_2\rangle = |L\rangle$. For simplicity assume that these states are orthonormal, so that the overlap matrix $B_{jl} = \langle \phi_j | \phi_l \rangle = \delta_{jl}$ is the identity. The eigenenergies and eigenstates of the full problem are then approximated by the eigenvalues and eigenvectors of the 2×2 matrix

$$\begin{pmatrix} H_{11} & H_{12} \\ H_{21} & H_{22} \end{pmatrix} \equiv \begin{pmatrix} \langle R|H|R \rangle & \langle R|H|L \rangle \\ \langle L|H|R \rangle & \langle L|H|L \rangle \end{pmatrix} = \begin{pmatrix} E_R & -\Delta \\ -\Delta^* & E_L \end{pmatrix}. \tag{C.38}$$

The eigenvalues of the matrix (C.38) are

$$\bar{E}_{\substack{1 \\ 2}} = \frac{E_L + E_R}{2} \mp \sqrt{\left(\frac{E_L - E_R}{2}\right)^2 + |\Delta|^2}. \tag{C.39}$$

The corresponding ground $|\bar{E}_1\rangle$ and excited $|\bar{E}_2\rangle$ eigenkets can be written as

$$\left|\bar{E}_{\substack{1 \\ 2}}\right\rangle = \frac{\Delta}{|\Delta|} \sqrt{\frac{1}{2}\left(1 \mp \frac{u}{\sqrt{1+u^2}}\right)} |L\rangle \pm \sqrt{\frac{1}{2}\left(1 \pm \frac{u}{\sqrt{1+u^2}}\right)} |R\rangle, \tag{C.40}$$

assuming $E_R \le E_L$; in Eq. (C.40) we have introduced

$$u = \frac{E_L - E_R}{2|\Delta|}, \tag{C.41}$$

measuring the relative importance of the diagonal energy difference to the off-diagonal coupling strength.

The eigenvalues are sketched in Fig. C.2. We see that the two eigenenergies are centered around $(E_L + E_R)/2$, due to the conservation of the matrix trace upon diagonalization. The eigenenergies never come closer than the smaller of $|E_L - E_R|$ and $|2\Delta|$. The off-diagonal coupling Δ induces a sort of mutual "repulsion" between the energy levels.

In the special limit $E_L = E_R$, i.e. $u = 0$, the two eigenvalues acquire a minimum separation $|2\Delta|$. The corresponding kets turn into symmetric $|\bar{E}_1\rangle = 2^{-1/2}(|L\rangle + |R\rangle)$ and antisymmetric $|\bar{E}_2\rangle = 2^{-1/2}(|L\rangle - |R\rangle)$ combinations of the original states. For increasing u, the eigenkets (C.40) deviate more and more from these simple symmetric and antisymmetric combinations: the ground state $|\bar{E}_1\rangle$ acquires a prevalent $|R\rangle$ character; correspondingly $|\bar{E}_2\rangle$ acquires a mainly $|L\rangle$ character.

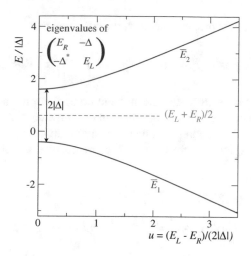

Fig. C.2 The eigenvalues, Eq. (C.39), of the 2×2 matrix (C.38), as a function of the asymmetry ratio u. The symmetric case $u = 0$ describes the splitting of degenerate levels by a off-diagonal "perturbation". This splitting determines, e.g., the bonding-antibonding splitting in the electronic structure of homonuclear diatomic molecules. For large $u \gg 1$, $\bar{E}_1 \simeq E_R$ and $\bar{E}_2 \simeq E_L$. This limit describes, e.g., the electronic bonding/antibonding states of strongly ionic dimers

C.6 The Schrödinger Equation in Real Space

The variational method outlined above provides also a useful reformulation of Schrödinger's equation, by adopting the basis of position eigenstates $|\mathbf{r}'\rangle$ for the set of kets $|\phi_l\rangle$. Because for the translational degrees of freedom the set of all $|\mathbf{r}'\rangle$ is complete, this method yields an exact mapping of the Schrödinger equation (C.30) for the translational degrees of freedom. This mapping to a differential equation for the wavefunction $\psi(\mathbf{r}')$ takes the name of "position representation".

To obtain this mapping, we need to represent the momentum operator on the position basis. Applying Eq. (C.21) on an arbitrary state $|\alpha\rangle$, we have

$$
\left(I - i\hbar^{-1} \mathbf{P} \cdot \mathbf{x} \right) |\alpha\rangle = \mathcal{T}(\mathbf{x})|\alpha\rangle = \mathcal{T}(\mathbf{x}) \int d^3 r' \, |\mathbf{r}'\rangle \langle \mathbf{r}'|\alpha\rangle
$$

$$
= \int d^3 r' \, \mathcal{T}(\mathbf{x})|\mathbf{r}'\rangle \langle \mathbf{r}'|\alpha\rangle = \int d^3 r' \, |\mathbf{r}' + \hat{\mathbf{x}}\,\delta x\rangle \langle \mathbf{r}'|\alpha\rangle
$$

$$
= \int d^3 r' \, |\mathbf{r}'\rangle \langle \mathbf{r}' - \hat{\mathbf{x}}\,\delta x|\alpha\rangle
$$

$$
= \int d^3 r' \, |\mathbf{r}'\rangle \left(\langle \mathbf{r}'|\alpha\rangle - \mathbf{x} \cdot \nabla_{r'} \langle \mathbf{r}'|\alpha\rangle \right)
$$

$$
= |\alpha\rangle - \mathbf{x} \cdot \int d^3 r' \, |\mathbf{r}'\rangle \nabla_{r'} \langle \mathbf{r}'|\alpha\rangle .
$$

By identifying the terms proportional to $\mathbf{x}$, we conclude that

$$\mathbf{P}|\alpha\rangle = -i\hbar \int d^3r' \, |\mathbf{r}'\rangle \boldsymbol{\nabla}_{r'}\langle \mathbf{r}'|\alpha\rangle = -i\hbar \int d^3r' \, |\mathbf{r}'\rangle \boldsymbol{\nabla}_{r'}\psi_\alpha(\mathbf{r}') . \qquad (C.42)$$

This expression indicates that, after the momentum operator has acted on a state $|\alpha\rangle$, its wavefunction becomes $-i\hbar \times$ the gradient of the wavefunction $\psi_\alpha(\mathbf{r}')$ of the state. In this sense, the position representation of the momentum operator is $-i\hbar \boldsymbol{\nabla}_{r'}$.

The following expressions for the matrix elements of $\mathbf{P}$ are direct consequences of Eq. (C.42):

$$\langle \mathbf{r}'|\mathbf{P}|\mathbf{r}''\rangle = -i\hbar \boldsymbol{\nabla}_{r'}\delta(\mathbf{r}' - \mathbf{r}'') \qquad (C.43)$$

$$\langle \beta|\mathbf{P}|\alpha\rangle = \int d^3r' \, \psi_\beta^*(\mathbf{r}') \left(-i\hbar \boldsymbol{\nabla}_{r'}\right) \psi_\alpha(\mathbf{r}') . \qquad (C.44)$$

For a particle with mass m moving under the action of conservative forces described by a potential energy function $V(\mathbf{r})$, the Hamiltonian is

$$H = H_{\text{kin}} + V(\mathbf{R}) = \frac{|\mathbf{P}|^2}{2m} + V(\mathbf{R}) . \qquad (C.45)$$

The expression (C.43) for the momentum matrix operator allows us to recognize that on the position basis the matrix of the $|\mathbf{P}|^2$ operator is diagonal:

$$\langle \mathbf{r}'| \, |\mathbf{P}|^2 \, |\mathbf{r}''\rangle = -\hbar^2 \nabla_{r'}^2 \, \delta(\mathbf{r}' - \mathbf{r}'') . \qquad (C.46)$$

Also the potential-energy term is of course diagonal on this basis:

$$\langle \mathbf{r}'|V(\mathbf{R})|\mathbf{r}''\rangle = V(\mathbf{r}') \, \delta(\mathbf{r}' - \mathbf{r}'') . \qquad (C.47)$$

With these results in mind, evaluate now the matrix elements in Eq. (C.36) on the position basis: $|\phi_j\rangle = |\mathbf{r}'\rangle$ and $|\phi_l\rangle = |\mathbf{r}''\rangle$. Due to the Dirac deltas in Eqs. (C.46) and (C.47), the $\sum_l$ (here an $\int d^3r''$) drops out, resulting in a substantial simplification. Therefore, the Schrödinger equation (C.30) involves a single position variable $\mathbf{r}'$:

$$\left[-\frac{\hbar^2}{2m} \, \nabla_{r'}^2 + V(\mathbf{r}') \right] \psi_{\mathcal{E}}(\mathbf{r}') = \mathcal{E} \, \psi_{\mathcal{E}}(\mathbf{r}') . \qquad (C.48)$$

This is a second-order differential secular equation in 3D space. Simpler 1D and 2D versions can be formulated whenever the motion in the remaining dimensions is either impossible or trivial. For several interacting particles moving simultaneously, under the action of a many-body Hamiltonian

$$H = H_{\text{kin}} + V(\mathbf{R}_1, \mathbf{R}_2, \mathbf{R}_3, \ldots) = \sum_\alpha \frac{|\mathbf{P}_\alpha|^2}{2m} + V(\mathbf{R}_1, \mathbf{R}_2, \mathbf{R}_3, \ldots) , \qquad (C.49)$$

the equation in real space generalizes to

$$\left[-\frac{\hbar^2}{2m} \sum_\alpha \nabla^2_{r'_\alpha} + V(\mathbf{r}'_1, \mathbf{r}'_2, \mathbf{r}'_3, \ldots) \right] \psi_E(\mathbf{r}'_1, \mathbf{r}'_2, \mathbf{r}'_3, \ldots) = E \, \psi_E(\mathbf{r}'_1, \mathbf{r}'_2, \mathbf{r}'_3, \ldots) .$$

$$(C.50)$$

For most practical problems, the real-space formulation (C.48) and (C.50) is far too complicated to solve analytically. In these cases one usually goes back to approximate numerical methods based on the mapping of Eq. (C.30) onto an algebraic matrix problem, as described in Sect. C.5. However, the real-space formulation of the Schödinger problem can be solved exactly in a few simple cases.

C.7 Exact Solutions of Schrödinger's Equation

In the following we report three simple but physically fundamental examples of exactly-solvable mechanical systems. A few other examples, e.g. the one-electron atom, are mentioned elsewhere in the present book.

C.7.1 A Free Particle

When in Eq. (C.48) the potential energy $V(\mathbf{r}') = 0$, the motion of the particle is free. One can separate the motion in the three Cartesian directions by decomposing the solution

$$\psi_{\mathcal{E}}(\mathbf{r}') = \psi_{\mathcal{E}_x}(r'_x) \, \psi_{\mathcal{E}_y}(r'_y) \, \psi_{\mathcal{E}_z}(r'_z) , \qquad (C.51)$$

with $\mathcal{E}_u$ representing the contribution of the u-component of the motion to the total kinetic energy $\mathcal{E} = \mathcal{E}_x + \mathcal{E}_y + \mathcal{E}_z$. The equation for each component is

$$-\frac{\hbar^2}{2m} \nabla^2_{r'_u} \psi_{\mathcal{E}_u}(r'_u) = \mathcal{E}_u \, \psi_{\mathcal{E}_u}(r'_u) . \qquad (C.52)$$

Any kind of harmonic-type functions such as, e.g., $\psi(r'_u) = \exp(\kappa r'_u)$ or $\psi(r'_u) = \sin(\kappa r'_u)$ solve Eq. (C.52) for a suitable $\mathcal{E}_u$. However, not all solutions are equally acceptable: in these two examples, whenever $\Re(\kappa) \neq 0$ the exponential one would diverge at large r'_u, and for $\Im(\kappa) \neq 0$ the trigonometric one would have similar shortcomings. In practice, solutions of the type

$$\psi_{k_u}(r'_u) = (2\pi)^{-1/2} \exp(i \, k_u \, r'_u) \qquad (C.53)$$

are usually adopted, with real k_u. The factor $(2\pi)^{-1/2}$ yields a convenient normalization for a wavefunction defined in the whole range $-\infty < r'_u < \infty$. In such a case, all real values of k_u are possible, and provide an energy contribution $\mathcal{E}_u = \hbar^2 k_u^2/(2m)$, as one can immediately verify by inserting the solution (C.53) into Eq. (C.52).

The vector $\mathbf{k}$ with components $\{k_x, k_y, k_z\}$ represents the eigenvalue of the vector operator $\mathbf{K}$ introduced in Eq. (C.13). Indeed, the states defined by the plane-wave function of Eq. (C.53) are eigenstates $|\mathbf{k}\rangle$ of $\mathbf{K}$ and therefore also of the translation operator $\mathcal{T}(\mathbf{x})$. According to Eq. (C.20), these $|\mathbf{k}\rangle$ states are also eigenstates of the momentum operator $\mathbf{P}$ with eigenvalue $\mathbf{p}' = \hbar\mathbf{k}$. The ordinary relations of energy and momentum hold for these eigenvalues:

$$\mathcal{E} = \mathcal{E}_x + \mathcal{E}_y + \mathcal{E}_z = \frac{\hbar^2 k_x^2}{2m} + \frac{\hbar^2 k_y^2}{2m} + \frac{\hbar^2 k_z^2}{2m} = \frac{p'^2_x}{2m} + \frac{p'^2_y}{2m} + \frac{p'^2_z}{2m} = \frac{|\mathbf{p}'|^2}{2m}.$$
(C.54)

It is often convenient to express the free-particle wavefunctions in a finite—rather than infinite—region of space, with volume $V = L \times L \times L$. Periodic boundary conditions are usually applied across this cube. As a consequence, not all values of wave vector are allowed, but only those with components

$$k_u = \frac{2\pi}{L} n_u, \quad n_u = 0, \pm 1, \pm 2, \pm 3, \ldots$$
(C.55)

In this $L \times L \times L$ cube, the plane-wave eigenfunctions are normalized as follows:

$$\psi_{\mathbf{k}}(\mathbf{r}) = L^{-3/2} \exp(i\,\mathbf{k}\cdot\mathbf{r}).$$
(C.56)

The corresponding translational kinetic energy takes on the discrete values

$$\mathcal{E}_{\mathbf{n}} = \frac{|\mathbf{p}'_{\mathbf{n}}|^2}{2m} = \frac{\hbar^2 |\mathbf{k}_{\mathbf{n}}|^2}{2m} = \frac{(2\pi\hbar)^2}{2mL^2}(n_x^2 + n_y^2 + n_z^2).$$
(C.57)

In the limit of infinite size $L \to \infty$ these discrete kinetic-energy eigenvalues turn into the same continuum of positive energies as as those of Eq. (C.54).

C.7.2 A Particle Confined in an Infinitely Deep Square Well

A particle confined in a finite region of space is particularly illuminating for the implications of the Heisenberg uncertainty relation for the energy spectrum.

Consider the potential energy

$$V(r_x) = \begin{cases} 0 & \text{if } 0 \le r_x \le L \\ +\infty & \text{elsewhere} \end{cases},$$
(C.58)

confining a particle to an interval of width L in 1D. Inside the confinement region, the particle moves freely: its wavefunction follows Eq. (C.52). The confining potential forces the wavefunction to vanish at $r_x \leq 0$ and $r_x \geq L$ (otherwise the infinite repulsion would make $\mathcal{E}_x$ diverge). This requirement imposes a vanishing boundary condition to the eigenfunctions of Eq. (C.52). The general solution

$$\psi_{k_x}(r_x) = \sqrt{\frac{2}{L}} \sin(k_x r_x), \quad k_x = \frac{\pi n}{L}, \quad n = 1, 2, 3, \ldots \quad (C.59)$$

exhibits $n - 1$ nodes. At each node, $\psi_{k_x}(r_x)$ changes sign.

The corresponding (kinetic) energy eigenvalues are

$$\mathcal{E}_n = \frac{\hbar^2 k_x^2}{2m} = \frac{\pi^2 \hbar^2 n^2}{2mL^2}, \quad n = 1, 2, 3, \ldots . \quad (C.60)$$

Compare this result with Eq. (4.39): note in particular that the ground-state energy of the free particle vanishes, while that of the confined particle, obtained by substituting $n = 1$ in Eq. (C.60), is nonzero. Its value increases with the inverse of the particle mass and the inverse square of the confinement size L. This a fundamental consequence of Heisenberg's uncertainty relation: the more restrictively a quantum particle is confined in position space, the fuzzier its momentum becomes, resulting in a larger and larger average kinetic energy.

C.7.3 The Linear Harmonic Oscillator

A third basic problem of mechanics is that of a mass attached to a fixed point through an elastic spring, which we address for simplicity in 1D. The restoring force $-Kr_x$ is represented by the potential energy

$$V(r_x) = \frac{1}{2} K r_x^2 . \quad (C.61)$$

As it is straightforward to verify by direct substitution, the eigenfunctions of Schrödinger's equation

$$\left(-\frac{\hbar^2}{2m} \nabla_{r_x}^2 + \frac{1}{2} K r_x^2 \right) \psi_{\mathcal{E}}(r_x) = \mathcal{E} \psi_{\mathcal{E}}(r_x) \quad (C.62)$$

are

$$\psi_n(r_x) = \left(\frac{1}{\pi^{1/2} 2^n n! \, x_0} \right)^{1/2} H_n \left(\frac{r_x}{x_0} \right) \exp \left(-\frac{r_x^2}{2x_0^2} \right), \quad n = 0, 1, 2, 3, \ldots , \quad (C.63)$$

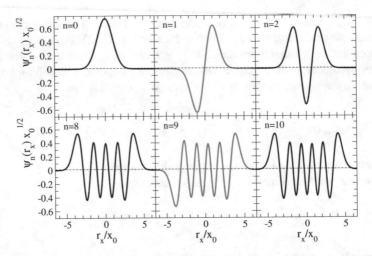

Fig. C.3 The eigenfunctions for the ground state and for a few excited states of the linear harmonic oscillator. Note that the number of nodes (zeroes of the eigenfunctions) equals the quantum number n

where

$$x_0 = \hbar^{1/2} (mK)^{-1/4}, \qquad H_n(\xi) = (-1)^n \exp\left(\xi^2\right) \frac{d^n}{d\xi^n} \exp\left(-\xi^2\right). \qquad (C.64)$$

The functions $H_n(\xi)$ are (Hermite) polynomials of degree n, e.g. $H_0(\xi) = 1$, $H_1(\xi) = 2\xi$, $H_2(\xi) = 4\xi^2 - 2$, ... As illustrated in Fig. C.3, $\psi_n(r_x)$ exhibits n nodes, at the points where $H_n(r_x/x_0)$ changes sign.

Introducing the angular frequency $\omega_0 = (K/m)^{1/2}$ of oscillation, the energy eigenvalues

$$\mathcal{E}_n = \hbar\omega_0 \left(n + \frac{1}{2}\right) \qquad (C.65)$$

form an equally-spaced ladder, with spacing $\hbar\omega_0$. The energy $\mathcal{E}_n$ of each eigenstate consists of 50% kinetic plus 50% potential contributions.

The ground-state position probability distribution $P_0(r_x) = |\psi_0(r_x)|^2$ is a Gaussian with standard deviation $2^{-1/2}x_0$, measuring the amplitude of the zero-point motion. Like for the infinite square well, the "zero-point" ground-state energy $\mathcal{E}_0 = \hbar\omega_0/2$ is positive, as a consequence of confinement and Heisenberg's uncertainty relation. The 50% kinetic and 50% potential energy composition of $\mathcal{E}_0$ contrasts the 100% kinetic energy $\mathcal{E}_1$ of the ground-state of the infinite square well, Eq. (C.60).

The quantum harmonic oscillator is a fundamental model for several physical phenomena, including:

- the vibrational degree of freedom of a diatomic molecule. In this context, the symbol v often replaces the vibrational quantum number n, see Eq. (3.21);

- each normal mode of vibration of a polyatomic molecule, see Fig. 3.12;
- each normal mode of vibration of a solid, see Sect. 5.3;
- each normal oscillatory mode of the electromagnetic fields in a cavity, see Sect. 4.3.2.2.

C.8 Angular Momentum

In Sect. C.2 we introduced linear momentum $\mathbf{P}$ as the generator of translations. Likewise, angular momentum generates rotations. The orbital angular momentum operator $\mathbf{L} = \mathbf{R} \times \mathbf{P}$ generates the rotations of the position degrees of freedom of a particle around the origin of the reference system. Starting from the commutation relations (C.22) for the $\mathbf{R}$ and $\mathbf{P}$ operators, it is straightforward to realize that $[L_x, L_y] = i \hbar L_z$, and analogous relations obtained by cyclic permutations of the components. These commutation relations are fundamental in nature. They can be extended to the total angular momentum $\mathbf{J}$, by requiring that $\mathbf{J}$ is related to the operator $\mathcal{D}(\hat{\mathbf{n}}, \delta\varphi)$ implementing a rotation by an infinitesimal angle $\delta\varphi$ around the direction $\hat{\mathbf{n}}$, as follows:

$$\mathcal{D}(\hat{\mathbf{n}}, \delta\varphi) = \mathcal{I} - i\hbar^{-1}\mathbf{J} \cdot \hat{\mathbf{n}} \, \delta\varphi. \tag{C.66}$$

One can prove that $\mathbf{J} \equiv \mathbf{L}$ for a spinless particle. For particles with spin, such as electrons, instead, a rotation must include spin as well. As a result, an extra contribution has to be included in the *total angular momentum*

$$\mathbf{J} = \mathbf{L} + \mathbf{S}. \tag{C.67}$$

Spin fulfills the same fundamental commutation relations $[S_x, S_y] = i \hbar S_z$, etc. as the L_u components. Moreover, spin acts in a space different from the one of translations, therefore $[L_u, S_v] = 0$. For systems formed by several particles the rotations of all must be included, so that

$$\mathbf{J} = \mathbf{J}_1 + \mathbf{J}_2 + \mathbf{J}_3 + \ldots \tag{C.68}$$

The angular momentum commutation relations

$$[J_x, J_y] = i \hbar J_z, \text{ etc.} \tag{C.69}$$

are all that it takes to determine the properties (eigenvalues, eigenvectors, matrix elements) of the angular momentum operators, irrespective of their spin or orbital nature [17].

One can show that the operator $|\mathbf{J}|^2$ commutes with any component operator, e.g. J_z. As a consequence, $|\mathbf{J}|^2$ and J_z are compatible observables and can be diagonalized simultaneously, see Sect. C.2.1. Call $|J, M_J\rangle$ the basis of common eigenstates of $|\mathbf{J}|^2$

and J_z. The expressions for their eigenvalues are [17]:

$$|\mathbf{J}|^2|J, M_J\rangle = \hbar^2 J(J+1)|J, M_J\rangle, \quad \text{with } J = 0, \ 1/2, \ 1, \ 3/2, \ 2, \ 5/2, \ldots \quad (C.70)$$

$$J_z|J, M_J\rangle = \hbar M_J|J, M_J\rangle, \qquad\qquad \text{with } M_J = -J, \ -J+1, \ldots J-1, \ J. \quad (C.71)$$

For given J, the projection quantum number M_J can take one of the $2J + 1$ values listed above.

A rotationally-invariant QM system, e.g. an atom, has $[H, \mathcal{D}(\hat{\mathbf{n}}, \delta\varphi)] = 0$ for any $\hat{\mathbf{n}}$ and $\delta\varphi$, see Eq. (C.34). Due to Eq. (C.66), this implies $[H, \mathbf{J}] = 0$, indicating that the operators H, $|\mathbf{J}|^2$, and J_z are compatible, and can therefore be diagonalized simultaneously. As a consequence, the energy eigenstates and eigenvalues of a rotationally invariant system can be labeled with the J and M_J quantum numbers.

C.8.1   The Coupling of Angular Momenta

One often needs to combine several angular momenta, e.g. those carried by several particles or by the orbital and spin degrees of freedom of the same particle. The main question is: what are the allowed eigenvalues of the total angular momentum $|\mathbf{J}|^2$, for given eigenvalues of the combined angular momenta?

Here we provide the answer to this problem when the combining angular momenta are just two. We combine, say, the orbital angular momentum $\mathbf{L}$ and the spin one $\mathbf{S}$. The starting observation is that one can identify two sets of 4 mutually commuting operators $(|\mathbf{L}|^2, |\mathbf{S}|^2, L_z, S_z)$ and $(|\mathbf{L}|^2, |\mathbf{S}|^2, |\mathbf{J}|^2, J_z)$. Each of these sets can be diagonalized simultaneously to produce a basis of orthonormal states. Basis one and basis two, obtained diagonalizing the first and second set respectively, do not coincide. However, observe that two out of four operators coincide: $|\mathbf{L}|^2$ and $|\mathbf{S}|^2$. They can then be taken diagonal in both bases. It is then possible to fix the corresponding quantum numbers l and s.

In basis one, of states $|l, s, m_l, m_s\rangle$, L_z and S_z are diagonal. For fixed l and s, this basis is formed by $d = (2l + 1) \times (2s + 1)$ states labeled by all possible values $m_l = -l, \ -l+1, \ \ldots l-1, \ l$ and $m_s = -s, \ -s+1, \ \ldots s-1, \ s$ of the individual $\hat{\mathbf{z}}$-projections. Alternatively, to span this d-dimensional space of states, one can select the second basis of states $|l, s, j, m_j\rangle$, that has diagonal $|\mathbf{J}|^2$ and J_z. The coupling of these angular momenta provides the change of basis from basis one to basis two.

To identify the $|l, s, j, m_j\rangle$ states we need to determine the values of j compatible with the given l and s. The commutation relations are sufficient to answer completely this question [17] and also to express the "coupled states" $|l, s, j, m_j\rangle$ in terms of the original L_z and S_z eigenstates $|l, s, m_l, m_s\rangle$. For our purposes, it suffices to retain the main result of this calculation, namely that the allowed values for j are

Fig. C.4 An intuitive mnemonic sketch for the rule of angular-momentum composition, Eq. (C.72)

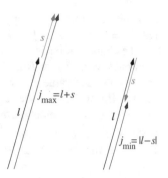

$$j = |l - s|, \ |l - s| + 1, \ \dots \ l + s - 1, \ l + s. \tag{C.72}$$

The extremal values recall the classical picture of vector composition (Fig. C.4); the discrete values are characteristic of the QM of angular momentum. One should check that the number d of states in basis one and basis two is the same, i.e. that

$$(2l + 1) \cdot (2s + 1) = 2(|l - s|) + 1 + 2(|l - s| + 1) + 1 + \dots$$
$$+ 2(l + s - 1) + 1 + 2(l + s) + 1.$$

A unitary transformation in this d-dimensional space implements the coupling of angular-momentum eigenstates. In practice this transformation can be written as a multiplication of the basis states by a unitary $d \times d$ matrix:

$$|l, s, j, m_j\rangle = \sum_{m_l, m_s} C^{j \, m_j}_{l \, m_l \, s \, m_s} |l, s, m_l, m_s\rangle. \tag{C.73}$$

The coefficients $C^{j \, m_j}_{l \, m_l \, s \, m_s}$ defining the linear basis transformation (C.73) are named Clebsch–Gordan coefficients, are taken real by convention, and are tabulated in specific books [49, 50].

A priori, either basis one or basis two are equally suited to describe a particle's orbital and spin dynamics. The difference is that the $|m_l, m_s\rangle$ basis (hence, for compactness' sake, we drop the fixed l and s labels) emphasizes the invariance against separate position-space and spin-space rotation, while the $|j, m_j\rangle$ basis emphasizes the invariance for global rotations (equal rotations for position and spin).

Consider the concrete example of an electron with its spin $s = 1/2$. With the basic nonrelativistic Hamiltonian (1.1) all the $d = (2l + 1) \times 2$ states within the $(l, 1/2)$ multiplet have the same energy: this degeneracy occurs whether we describe them in terms of the $|m_l, m_s\rangle$ basis or of the $|j, m_j\rangle$ basis. When H includes relativistic interactions, which are invariant under global rotations, but not under separate orbital and spin rotations, the coupled $|j, m_j\rangle$ basis becomes the basis of choice, because $[H, \mathbf{J}] = 0$, but $[H, L_z] \neq 0$ and $[H, S_z] \neq 0$.

The change of basis dictates how the orbital angular momentum l of an electron combines with its spin. Rule (C.72) yields the appropriate j values. For $l = 0$—i.e. s states—a single value $j \equiv s = 1/2$ is obtained; for any $l = 1, 2, 3 \ldots$—i.e. p, d, f... states—Eq. (C.72) foresees the two values $j = l - 1/2$ and $j = l + 1/2$. For s states, the transformation matrix from the $|m_l = 0, m_s\rangle$ basis to the $|j = 1/2, m_j\rangle$ basis is trivially the 2×2 identity $C_{00\,1/2 m_s}^{1/2 m_j} = \delta_{m_j m_s}$, since, for $l = 0$, $\mathbf{J}$ coincides with $\mathbf{S}$. For $l \geq 1$, the spin-orbital "maximally aligned" components are simply

$$\left|j = l + 1/2, m_j = \pm(l + 1/2)\right\rangle = |m_l = \pm l, m_s = \pm 1/2\rangle.$$

Each of the remaining coupled states is expressed in terms of two uncoupled states only, namely those whose $J_z = L_z + S_z$ component match:

$$\left|j = l + 1/2, m_j\right\rangle = c_1 \left|m_l = m_j + 1/2, m_s = -1/2\right\rangle + c_2 \left|m_l = m_j - 1/2, m_s = +1/2\right\rangle.$$

The orthogonal ket is

$$\left|j = l - 1/2, m_j\right\rangle = c_2 \left|m_l = m_j + 1/2, m_s = -1/2\right\rangle - c_1 \left|m_l = m_j - 1/2, m_s = +1/2\right\rangle.$$

Here

$$c_1 = C_{l\,m_j+1/2\,1/2\,-1/2}^{l+1/2 m_j} = \sqrt{\frac{l + 1/2 - m_j}{2l + 1}} \text{ and } c_2 = C_{l\,m_j-1/2\,1/2\,1/2}^{l+1/2 m_j} = \sqrt{\frac{l + 1/2 + m_j}{2l + 1}}$$

are the matrix elements of the basis-change transformation. Clearly, $c_1^2 + c_2^2 = 1$, because this transformation is unitary.

Even more concretely, for a p ($l = 1$) orbital triplet, the explicit 6×6 transformation matrix between the $|m_l, m_s\rangle$ basis and the coupled $|j, m_j\rangle$ basis acts as follows:

$$
\begin{pmatrix} \left|j = \frac{1}{2}, m_j = \frac{1}{2}\right\rangle \\ \left|j = \frac{1}{2}, m_j = -\frac{1}{2}\right\rangle \\ \left|j = \frac{3}{2}, m_j = \frac{3}{2}\right\rangle \\ \left|j = \frac{3}{2}, m_j = \frac{1}{2}\right\rangle \\ \left|j = \frac{3}{2}, m_j = -\frac{1}{2}\right\rangle \\ \left|j = \frac{3}{2}, m_j = -\frac{3}{2}\right\rangle \end{pmatrix}
=
\begin{pmatrix}
0 & -\sqrt{\frac{1}{3}} & 0 & \sqrt{\frac{2}{3}} & 0 & 0 \\
0 & 0 & -\sqrt{\frac{2}{3}} & 0 & \sqrt{\frac{1}{3}} & 0 \\
1 & 0 & 0 & 0 & 0 & 0 \\
0 & \sqrt{\frac{2}{3}} & 0 & \sqrt{\frac{1}{3}} & 0 & 0 \\
0 & 0 & \sqrt{\frac{1}{3}} & 0 & \sqrt{\frac{2}{3}} & 0 \\
0 & 0 & 0 & 0 & 0 & 1
\end{pmatrix}
\cdot
\begin{pmatrix} \left|m_l = 1, m_s = +\frac{1}{2}\right\rangle \\ \left|m_l = 0, m_s = +\frac{1}{2}\right\rangle \\ \left|m_l = -1, m_s = +\frac{1}{2}\right\rangle \\ \left|m_l = 1, m_s = -\frac{1}{2}\right\rangle \\ \left|m_l = 0, m_s = -\frac{1}{2}\right\rangle \\ \left|m_l = -1, m_s = -\frac{1}{2}\right\rangle \end{pmatrix},
$$

where we list the $j = 1 - 1/2 = 1/2$ doublet followed by the $j = 1 + 1/2 = 3/2$ quartet. In a pictorial "box" notation where ▦ stands for the uncoupled basis state $|m_l = 1, m_s = +1/2\rangle$, we can write e.g., the two bottom rows of the matrix equality above as

$$\left| j = 3/2,\ m_j = -1/2 \right\rangle = \sqrt{1/3}\ \boxed{\uparrow} + \sqrt{2/3}\ \boxed{\downarrow}$$

and

$$\left| j = 3/2,\ m_j = -3/2 \right\rangle = \boxed{\downarrow}.$$

In summary, when we measure the *total* angular momentum $|\mathbf{J}|^2$ of a one-electron atom, then each multiplet of states with given $l \geq 1$ yields *two* groups of states characterized by different values of j, namely $j = l - 1/2$ and $j = l + 1/2$. Unless spherical symmetry is broken (e.g. by some external field), all states with given j and different m_j have the same energy. Multiplets of states with different j would be degenerate if relativistic effects were ignored. In Sect. 2.1.7 we express the relativistic energy corrections on the coupled basis. We show that these terms are diagonal on the $|j, m_j\rangle$ basis, and that they are associated to different energies for different j, thus clarifying the need for the coupled basis.

The angular-momentum composition rules discussed here are more general than the composition of the orbital and spin angular momenta of a single electron. These rules are purely algebraic: they apply equally well to any kind of angular momentum. We rely on this same formalism for combining the angular momenta of many-electron atoms, especially in Sects. 2.2.4 and 2.2.9.3.

C.8.2 Coupled Magnetic Moments

The magnetic properties of a rotating charge are determined by angular-momentum properties, Eq. (2.56). Here we evaluate the matrix elements of the magnetic-moment operator on the uncoupled basis $|m_l,\ m_s\rangle$ and on the spin-orbit coupled basis $|j,\ m_j\rangle$. In the uncoupled basis, the μ_z operator is diagonal, with eigenvalues

$$\langle m_l,\ m_s | \mu_z | m_l,\ m_s \rangle = -\mu_B \langle m_l,\ m_s | \frac{L_z + 2S_z}{\hbar} | m_l,\ m_s \rangle = -\mu_B\,(m_l + 2m_s).$$

$$(C.74)$$

The matrix elements of the μ_x and μ_y components too have explicit expressions, which can be obtained from the (nondiagonal) matrix elements of $L_{x/y}$ and $S_{x/y}$ [17].

In principle one could obtain the matrix elements of μ in the coupled basis $|j,\ m_j\rangle$ by using explicitly the Clebsch–Gordan transformation (C.73). However, a simpler and more instructive method yields these matrix elements within each subspace at fixed total angular momentum j. The key point is a symmetry argument: on average, *all vector quantities* characterizing a spherically symmetric object freely rotating in space are *proportional to its total angular momentum*. This means in particular that

$$\langle \boldsymbol{\mu} \rangle \propto \langle \mathbf{J} \rangle, \quad \langle \mathbf{L} \rangle \propto \langle \mathbf{J} \rangle, \quad \text{and} \quad \langle \mathbf{S} \rangle \propto \langle \mathbf{J} \rangle.$$

The matrix elements of $\mathbf{J}$ are well known. The only unknown quantities are the individual proportionality constants. To obtain the relevant ones, note that by definition

$$\langle j, m_j | \boldsymbol{\mu} | j, m_j \rangle = -\frac{\mu_B}{\hbar} \langle j, m_j | \mathbf{L} + 2\mathbf{S} | j, m_j \rangle = -\frac{\mu_B}{\hbar} \langle j, m_j | \mathbf{J} + \mathbf{S} | j, m_j \rangle$$

$$\tag{C.75}$$

$$= -(1 + \gamma) \frac{\mu_B}{\hbar} \langle j, m_j | \mathbf{J} | j, m_j \rangle = -g_j \frac{\mu_B}{\hbar} \langle j, m_j | \mathbf{J} | j, m_j \rangle,$$

where we introduce the ratio γ between $\langle j, m_j | \mathbf{S} | j, m_j \rangle$ and $\langle j, m_j | \mathbf{J} | j, m_j \rangle$. We determine γ by observing that the same ratio is involved when we take the scalar product with $\mathbf{J}$:

$$\langle j, m_j | \mathbf{S} | j, m_j \rangle = \gamma \langle j, m_j | \mathbf{J} | j, m_j \rangle,$$
$$\langle j, m_j | \mathbf{J} \cdot \mathbf{S} | j, m_j \rangle = \gamma \langle j, m_j | \mathbf{J} \cdot \mathbf{J} | j, m_j \rangle. \tag{C.76}$$

γ can be extracted from Eq. (C.76) by replacing the scalar product $\mathbf{J} \cdot \mathbf{S}$ with the expression

$$\mathbf{J} \cdot \mathbf{S} = \frac{|\mathbf{J}|^2 + |\mathbf{S}|^2 - |\mathbf{L}|^2}{2}, \tag{C.77}$$

obtained by squaring $(\mathbf{J} - \mathbf{S}) = \mathbf{L}$. We obtain the proportionality constant

$$\gamma = \frac{\langle j, m_j | \mathbf{J} \cdot \mathbf{S} | j, m_j \rangle}{\langle j, m_j | |\mathbf{J}|^2 | j, m_j \rangle} = \frac{\langle j, m_j | |\mathbf{J}|^2 + |\mathbf{S}|^2 - |\mathbf{L}|^2 | j, m_j \rangle}{2 j(j+1) \hbar^2}$$

$$= \frac{j(j+1) + s(s+1) - l(l+1)}{2j(j+1)}.$$

This result allows us to evaluate the proportionality constant g_j between $\boldsymbol{\mu}/\mu_B$ and $-\mathbf{J}/\hbar$ introduced in Eq. (C.75):

$$g_j = 1 + \gamma = 1 + \frac{j(j+1) + s(s+1) - l(l+1)}{2j(j+1)} \tag{C.78}$$

called *Landé g-factor*. g_j measures (in units of μ_B) the *effective* atomic magnetic moment resulting from the combined orbital and spin contributions, as seen within a given fixed-j multiplet. In the spectroscopy of atoms the states of such a j-multiplet split in a magnetic field as if they had a component of $\langle \boldsymbol{\mu} \rangle$

$$\langle j, m_j | \mu_z | j, m_j \rangle = -g_j \mu_B m_j \tag{C.79}$$

in the $\hat{\mathbf{z}}$ direction of the field. Be warned that not all off-diagonal matrix elements of S_z (and thus of μ_z) vanish in the coupled basis $| j, m_j \rangle$. Note also that, contrary to what one might expect for a combination of two magnetic moments with $g_l = 1$ and $g_s = 2$, the values of g_j are *not* restricted to the range $1 \leq g_j \leq 2$.

C.9 Perturbation Theory

Real QM system are sometimes similar to simple systems for which exact analytic solutions are available, as in the examples of Sect. C.7. We express this resemblance by

$$H = H_0 + V ,$$ (C.80)

where H_0 is the "simple-system" Hamiltonian, and the operator $V \equiv H - H_0$ is the difference between the full Hamiltonian for the actual problem and the simple Hamiltonian. Consider now

$$H^{(\lambda)} = H_0 + \lambda V ,$$ (C.81)

where $0 \leq \lambda \leq 1$ is a real parameter which tunes continuously the Hamiltonian system from the simple H_0 to the full H. If V is sufficiently "small" (meaning that H_0 is really similar to H), then the set of eigenenergies and eigenkets should evolve continuously as follows:

$$
\begin{array}{ccc}
\lambda & : & 0 \to 1 \\
H^{(\lambda)} & : & H_0 \to H \\
E_n^{(\lambda)} & : & E_n^{(0)} \to E_n \\
|n^{(\lambda)}\rangle & : & |n^{(0)}\rangle \to |n\rangle .
\end{array}
$$ (C.82)

It should be possible to Taylor-expand the functional dependence of the eigenenergy and eigenkets:

$$E_n^{(\lambda)} = E_n^{(0)} + \lambda E_n^{(1)} + \lambda^2 E_n^{(2)} + \dots$$ (C.83)
$$|n^{(\lambda)}\rangle = |n^{(0)}\rangle + \lambda|n^{(1)}\rangle + \lambda^2|n^{(2)}\rangle + \dots$$ (C.84)

It is possible to prove that the 1st-order energy correction coefficient

$$E_n^{(1)} = \langle n^{(0)}|V|n^{(0)}\rangle .$$ (C.85)

The 2nd-order energy correction term

$$E_n^{(2)} = \sum_{k \neq n} \frac{|\langle n^{(0)}|V|k^{(0)}\rangle|^2}{E_n^{(0)} - E_k^{(0)}} .$$ (C.86)

The 1st-order correction to the eigenkets is

$$|n^{(1)}\rangle = \sum_{k \neq n} |k^{(0)}\rangle \frac{\langle n^{(0)}|V|k^{(0)}\rangle}{E_n^{(0)} - E_k^{(0)}} \,. \tag{C.87}$$

The corrections of Eqs. (C.85)–(C.87) are used to generate approximate eigenvalues and eigenkets by substitution into Eqs. (C.83) and (C.84), setting $\lambda = 1$. For example, the 1st-order approximation to the eigenvalues of H is given by

$$E_n \simeq E_n^{(0)} + E_n^{(1)} = E_n^{(0)} + \langle n^{(0)}|V|n^{(0)}\rangle \,. \tag{C.88}$$

Note that:

- Expression (C.85)–(C.88) involve *unperturbed* energies and matrix elements of the perturbing Hamiltonian V over *unperturbed* states. All these quantities involving the simple system H_0 can either be evaluated as analytic expressions, or be computed by numerical integration.
- Equations (C.86) and (C.87) involve energy denominators $E_n^{(0)} - E_k^{(0)}$ which vanish for degenerate H_0-eigenstates. In such degenerate cases, Eqs. (C.86) and (C.87) do not hold as such. However, in a degenerate situation, one is free to chose any linear combination of the unperturbed basis states within each degenerate space. One can then select suitable combinations for zeroing all off-diagonal matrix elements $\langle n^{(0)}|V|k^{(0)}\rangle$ at the numerators above the vanishing denominators. This means that one must first diagonalize the perturbation operator within all degenerate subspaces of the H_0 system. After this diagonalization, Eqs. (C.86) and (C.87) hold again, with the provision that the 0/0 terms are set to 0.
- The kets in any truncated version of Eq. (C.84) are generally unnormalized. It is straightforward to normalize each approximate eigenket by dividing it by its norm.

An important consequence of Eq. (C.88) is that, even in the plausible condition that the diagonal and the off-diagonal matrix elements of V have similar magnitude ($|\langle n^{(0)}|V|n^{(0)}\rangle| \simeq |\langle n^{(0)}|V|k^{(0)}\rangle|$), the diagonal ones, acting at 1st order, are responsible for the largest corrections to the eigenvalues. Each off-diagonal element starts to act at 2nd order, Eq. (C.86): its correction to the eigenvalue involves the ratio of the square of the off-diagonal matrix element itself divided by the difference of unperturbed energies. This quadratic dependence often generates *very* small corrections. For example, if $E_n^{(0)} - E_k^{(0)} \simeq 1$ eV and $|\langle n^{(0)}|V|k^{(0)}\rangle| \simeq 1$ meV, then the lowest-order correction due to off-diagonal elements is $E_n^{(2)} \simeq (1 \text{ meV})^2/(1 \text{ eV}) \simeq 1 \text{ μeV}$. This tiny correction should be compared with the much larger correction due to the diagonal matrix element, $E_n^{(1)} = \langle n^{(0)}|V|n^{(0)}\rangle \simeq 1$ meV.

The degenerate case $E_n^{(0)} - E_k^{(0)} = 0$ marks an apparent exception to the above observation. As noted above, this singular condition must be treated by pre-diagonalizing the perturbation operator in the degenerate space. After this unitary rotation, the off-diagonal elements of V transform into purely diagonal ones: Eq. (C.85) applies, and predicts 1st-order corrections, of the same order as the perturbation matrix elements.

It is an instructive exercise to apply the formulas of perturbation theory to the 2-states problem of Eq. (C.38), assuming that the off-diagonal Hamiltonian acts as a small perturbation to the diagonal part, i.e. Δ is very small. One can verify that perturbation theory provides indeed the correct Taylor expansion to the eigenvalues, Eq. (C.39). Specifically, in the nondegenerate case $E_L \neq E_R$, one obtains the precise 2nd-order correction to the eigenvalues, proportional to $|\Delta|^2/(E_L - E_R)$. In contrast, the degenerate case $E_L = E_R$ requires the pre-diagonalization of the perturbation, which here provides the 1st-order corrections $\pm|\Delta|$.

C.10 Interaction of Charged Particles and Electromagnetic Radiation

A charged particle such an electron or a proton interacts with the electromagnetic fields. The form of this interaction is

$$H = \frac{1}{2m} (\mathbf{P} - q\mathbf{A})^2 + q\phi + V , \tag{C.89}$$

where m and q are the particle's mass and charge, $\mathbf{A}$ and ϕ are the electromagnetic vector and scalar potentials, respectively, and V is the potential energy describing non-electromagnetic forces acting on the particle. The potentials are functions of position, and often of time, as well. In QM the particle position $\mathbf{R}$ is an operator. Therefore the electromagnetic potentials are operators, and precisely they are functions of the position operator $\mathbf{R}$. This implies that while $[A_u(\mathbf{R}), R_v] = 0$, in general $[A_u(\mathbf{R}), P_v] \neq 0$, so the potentials and the particle momentum are not compatible observables. The meaning of the square in Eq. (C.89) is then

$$(\mathbf{P} - q\mathbf{A})^2 = |\mathbf{P}|^2 - q(\mathbf{P} \cdot \mathbf{A} + \mathbf{A} \cdot \mathbf{P}) + q^2 |\mathbf{A}|^2. \tag{C.90}$$

Of these three terms, the first one generates the standard kinetic energy of the uncoupled particle, the others describe the particle-fields coupling. For static fields, one can represent the electric fields in terms of the ϕ potential, and the magnetic field in terms of $\mathbf{A}$. When the fields change in time, as in the presence of electromagnetic radiation, a suitable gauge choice must be adopted.

As long as the electromagnetic fields are not too intense, this interaction is usually addressed by *time-dependent perturbation theory*, a rather advanced topic whose mathematical subtleties far exceed the level of the present Appendix. To our purposes it suffices to retain that the spontaneous emission rate in the electric-dipole approximation, Eq. (2.45), is obtained by a linear-response perturbative treatment of the term proportional to q in Eq. (C.90). The elastic scattering of radiation, e.g. in the X-ray diffraction investigation of the structure of solids, Sect. 5.1.3, in terms of Eq. (C.90) is described as a first-order contribution of the q^2 term plus the second-order effect of the q term.

Solutions to Problems

Problems of Chap. 2

2.1 5 components; 25.4 mm

2.2 $1s^2 2s^2 2p^5 3s\ ^1P_1$, $1s^2 2s^2 2p^5 4s\ ^1P_1$; both split into 3 components separated by 58 μeV

2.3 5145 eV; 2827 eV, 26835 eV, 31199 eV

2.4 $\gamma_{3s\text{-}1s} = 0$; $\gamma_{2p\text{-}1s} = 6.27 \times 10^8$ s^{-1}

2.5 Max z projection of μ: $7\,\mu_B = 6.492 \times 10^{-23}$ A m^2; $|\mu| = \sqrt{61}\,\mu_B = 7.243 \times 10^{-23}$ A m^2; 5 components

2.6 4 lines; 3 sub-lines for each line with $\Delta J = 0$, 6 sub-lines for the lines $\Delta J = 1$ and $\Delta J = -1$

2.7 $^3G_{3/4/5}$, $^3F_{2/3/4}$, $^3D_{1/2/3}$; $^3G_3$, $^3F_3$, $^3D_3$

2.8 1.606×10^{-23} A m^2; 1.606×10^{-23} A m^2; 5.487×10^{-23} A m^2

2.9 24350 eV (K); 3604 eV (L_I); 3330 eV (L_{II}); 3173 eV (L_{III}); the K shell; 42.30

2.10 21.25%; 0.18°

2.11 L, 5400 eV; M, 1000 eV

2.12 $\frac{2}{5}E_{Ha}(r_m/a_0)^2 = 3.15$ neV

2.13 90.3 meV; 116.1 meV; 129.3 meV. Zeeman splittings, since $\mu_B B = 0.17$ meV $\ll \xi = 12.9$ meV

N. Manini, *Introduction to the Physics of Matter*, Undergraduate Lecture Notes in Physics, https://doi.org/10.1007/978-3-030-57243-3

Problems of Chap. 3

3.1 $l = 4$; 2581.3 cm^{-1} and 2735.9 cm^{-1}

3.2 408.6 pN; 0.0183 J mol^{-1} K^{-1}

3.3 380 kg/s^2; 1.46 × 10^{-10} m; 280 K

3.4 14 levels; $Z = 3.885$; bound fraction = 97.25%

3.5 2.297 × 10^{-9} N; 1.508 meV

3.6 62964 GHz; 50.6 meV (in the harmonic approximation)

3.7 21.459 J mol^{-1} K^{-1}

3.8 1.3575 × 10^{-18} J; 1.0134 × 10^{-10} m

3.9 9.42110^{13} Hz

3.10 $l = 4$ [taking the following effects into account: (i) the decrease of the zero-point energy, (ii) the increase of V_{ad}, and (iii) the decrease of the $V_{centrif}$ term. Neglecting effects (i) and (ii) the result would be $l = 8$]

3.11 74.08 pm; 6.19 pm; 0.084

3.12 343.7 nm

3.13 474 meV

3.14 Verify that $\varepsilon_{\pi^*} < \varepsilon_{\sigma^*}$ for $R = R_M$; R(degeneracy of π^* and σ^*) = 99.8 pm

3.15 The 6 line assignments are: $l = 2 \rightarrow 3$, $l = 1 \rightarrow 2$, $l = 0 \rightarrow 1$, $l = 1 \rightarrow 0$, $l = 2 \rightarrow 1$, and $l = 3 \rightarrow 2$, respectively. $\Delta E_{vib} = 1.410$ eV

Problems of Chap. 4

4.1 $16\pi^2 \hbar \nu^3 c^{-3} \rho_\nu^{-1} = 5.297 \times 10^{-21}$

4.2 $P = \hbar c (N/V)^{4/3} (3\pi^2)^{1/3}/4 = 5.712 \times 10^{24}$ Pa

4.3 1.401 × 10^{10} Pa; 1.633 × 10^6 Pa

4.4 Percentage$(J = 1/2) = 30.81\%$; 2.07 × 10^{-23} J/K; 3.963 × 10^{-26} J/K

4.5 1.70

4.6 0.42102 J mol^{-1} K^{-1}; 0.0327

4.7 1.372×10^{-6} J/(kg K)

4.8 15.625; 39.06

4.9 1.046×10^{6} m/s; 2.090×10^{9} Pa

4.10 2046 K; 14.372

4.11 8953 K

4.12 944 W; 302.1 K

4.13 4.96 GPa; $\Delta P = 1.76$ GPa

4.14 8.0 mW

Problems of Chap. 5

5.1 1.519×10^{6} m/s

5.2 $\alpha - 3$; $A = 1.831 \times 10^{-5}$ J mol^{-1} K^{-4}; 23.9 K

5.3 427.5 nm

5.4 Gap $\simeq E = 6.8$ eV; insulator; transparent

5.5 A$\rightarrow$ 512.6 pm; B$\rightarrow$ 887.8 pm; C$\rightarrow$ 724.9 pm; A, with density 1479 kg m^{-3}

5.6 3704 m/s

5.7 $w(k) = 2\sqrt{C/M_{\text{Fe}}} \, \sin(ka/2)$; 8.26 THz

5.8 20.7 μs

5.9 24.35°

5.10 $\Theta_{\text{D}} = 360$ K; $v_{\text{s}} = 3420$ m/s; approximately $6R \simeq 50$ J mol^{-1} K^{-1}

5.11 (a) 2.14 eV; (b) 2.743×10^{-29} kg $= 30.1 \, m_e$

5.12 461.7 nm

5.13 2.134×10^{-5} K^{-1}

5.14 $C_V(1 \text{ K}) = 1.53 \times 10^{-8}$ J K^{-1}; $C_V(1000 \text{ K}) = 1.843 \times 10^{-3}$ J K^{-1}

References

1. R. Eisberg, R. Resnick, *Quantum Physics*, 2nd edn. (Wiley, New York, 1974)
2. J. Brehm, W. Mullin, *Introduction to the Structure of Matter: A Course in Modern Physics* (Wiley, New York, 1989)
3. M. Alonso, E. Finn, *Fundamental University Physics: III* (Quantum and Statistical Physics (Addison Wesley, Reading, 1968)
4. A. Rigamonti, P. Carretta, *Structure of Matter, An Introductory Course with Problems and Solutions* (Springer, Milan, 2007)
5. E. Condon, H. Odabaşi, *Atomic Structure* (Cambridge University Press, London, 1980)
6. B. Bransden, C.J. Joachain, *Physics of Atoms and Molecules* (Prentice Hall, Englewood Cliffs, 2003)
7. R. Cowan, *The Theory of Atomic Structure and Spectra* (University of California Press, Berkeley, 1981)
8. E.B. Wilson, J. Decius, P. Cross, *Molecular Vibrations: The Theory of Infrared and Raman Vibrational Spectra* (McGraw-Hill, New York, 1955)
9. G. Herzberg, *Molecular Spectra and Molecular Structure*, vol. III (Van Nostrand Reinhold, New York, 1966)
10. N. Ashcroft, M. Mermin, *Solid State Physics* (Holt-Saunders, Philadelphia, 1976)
11. C. Kittel, *Introduction to Solid State Physics* (Wiley, New York, 2005)
12. G. Grosso, G. Pastori Parravicini, *Solid State Physics* (Academic, Oxford, 2014)
13. J. Slater, *Solid-State and Molecular Theory: A Scientific Biography* (Wiley, New York, 1975)
14. The web site https://physics.nist.gov/cuu/Constants/ maintains a table of physical constants
15. L. Schiff, *Quantum Mechanics* (McGraw-Hill, Singapore, 1955)
16. P. Caldirola, R. Cirelli, G.M. Prosperi, *Introduzione alla Fisica Teorica* (UTET, Torino, 1982)
17. J. Sakurai, *Modern Quantum Mechanics* (Benjamin, Menlo Park, 1985)
18. T. Hänsch, I. Shahin, E. Schawlow, Nat. Phys. Sci. **235**, 63 (1972)
19. B. Cordero, V. Gómez, A. Platero-Prats, M. Revés, J. Echeverría, E. Cremades, F. Barragán, S. Alvarez, Dalton Trans. **2008**, 2832 (2008)
20. The web site https://www.nist.gov/pml/atomic-spectra-database provides a database of atomic spectroscopy levels and transitions
21. J. Slater, *Quantum Theory of Matter* (McGraw-Hill, New York, 1951)
22. Several international facilities currently run sources of X-ray photons for research and applications. A list is maintained at https://en.wikipedia.org/wiki/List_of_synchrotron_radiation_facilities
23. R.M. Dreizler, E. Gross, *Density Functional Theory: An Approach to the Quantum Many-Body Problem* (Springer, Berlin, 1990)

© The Editor(s) (if applicable) and The Author(s), under exclusive license
to Springer Nature Switzerland AG 2020
N. Manini, *Introduction to the Physics of Matter*, Undergraduate Lecture Notes
in Physics, https://doi.org/10.1007/978-3-030-57243-3

24. J. Kielkopf, K. Myneni, F. Tomkins, J. Phys. B: At. Mol. Opt. Phys. **23**, 251 (1990)
25. 3D structures displayed in several illustrations of the present textbook are available at http://materia.fisica.unimi.it/manini/dida/structures.html as xyz files
26. Z. Yan, J. Babb, A. Dalgarno, G.W.F. Drake, Phys. Rev. A **54**, 2824 (1996)
27. J.L. Gardner, J.A.R. Samson, J. Chem. Phys. **62**, 1447 (1975)
28. R.E. Grisenti, W. Schöllkopf, J.P. Toennies, G.C. Hegerfeldt, T. Köhler, M. Stoll, Phys. Rev. Lett. **85**, 2284 (2000)
29. G. Wannier, *Statistical Physics* (Wiley, New York, 1966)
30. R. Balescu, *Equilibrium and Nonequilibrium Statistical Mechanics* (Wiley, New York, 1975)
31. G. Turrell, *Mathematics for Chemistry and Physics* (Academic, London, 2002)
32. W.H. Lien, N.E. Phillips, Phys. Rev. **133**, A1370 (1964)
33. H. Kleinke, Eur. J. Inorg. Chem. **1998**, 1369 (1998)
34. L. Pitaevskii, S. Stringari, *Bose-Einstein Condensation and Superfluidity* (Oxford University Press, Oxford, 2016)
35. R. Bini, L. Ulivi, H. Jodl, P. Salvi, J. Chem. Phys. **103**, 1353 (1995)
36. The web page http://materia.fisica.unimi.it/manini/scripts.html provides a set of tools to manipulate xyz files and to simulate diffraction patterns
37. K. Koyama, T. Kajitani, Y. Morii, H. Fujii, M. Akayama, Phys. Rev. B **55**, 11414 (1997)
38. E. Wollan, C. Shull, Phys. Rev. **73**, 830 (1948)
39. H.J. Fritzsche, J. Phys. Chem. Solids **6**, 69 (1958)
40. The web site http://britneyspears.ac/lasers.htm maintains a detailed, mostly correct, and entertaining "hands on" approach to semiconductors physics and technology
41. F.M.F. de Groot, M.H. Krisch, J. Vogel, Phys. Rev. B **66**, 195112 (2002)
42. E. Palik, *Handbook of Optical Constants of Solids* (Academic, San Diego, 1998)
43. M.A. Green, Sol. Energ. Mat. Sol. Cells **92**, 1305 (2008)
44. G. Onida, L. Reining, A. Rubio, Rev. Mod. Phys. **74**, 601 (2002)
45. K.K. Rao, T.J. Moravec, J.C. Rife, R.N. Dexter, Phys. Rev. B **12**, 5937 (1975)
46. D.W. Berreman, Phys. Rev. **130**, 2193 (1963)
47. The Nobel prize for the blue LED to I. Akasaki, H. Amano, S. Nakamura, https://www.nobelprize.org/prizes/physics/2014/press-release/ (2014)
48. J.P. Elliott, P.G. Dawber, *Symmetry in Physics* (McMillan, London, 1979)
49. P. Butler, *Point Group Symmetry Applications* (Plenum, New York, 1981)
50. A.R. Edmonds, *Angular Momentum in Quantum Mechanics* (Princeton University Press, Princeton, 1974)

Index

THE PERIODIC TABLE OF ELEMENTS

Legend: Atomic # / Symbol / Name / Mass

- **B** Solid
- **Br** Liquid
- **N** Gas
- **Rf** Unknown

Metals: Alkali metals · Alkaline earth metals · Transition metals · Post-transition metals · Lanthanides · Actinides

Nonmetals: Semiconductors / poor metals · Nonmetals · Noble gases

Pnictogens · Chalcogens · Halogens

For elements with no stable isotopes, the mass number of the isotope with the longest half-life is in parentheses

Group	1	2	3	4	5	6	7	8	9	10	11	12	13	14	15	16	17	18
1	1 **H** Hydrogen 1.008																	2 **He** Helium 4.0026
2	3 **Li** Lithium 6.94	4 **Be** Beryllium 9.0122											5 **B** Boron 10.81	6 **C** Carbon 12.011	7 **N** Nitrogen 14.007	8 **O** Oxygen 15.999	9 **F** Fluorine 18.998	10 **Ne** Neon 20.180
3	11 **Na** Sodium 22.990	12 **Mg** Magnesium 24.305											13 **Al** Aluminium 26.982	14 **Si** Silicon 28.085	15 **P** Phosphorus 30.974	16 **S** Sulfur 32.06	17 **Cl** Chlorine 35.45	18 **Ar** Argon 39.948
4	19 **K** Potassium 39.098	20 **Ca** Calcium 40.078	21 **Sc** Scandium 44.956	22 **Ti** Titanium 47.867	23 **V** Vanadium 50.942	24 **Cr** Chromium 51.996	25 **Mn** Manganese 54.938	26 **Fe** Iron 55.845	27 **Co** Cobalt 58.933	28 **Ni** Nickel 58.693	29 **Cu** Copper 63.546	30 **Zn** Zinc 65.38	31 **Ga** Gallium 69.723	32 **Ge** Germanium 72.630	33 **As** Arsenic 74.922	34 **Se** Selenium 78.971	35 **Br** Bromine 79.904	36 **Kr** Krypton 83.798
5	37 **Rb** Rubidium 85.468	38 **Sr** Strontium 87.62	39 **Y** Yttrium 88.906	40 **Zr** Zirconium 91.224	41 **Nb** Niobium 92.906	42 **Mo** Molybdenum 95.95	43 **Tc** Technetium (98)	44 **Ru** Ruthenium 101.07	45 **Rh** Rhodium 102.91	46 **Pd** Palladium 106.42	47 **Ag** Silver 107.87	48 **Cd** Cadmium 112.41	49 **In** Indium 114.82	50 **Sn** Tin 118.71	51 **Sb** Antimony 121.76	52 **Te** Tellurium 127.60	53 **I** Iodine 126.90	54 **Xe** Xenon 131.29
6	55 **Cs** Caesium 132.91	56 **Ba** Barium 137.33	57–71	72 **Hf** Hafnium 178.49	73 **Ta** Tantalum 180.95	74 **W** Tungsten 183.84	75 **Re** Rhenium 186.21	76 **Os** Osmium 190.23	77 **Ir** Iridium 192.22	78 **Pt** Platinum 195.08	79 **Au** Gold 196.97	80 **Hg** Mercury 200.59	81 **Tl** Thallium 204.38	82 **Pb** Lead 207.2	83 **Bi** Bismuth 208.98	84 **Po** Polonium (209)	85 **At** Astatine (210)	86 **Rn** Radon (222)
7	87 **Fr** Francium (223)	88 **Ra** Radium (226)	89–103	104 **Rf** Rutherfordium (267)	105 **Db** Dubnium (268)	106 **Sg** Seaborgium (269)	107 **Bh** Bohrium (270)	108 **Hs** Hassium (277)	109 **Mt** Meitnerium (278)	110 **Ds** Darmstadtium (281)	111 **Rg** Roentgenium (282)	112 **Cn** Copernicium (285)	113 **Nh** Nihonium (286)	114 **Fl** Flerovium (289)	115 **Mc** Moscovium (290)	116 **Lv** Livermorium (293)	117 **Ts** Tennessine (294)	118 **Og** Oganesson (294)

Lanthanides (Period 6):

57 **La** Lanthanum 138.91	58 **Ce** Cerium 140.12	59 **Pr** Praseodymium 140.91	60 **Nd** Neodymium 144.24	61 **Pm** Promethium (145)	62 **Sm** Samarium 150.36	63 **Eu** Europium 151.96	64 **Gd** Gadolinium 157.25	65 **Tb** Terbium 158.93	66 **Dy** Dysprosium 162.50	67 **Ho** Holmium 164.93	68 **Er** Erbium 167.26	69 **Tm** Thulium 168.93	70 **Yb** Ytterbium 173.05	71 **Lu** Lutetium 174.97

Actinides (Period 7):

89 **Ac** Actinium (227)	90 **Th** Thorium 232.04	91 **Pa** Protactinium 231.04	92 **U** Uranium 238.03	93 **Np** Neptunium (237)	94 **Pu** Plutonium (244)	95 **Am** Americium (243)	96 **Cm** Curium (247)	97 **Bk** Berkelium (247)	98 **Cf** Californium (251)	99 **Es** Einsteinium (252)	100 **Fm** Fermium (257)	101 **Md** Mendelevium (258)	102 **No** Nobelium (259)	103 **Lr** Lawrencium (266)